PROBABILITY AND STATISTICS
FOR PHYSICAL SCIENCES

PROBABILITY AND STATISTICS FOR PHYSICAL SCIENCES

SECOND EDITION

BRIAN R. MARTIN
MARK F. HURWITZ

ACADEMIC PRESS
An imprint of Elsevier

ELSEVIER

Academic Press is an imprint of Elsevier
125 London Wall, London EC2Y 5AS, United Kingdom
525 B Street, Suite 1650, San Diego, CA 92101, United States
50 Hampshire Street, 5th Floor, Cambridge, MA 02139, United States
The Boulevard, Langford Lane, Kidlington, Oxford OX5 1GB, United Kingdom

Notices

Knowledge and best practice in this field are constantly changing. As new research and experience broaden our understanding, changes in research methods, professional practices, or medical treatment may become necessary.

Practitioners and researchers must always rely on their own experience and knowledge in evaluating and using any information, methods, compounds, or experiments described herein. In using such information or methods they should be mindful of their own safety and the safety of others, including parties for whom they have a professional responsibility.

To the fullest extent of the law, neither nor the Publisher, nor the authors, contributors, or editors, assume any liability for any injury and/or damage to persons or property as a matter of products liability, negligence or otherwise, or from any use or operation of any methods, products, instructions, or ideas contained in the material herein.

ISBN: 978-0-443-18969-2

For information on all Academic Press publications visit our website at
https://www.elsevier.com/books-and-journals

Acquisitions Editor: Stephanie Cohen
Publisher: Katey Birtcher
Editorial Project Manager: Mason Malloy
Publishing Services Manager: Shereen Jameel
Senior Project Manager: Manikandan Chandrasekaran
Cover Designer: Mark Rogers

Printed in the United States of America

Last digit is the print number: 9 8 7 6 5 4 3 2 1

Working together to grow libraries in developing countries

www.elsevier.com • www.bookaid.org

Contents

Preface

The words 'physical sciences' in the title of this book embrace a wide range of disciplines, including physics, astronomy, chemistry, earth science, engineering, and others. Practitioners in these various fields all use the methods of statistics[1] at some time. This is often initially during their undergraduate studies but is seldom via a full lecture course. Some lectures on statistics are usually given as part of a general mathematical methods course, or as part of a laboratory course. Neither route is entirely satisfactory. The student learns a few techniques, typically the use of the unconstrained linear least-squares method to fit data, and the analysis of errors, but without necessarily the theoretical background that justifies the methods and allows one to appreciate their limitations. On the other hand, physical scientists, particularly undergraduates, rarely have the time, and possibly the inclination, to study statistics in detail. What we have tried to do in this book is therefore to steer a path between the extremes of a recipe of methods with a collection of useful formulas, and a detailed account of mathematical statistics, while at the same time developing the subject in a reasonably logical way, so that it may be used by all physical scientists as defined above. Proofs of some of the more important results stated have been included in those cases where they are short, but this book has been written by two physical scientists (a physicist and an engineer) for other physical scientists and there is no pretense of

mathematical rigor. The proofs are useful for showing how the definitions of certain statistical quantities and their properties may be used. Nevertheless, a reader uninterested in the proofs can easily skip over these, hopefully to come back to them later. Above all, the size of the book has been contained so that it can be read in its entirety by anyone with a basic exposure to mathematics, principally calculus and matrices, at the level of a first-year undergraduate student of a physical science.

The structure of the book follows closely that of the first edition written by one of us (BM), but with several changes and additions. In this edition there is a short discussion of experimental design; a clearer discussion of the *outcomes* of an experiment and their relation to *events*; more details about probability distribution functions and why a particular function is used; additional examples and problems; and a brief discussion of the use of simulation techniques to test the robustness of data and experimental design. In addition, we have corrected those typographical and other errors in the first edition that we are aware of. We would be grateful to have any others that have been missed brought to our attention.

Statistics in physical science is principally concerned with the analysis of numerical data obtained from experiments, so in Chapter 1 there is a review of what is meant by an experiment and how the data that it produces are displayed and characterized by

---

[1] The word 'statistics' is used here as the name of the subject; it is a collective noun and hence singular. In Section 1.3 we will introduce another meaning of the word to describe a function (or functions) of the data.

a few simple numbers. This leads naturally to a discussion of the vexed question of probability — what we mean by this term and how it is calculated. In Chapter 2 we discuss two interpretations, the frequency interpretation and the subjective interpretation, the latter also called the Bayesian interpretation, in recognition of the foundational work done by its founder Thomas Bayes, an 18th-century English clergyman. Neither interpretation is without its critics, but the great majority of physical scientists in practice use the frequency interpretation. For this reason, we have concentrated on this approach when discussing methods used in statistical analyses. But in recent years there has been an increasing interest in the Bayesian approach, and so we have included brief examples of this where appropriate.

There then follow two chapters on probability distributions: Chapter 3 reviews some basic concepts, and in Chapter 4 there is a discussion of the properties of several theoretical distributions commonly met in the physical sciences. In practice, scientists rarely have access to the whole population of events, that is, all events that could possibly exist, but instead must rely on a sample from which to deduce inferences about the population. Therefore in Chapter 5 the basic ideas involved in obtaining suitable samples, and how these samples can be used to produce reliable estimates, are discussed. This chapter also has a short description of the Monte Carlo method for evaluating complex multidimensional integrals often encountered in estimation problems, and its use in the important topic of simulation of data using computers. This is followed in Chapter 6 by a review of some sampling distributions associated with the important and ubiquitous normal distribution, the latter more familiar to physical scientists as the Gaussian function. The procedure whereby estimates are inferred for the individual parameters of

a population from sample statistics is called 'parameter estimation', or sometimes 'point estimation'. Chapters 7 and 8 explore this for some practical techniques. Parameter estimation is generalized in Chapter 9 by considering how to obtain precise estimates for the range of values within which an estimate may lie. This is called 'interval estimation'.

In the final two chapters, Chapters 10 and 11, methods for testing hypotheses about statistical data are discussed. In the first of these the emphasis is on hypotheses about individual parameters, and in the second some other hypotheses are discussed, such as whether a sample comes from a given population distribution, and 'goodness-of-fit' tests, that is, how well a mathematical model describes a set of experimental data. Chapter 11 also briefly describes tests that can be made in the absence of any information about the underlying population distribution.

All the chapters contain examples, with solutions for each given immediately following them. Most numerical statistical analyses are of course carried out using computers, and several statistical packages exist to enable this. But the aim of the present book is to introduce the ideas of statistics, and so the examples are simple and illustrative only, and any numerical calculations that are needed can easily be made using a simple spreadsheet. In an introduction to the subject there is an educational value in doing this, rather than simply entering a few numbers into a computer program and copying down the answer. The examples are an integral part of the text, and by working through them as the text is read, the reader's understanding of the material will be reinforced. There is also a set of problems at the end of each chapter and short answers to selected ones are given in Appendix D. Full solutions to all the problems are available on the book's website: https://educate.elsevier.com/book/details/9780443189692. The numbers of examples and problems have been increased in this edition but have still been kept relatively small to contain the size of the book. Numerous other problems, although not always relevant to physical sciences, may be found in the references given in the bibliography at the end of the book.

There are three other appendices: Appendix A gives an overview of some basic mathematics needed in the text, in case the reader needs to refresh their memory about details; Appendix B is about the principles of function optimization; and Appendix C contains a set of useful statistical tables, to complement the topics discussed in the chapters and to make the book reasonably self-contained.

Acknowledgments

Brian Martin thanks Andy Parker, Head of the Physics Department at Cambridge University, for a position as an Academic Visitor for the duration of writing this book, and the staff at the Betty and Gordon Moore Mathematics Library, the Raleigh Library of the Cavendish Laboratory, University of Cambridge, and the University of Cambridge Library for their dedication and co-operation in promptly supplying requested books in a safe way during the corona virus pandemic. Mark Hurwitz thanks his wife, Ingrid Kay McWilliams, for her patience and understanding on the many evenings and weekends occupied with work on this book.

Both authors thank Katey Birtcher, our editor at Elsevier, for her encouragement and patience during the long period of the corona virus pandemic, when lockdowns and other restrictions meant that progress was very slow, and Senior production manager Mani Chandrasekaran for his work liaising with the authors and co-ordinating the work of the production team.

June 2023

CHAPTER

1

Statistics, experiments, and data

Overview

This chapter is concerned with descriptive statistics. It starts by considering what is meant by the word 'statistics' and illustrates this with some examples. Basic concepts such as an 'experiment' and the 'observations' are then defined. The main part of the chapter is devoted to discussing how data are displayed and summarized using a few simple numbers, principally the mean, variance, and covariance. Examples are given to show how these are calculated, both for binned and unbinned data. The link to real situations is made in two ways. Firstly, there is a brief discussion of the usual form of the distribution of data when large numbers of measurements are made — the normal, or Gaussian — and secondly, there is an introduction to different types of error, distinguishing between random (statistical) and systematic errors.

Statistics dates back to the Renaissance, when Italian city states (probably the origin of the word 'statistics') found it useful to collect data on births, marriages, and deaths, the latter being of particular importance during the 16th and 17th centuries, when plague was endemic. This idea was adopted in other countries, and also expanded, so that by 1834, in the founding prospectus of the Statistical Society of London (later to become the Royal Statistical Society), statistics was defined very broadly as 'the ascertaining and bringing together of those facts which are calculated to illustrate the conditions and prospects of society'. It was not until the late 1800s that attention turned to drawing firm conclusions from the collections of data and making predictions about related areas of interest. The founders of this new aspect of statistics (Karl Pearson, Francis Galton, W. S. Gosset, Ronald Fisher, and others) developed pioneering statistical methods for applying statistics to a wide variety of problems in human and plant biology, drawing on the theory of probability that had been developed in the 17th century, mainly in France,[1] in the context of predicting gambling odds and life expectancies.

Statistical techniques now permeate numerous branches of modern life, including the physical, medical, and social sciences; engineering; and economics. In the context of modern physical sciences, statistics may be defined more narrowly as the branch of scientific method that deals with collecting data from experiments, describing the data (called *descriptive statistics*), and analyzing them to draw conclusions (called *inferential statistics*). Descriptive statistics allows repeated measurements to be characterized in a way that the application of inferential statistics gives answers to questions such as, 'What is the most likely result of a

---

[1]An introduction to probability is given in Chapter 2.

Probability and Statistics for Physical Sciences, Second Edition
https://doi.org/10.1016/B978-0-443-18969-2.00001-5

1

repeated measurement?' and 'Can two data sets be reasonably assumed as being measurements of the same quantity, despite there being small apparent differences between them?' There are other possible definitions of statistics that differ in their details, but all have the common elements of collections of data, which are the result of measurements, being described in some way and then used to make inferences. The following examples will illustrate a few of the many applications of statistics.

Consider a situation where several experimental groups claim to have discovered a new elementary particle, but all of them have very few examples to support their claim. Statistics tells us how to test whether the various experiments are consistent with each other and, if they are, how to combine their results so that one can be more confident about the claims. Another example concerns the efficacy of a medical treatment, such as a drug. A new drug is never licensed on the basis of its effect on a single patient. Regulatory authorities rightly require testing on a large number of patients having different ages, genders, comorbidities, and possibly other characteristics, to study its effect. But both time and cost limit how many people can be tested, so in practice, smaller groups of patients are used. Statistics specifies how such groups are best chosen to ensure that any inferences drawn from the data are meaningful for the wider population.

A different class of situations is where the predictions from a theory, or a law of nature, depend on one or more parameters, such as the electromagnetic force between two charged particles that depends on the strength of their electric charges. Such parameters can be determined from experiment by fitting data with a mathematical function that includes the unknown parameters and varying the latter to produce different predictions. Statistics specifies ways of doing this that lead to precise statements about the best values of the parameters. A related situation is where there are competing theories with different predictions for some phenomenon. Statistical analysis can use experimental data to test the predictions in a way that leads to precise statements about the relative likelihoods of the different theories being correct.

Both descriptive and inferential statistics have a bearing on *experimental design*. The aim of the latter is to achieve the optimal conditions in an experiment so that descriptive and inferential statistics may be used to draw quantifiable and reliable conclusions from the experimental results. Using statistical methods, experiments can be designed to increase the likelihood of obtaining a meaningful result. In some fields, such as market research, it is possible to codify techniques for optimizing experiments, for example, by deciding which questions should be asked and what their purpose is, and how many questionnaires are needed. How optimization is accomplished in physical science depends very much on the precise nature of the experiments and the sources and magnitudes of both random and nonrandom uncertainties, called 'errors' in physical science.[2]

An example of the importance of good experimental design in physics is the problem of measuring the lifetime of a particular radioactive isotope of an element that has several decay

---

[2]This is an unfortunate name, because in everyday usage an error is usually thought of as something that could, with forethought and care, be avoided. Nonrandom errors are in principle avoidable, or at least minimized, by careful design, but random errors are intrinsic to the process of measuring and cannot be totally avoided, although they can also be reduced by the use of good design, but mainly by repeated measurement, as we shall see in later chapters.

modes. A well-designed experiment here should be able to unambiguously identify the decay mode of interest in the presence of other decay modes. Again, how these goals are achieved depends on the nature of the investigation and on the sources and magnitudes of the various possible errors. To test the robustness of the experimental design, it is common practice to produce computer simulations of different configurations and to analyze 'pseudodata' produced by the different operating conditions. These can also be used to test the appropriateness of particular statistical techniques used to analyze the data produced.[3] Because experimental design is a large and specialized subject in its own right, it will not be discussed in detail in this book, and only a few aspects of its applicability will be examined where they arise naturally in the discussion.

1.1 Experiments and observations

In this section we will consider some aspects of descriptive statistics, starting by defining what is meant by an experiment.[4] An *experiment* is a set of reproducible conditions that enable measurements to be made and which produce observations, called *outcomes* in statistics. The word 'reproducible' implies that in principle independent measurements of a given quantity can be made. In other words the result of a given measurement does not depend on the result of any other measurement. The possible outcomes are denoted O_i, with $i = 1, 2, 3, ..., N$. In principle, N could be infinitely large, even if only conceptually. For example, when measuring the length of a rod, there is no limit to the number of measurements that could in principle be made, and we could conceive of a hypothetical infinite number of outcomes. A subset of outcomes with $n \leq N$ defines an *event*, denoted by E, and the collection (also called the *set*) of all possible events is called the *population* in statistics. In physical sciences the population is the total set of measurable data points, and a subset of n data points is called a *sample of size n*. Because probabilities are defined in terms of events rather than outcomes, we will define a *sample space S* in terms of its events, thus, for example $S_n = \{E_1, E_2, ..., E_n\}$. The following examples will illustrate these definitions.

Example 1.1

An 'experiment' consists of a train traveling between two points, A and B. On the journey it must negotiate three sections consisting of a single track, each of which is governed by a set of traffic lights, and the state of these, either red (r = stop) or green (g = go), is recorded. What is the sample space for the experiment? Write an expression for the event corresponding to the train encountering a red signal at the second traffic light.

---

[3]This is often done using a general technique called the Monte Carlo method, which is described briefly in Section 5.5.

[4]One problem here is that statisticians and physical scientists use different words to describe the same things. We will usually use the statistical terminology when discussing formal mathematics but refer to that used in physical science where appropriate, such as in 'real-life' examples.

Solution: The basic data resulting from the experiment are sequences of three signals, each either red or green. The population therefore consists of combinations of three possible outcomes, and the sample space is denoted by eight events:

$$S = \{rrr, rrg, rgg, ggg, ggr, grr, rgr, grg\}.$$

The event E defined by the driver encountering a red signal at the second traffic light (and also possibly at other lights) consists of four events. Thus

$$E = \{rrr, rrg, grr, grg\}.$$

In this case the event is called *complex*, because it contains a number of *simple* events each containing a group of outcomes such as rrr, etc.

Example 1.2

The number of 'heads' obtained by tossing two coins simultaneously can assume the discrete values 0, 1, or 2. If we distinguish between the two coins, what is the sample space for the experiment? Show the content of its events.

Solution: The two possible outcomes of tossing each coin are a 'head' or a 'tail'. Denoting these by H and T, respectively, there are four possible events:

$$E_1 = (H, H); \quad E_2 = (H, T); \quad E_3 = (T, H); \quad E_4 = (T, T)$$

and the sample space is therefore $S = \{E_1, E_2, E_3, E_4\}$, which formally may be written as the *product space*

$$S = \{H, T\} \times \{H, T\} = (H, H), (H, T), (T, H), (T, T).$$

In these examples the observations are nonnumeric, but in physical science the outcomes are almost invariably numbers, or sets of numbers, and to relate these definitions to numerical situations, consider firstly an experiment of tossing a six-sided die. The sample space is $S = \{1, 2, 3, 4, 5, 6\}$ as only one of the six numbers on the face of a die can be observed at any time. A *simple* event is defined as the occurrence of one of the six numbers 1–6. The outcome of the experiment is thus the observation of one of the numbers that define the event, that is, the observation of one of the numbers 1–6, and is a discrete variable. If there were two dice, an example of a *complex* event would be the occurrence of a 2 on one die and a 3 on the other, and the outcome would be the observation of these pairs of numbers. Another example is an experiment to measure the heights of all students in a given class. In this case the outcomes are not discrete but continuous, and in practice, an event would be defined by an interval of heights. Then the occurrence of the event would be interpreted as the situation where a measured height fell within a specified range.

1.2 Random variables and sampling

In the examples above the outcomes are unambiguously defined, and the possibility of confusing the red and green traffic lights, or misreading the number on a die, has been ignored. Either possibility would introduce a bias in the results and render the conclusions unreliable. We will return to this in Section 1.6, where we discuss experimental errors, but here we will illustrate the potential problem of bias by considering how to test which of two lecturers is more effective in teaching their students. How to overcome the problem of bias is an example of good experimental design.

One possibility would be to assign an equal number of students to each lecturer, and each group would receive the same number of lectures and tutorial classes on the same subject. The outcomes would be the examination pass rates for the two groups, and these could be analyzed to see if there was a significant difference between them. But as mentioned earlier, to ensure a meaningful outcome, any experiment must be designed to take into account possible sources of error, particularly nonrandom errors. For example, if all the students assigned to one group had a higher grade average over many courses than those in the other group, we might expect this to bias the measured pass rates of the two groups, making it an unsatisfactory measure of the effectiveness of the teaching skills of the lecturers. Different biases, but possibly as significant, might be introduced if some of the members in one group had previously taken a course on a similar subject.

One way to eliminate bias would be to use the whole population of students in each group. However, this would only be possible if the two groups were taught different courses, which would introduce another, but different, bias. Also, it is often not practical to use an entire population in an experiment, so we must use another way to reduce the effects of bias as much as possible in our choices of the two samples. The simplest accepted way of ensuring this is to assign students to the two lecturers in such a way that all possible choices of members of the groups are equally likely, or, put another way, that if we had a very large population of students, then every possible sample of a particular size n has an equal chance of being selected.[5] This technique is called *simple random sampling*. In principle, this condition could be relaxed provided we could calculate the chance of each sample being selected. Samples obtained this way are called *random samples of size n*, and the outcomes are called *random variables*.[6] At first sight, these choices appear counterintuitive. We might think it would be better to assign students on the basis of ensuring equivalent grade averages, or perhaps the percentages of one gender in each group, but any alternative method of nonrandom selection usually results in one that is inherently biased toward some specific outcome, and not necessarily one the experimenter notices until the experiment is completed, and the results are analyzed and questioned. A random choice removes this bias. Note that the word 'chance' used above has anticipated the idea of probability that will be discussed in more detail in Chapter 2.

A question that naturally arises is: 'How are random samples chosen?' For small populations this is easy. One could assign a unique number to each member of the population and

---

[5]If more than one examiner marks all the examination papers, then similar considerations must be applied when allocating the completed exam papers to the examiners.

[6]Random variables are discussed in detail in Chapter 3.

write it on a ball. The balls could then be thoroughly shaken in a bag, or rotating drum, and n balls drawn. This is essentially how lottery numbers are decided, with one ball assigned to each integer, and drawing several balls sequentially and not replacing them. For large populations, simple methods like this are not practical and more complex methods have to be used. This problem will be considered in more detail when sampling is discussed more fully in Chapter 5.

Finally, we should mention that although in physical sciences 'sampling' almost invariably means simple random sampling, there are different types of sampling that are used in other fields. For example, pollsters often use *systematic sampling* (sometimes mistakenly presented as true random sampling), where every nth member of a population is selected, for example, every 100th name in a telephone directory. Another method is used when the population can be divided into mutually exclusive subpopulations. In this case simple random samples are selected from the subpopulations with a size proportional to the fraction of members that are in that subpopulation. For example, if the fractions of men and women that take a degree in physics are known, then simple random samples of sizes proportional to these fractions could be taken to make inferences about the populations of all physics students. This is called *stratified sampling* (or *blocking*) and is very efficient but not often applicable in physical sciences.

1.3 Displaying data

In physical sciences experiments almost invariably produce data as sets of numbers, so we will concentrate on numerical outcomes. The measurements could be a set of discrete numbers, that is, integers, like the numbers on the faces of a die, or a set of real numbers forming a continuous distribution, as in the case of the heights of the students mentioned above. We will start by describing how data are displayed.

Experimental results can be presented by simply drawing a vertical line on an axis at each data point, whose height is the value of the data point, but in practice, for both discrete and continuous data it is common to group the measurements into *intervals*, or *bins*, containing all the data for a range of values. The number of events in each bin is called the *frequency* of that bin. The binned data can be presented as a *frequency table*, or graphically. For discrete data, binning can be done exactly, and the results displayed in the form of a *bar chart*, where a vertical column is drawn, the height of which represents the number of events of a given type. The total of the heights of the columns is equal to the number of events. The width of the bins is arbitrary and sometimes for clarity a gap is left between one bin and the next. Both are matters of taste. We will also be interested in the frequency with which an outcome takes on values equal to, or less than, a stated value. This is called the *cumulative frequency*.

Example 1.3

The frequency table below shows the results of an experiment where 6 coins are simultaneously tossed 200 times and the number of 'heads' recorded.

Number of heads	0	1	2	3	4	5	6
Frequency	2	19	46	62	47	20	4

Display these data, and the cumulative frequency, as bar charts.

Solution: Figure 1.1(a) shows the data displayed as a bar chart. The total of the heights of the columns is 200. Figure 1.1(b) shows a plot of the cumulative frequency of the same data. The numbers on this plot are obtained by cumulatively summing entries on the bar chart of frequencies.

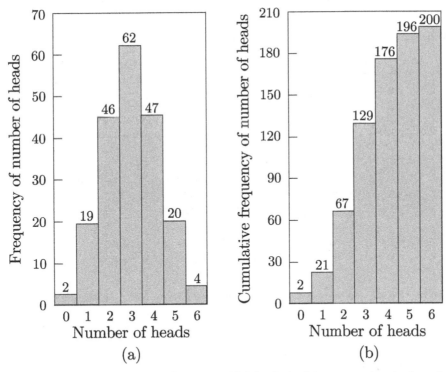

FIGURE 1.1 Bar charts showing (a) the frequency of heads obtained in an experiment where 6 coins were simultaneously tossed 200 times; and (b) the cumulative frequency of heads obtained in the same experiment.

For continuous data, the values of the edges of the bins have to be defined, and it is usual to choose bins of equal width, although this is not strictly essential. The raw data are then rounded to a specific accuracy, using the normal rules for rounding real numbers, and assigned to a particular bin. If a bin has lower and upper values of a and b, respectively, then a data point with value x in the range $a \leq x < b$ is assigned to this bin.[7] There is inevitably some loss of precision in binning data, although it will not be significant if the number of measurements is large. It is the price to be paid for putting the data in a useful visual form. The resulting plot is called a *histogram*. The only significant difference between this and a bar chart is that the number of events in a histogram is proportional to the area of the bins rather than their heights. The constant of proportionality is usually chosen so that the total area under the histogram is equal to the total number of events or is normalized to unity. The latter choice is useful if the histogram is interpreted as being generated by a mathematical form called a *probability density function (pdf)*. This is defined in Chapter 3 and is briefly illustrated in Section 1.5. In this case the associated cumulative plot is called the *empirical cumulative distribution function*, or just the *cumulative distribution*.

Pdfs and the associated distribution functions for both discrete and continuous variables are discussed in detail in Chapter 3.

Example 1.4

A data set x_i has 100 elements and an empirical distribution function $F_n(x)$ such that $F_{100}(2) = 0.8$. How many elements in the data set satisfy $x_i > 2$?

Solution: From the definition of $F_n(x)$, the number of elements that satisfy $x_i \leq 2$ is $100F_{100}(2) = 80$. Hence there are 20 elements that satisfy the condition $x_i > 2$.

The choice of bin width needs some care. If it is too narrow, there will be few events in each bin and fluctuations will be significant. If the bins are too wide, details can be lost by the data being spread over a wide range. Some authors have suggested empirical formulas for the optimal bin widths and number of bins, but they are very rarely used in physical science applications. In practice, the bin width is chosen so that the difference in the number of events from bin to bin is small, producing a gradual change in shape. About 10 events per bin over most of the range is often taken as a minimum when choosing bin widths, although this could be smaller at the end point.

Example 1.5

The table below shows data on the ages of a class of 230 university science students taking a course in statistics. Draw three histograms with bin sizes of 2 years, 1 year, and $\frac{1}{2}$ year, respectively, with each histogram normalized to a total area of unity. Which bin size is optimal?

---

[7]Some authors use the alternative convention $a < x \leq b$.

Age range	Student numbers
17.0–17.5	2
17.5–18.0	3
18.0–18.5	35
18.5–19.0	27
19.0–19.5	61
19.5–20.0	29
20.0–20.5	28
20.5–21.0	14
21.0–21.5	12
21.5–22.0	8
22.0–22.5	8
22.5–23.0	3

Solution: Figure 1.2 shows the three histograms. They have been normalized to a common area of unity by dividing the number of events in a bin by the product of the bin size and the total number of events. For example, with a bin size of 2 years, as shown in the left-hand histogram, the entry in the 19–21 age bin is $(61 + 29 + 28 + 14)/(2 \times 230) = 0.29$. The number of data is probably sufficient to justify a bin size of $1/2$ year because most of the bins contain a reasonable number of events. The other two histograms have lost significant detail because of the larger bin sizes.

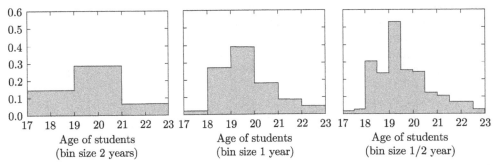

FIGURE 1.2 Normalized histograms of the ages of 230 university students taking a course on statistics, showing the effect of using different bin sizes.

If large samples can be obtained, smaller bin widths can be used and histograms can then frequently be represented by a smooth curve (often of a form known as the *normal*

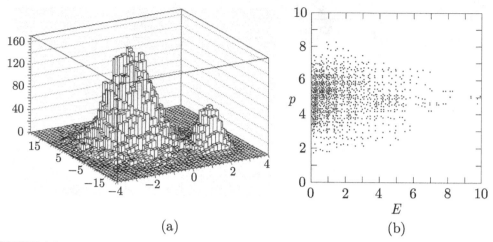

(a)   (b)

FIGURE 1.3   Examples of displays for data that depend on two variables: (a) Lego plot; (b) scatter plot.

distribution), and in the case of continuous data this can then be used to interpolate the data to intermediate values. This is illustrated in Section 1.4 below. But in cases where only limited data are available it may still be useful to have a smooth representation of the raw data that to some extent lessens the dependence on choice of bin width in histograms. Some specific functions have been suggested, but if all that is required is a mathematical representation to interpolate the data, any function that has the basic features of the histogram, for example, the observed number of maxima and minima, with a few parameters to represent their positions and widths, possibly combined with a low-order polynomial, could be used, although fitting this to the data may involve substantial computations. We return to this briefly in Section 1.5, and later in Chapter 8, where we will discuss methods that can be used to decide when a fit is acceptable.

Histograms can be extended to three dimensions for data with values that depend on two variables, in which case they are sometimes colloquially called *Lego plots*. Two-dimensional data, such as the energies (E) and momenta (p) of particles produced in a nuclear scattering reaction, can also be displayed in *scatter plots*, where points are plotted on a two-dimensional grid. The latter enable *correlations*[8] between the two variables to be easily seen. Examples of these types of displays are shown in Figure 1.3. Although there are other ways that data can be displayed, histograms, Lego plots, and scatter plots are by far the most common graphical representations of data used in physical sciences.

1.4 Summarizing data numerically

Although a histogram provides useful information about a set of measurements, it is inadequate for the purposes of making inferences, because many different histograms can be

---

[8]A nonzero correlation implies a relationship between variables. It is discussed in Section 1.4.3.

constructed from the same data set simply by changing the bin size. To make reliable inferences and to test the quality of such inferences, other quantities are needed that summarize the salient features of the data. A quantity constructed from a data sample is called a *statistic*[9] and is conventionally written using the Latin alphabet. The analogous quantity for a population is called a *parameter* and is written using the Greek alphabet. We will look firstly at statistics and parameters that describe frequency distributions.

1.4.1 Measures of location

The first measure of *location*, and the one most commonly used in physical sciences, is the *arithmetic mean*, usually simply called the *mean*. For a finite population of size N, the *population mean* is a parameter denoted by μ, and defined by

$$\mu = \frac{1}{N} \sum_{i=1}^{N} x_i. \tag{1.1}$$

The mean of a sample of size $n < N$ drawn from the population is the statistic, denoted $\bar{x}_n$, or often just $\bar{x}$, defined analogously by

$$\bar{x}_n = \frac{1}{n} \sum_{i=1}^{n} x_i, \tag{1.2a}$$

so that μ is the limit of the sample mean, when $n \to N$. Alternatively, if the values x_i in the sample occur with frequencies f_i, $(i = 1, 2, ..., k)$, respectively, where

$$\sum_{i=1}^{k} f_i = n \tag{1.2b}$$

is the total frequency, or sample size, then the sum in (1.2a) can be written as

$$\bar{x}_n = \frac{1}{n} \sum_{i=1}^{k} f_i x_i. \tag{1.3}$$

For binned data, an approximate value for $\bar{x}_n$ could be obtained quickly by interpreting x_i as the central value of a bin, and the quantity f_i as the frequency with which it appears in the bin. In this case the values of the mean calculated from (1.2a) and (1.3) will not be exactly the same, although the difference will be small for large samples divided into many bins.

Although the mean is the measure of location usually used in physical sciences, there are two other measures that are occasionally used. These are: the *mode*—the value of the quantity for which the frequency is a maximum—and the *median*—the value of the quantity that divides the cumulative frequency into two equal parts. In the coin-tossing experiment of

[9]This is the second use of this word, as mentioned in footnote 1 of the Preface.

Example 1.3 the mode is 3, with a frequency of 62. Both the 100th and 101st throws, arranged by order of size, fall in the class '3' and since the quantity in this example can only take on integer values between 0 and 6, the median value is also 3. In the case of a continuous quantity, such as that shown in Figure 1.2, the mode is the value of the (19–19.5) year bin and the median could be estimated by forming the cumulative frequency distribution from the raw data and using the 115th point on the plot to find the median age by interpolation.

The median is often quoted in situations where the distribution of events is very asymmetric, because it is less affected by events a long way from the mean. An example is the income of a population, where it is common practice to quote the median, because in relatively small populations, such as those in a single organization, the median is less influenced by the large incomes of a few very highly paid individuals and so yields an 'average' value that better represents the income that most employees receive. Note, however, that in using the median, a part of the distribution, the high-value tail in this case, is being rejected (called *trimming*) purely on the grounds of expediency. Similar considerations will arise later in Section 5.4 when we discuss what to do about experimental values that are very far from the mean, the so-called 'outliers'.

The median is an example of a more general measure of location called a *quantile*. This is the value of x below which a specific fraction of the observations falls. It is thus the inverse of the cumulative frequency. Commonly met quantiles are those that divide a set of observations into 100 equal intervals, ordered from smallest to largest. They are called *percentiles*. Thus, if the percentiles are denoted by $P_p(p = 0.01, 0.02, ..., 1)$, then $100p$ percent of the data are at, or fall below, P_p. The median therefore corresponds to the $P_{0.5}$ percentile. In practice, to find the percentile corresponding to a given value of p, the data are first ordered from lowest to highest and the quantity $q = np$ is calculated, where n is the sample size. Then if q is an integer, the values of the qth and the $(q+1)$th ordered values are averaged; if q is not an integer, the percentile is the kth ordered value, where k is found by rounding up q to the next integer.

Example 1.6

Use the sample data below to calculate the mean, the median, and the percentile $P_{0.81}$.

1.6	3.4	9.2	9.6	6.1	7.5	8.0	8.9	11.1	12.3
2.3	4.1	6.8	4.8	12.5	10.0	5.1	8.2	8.5	11.7

Solution: The mean is

$$\bar{x} = \frac{1}{20} \sum_{i=1}^{20} x_i = 7.585.$$

To find the median and $P_{0.81}$, we first order the data from lowest to highest:

1.6	2.3	3.4	4.1	4.8	5.1	6.1	6.8	7.5	8.0
8.2	8.5	8.9	9.2	9.6	10.0	11.1	11.7	12.3	12.5

The median is $P_{0.5}$, so $q = np = 20 \times 0.5 = 10$. As q is an integer, the median is the average of the 10th and 11th smallest data values, that is, 8.1. For $P_{0.81}$, $q = 16.2$, and because q is not an integer, $P_{0.81}$ is the 17th lowest data value, that is, 11.1.

Example 1.7

Two data sets $x_i (i = 1, n)$ and $y_i (i = 1, m)$ have means $\bar{x}_n$ and $\bar{y}_m$, respectively. In what circumstance is the average of the two means equal to the average of the sum of both data sets?

Solution: We require that

$$\frac{1}{2}(\bar{x}_n + \bar{y}_m) = \frac{1}{(n+m)} \left(\sum_{i=1}^{n} x_i + \sum_{i=1}^{n} y_i \right) = \frac{n\bar{x}_n + m\bar{y}_m}{(n+m)},$$

which is only true if $n = m$.

1.4.2 Measures of spread

The other most useful quantity to characterize a distribution measures the spread of the data about the mean and is called the *dispersion*. We might be tempted to use the average of the differences $d_i = x_i - \mu$ of x_i from the population mean, that is,

$$\bar{d} = \frac{1}{N} \sum_{i=1}^{N} (x_i - \mu), \tag{1.4a}$$

but from the definition of μ, (1.1), this quantity is identically zero. So instead, a quantity called the *variance* (sometimes abbreviated as *var*) is used that involves the squares of the differences d_i. The *population variance*, denoted by σ^2, is defined by

$$\sigma^2 = \frac{1}{N} \sum_{i=1}^{N} (x_i - \mu)^2, \tag{1.4b}$$

and the square root of the variance is called the *standard deviation* σ. The standard deviation is a measure of how spread out the distribution of the data is, with a larger value of σ meaning that the data have a greater spread about the mean. *The sample variance s^2 is defined for a sample of size n by*

1. Statistics, experiments, and data

$$s^2 \equiv \frac{1}{n-1} \sum_{i=1}^{k} f_i(x_i - \bar{x})^2,$$ (1.5)

where the values x_i in the sample occur with frequencies f_i, $(i = 1, 2, ..., k)$ respectively, as used in (2.1). Alternatively, if the sum is over individual data points,

$$s^2 = \frac{1}{n-1} \sum_{i=1}^{n} (x_i - \bar{x})^2 = \frac{1}{n-1} \left[\sum_{i=1}^{n} x_i^2 - \frac{1}{n} \left(\sum_{i=1}^{n} x_i \right)^2 \right]$$

$$= \frac{1}{n-1} \left[\sum_{i=1}^{n} x_i^2 - n\bar{x}^2 \right] = \frac{n}{n-1} \left(\overline{x^2} - \bar{x}^2 \right),$$ (1.6)

where an overbar is used to denote an average. For example, $\overline{x^2}$ is the mean value of x^2, defined by analogy with (1.2a) as

$$\overline{x^2} = \frac{1}{n} \sum_{i=1}^{n} x_i^2.$$ (1.7)

Equation (1.6) is useful when making numerical calculations. Just as for the mean, the sample variance calculated from (1.5) will not be exactly the same as that obtained from (1.6) if x_i is interpreted as the center of a bin, and frequency f_i means the number of values in the bin.

Example 1.8

Derive (1.6).

Solution: From (1.5), we have

$$s^2 = \frac{1}{n-1} \sum_{1}^{n} (x_i - \bar{x})^2 = \frac{1}{n-1} \left[\sum_{i} (x_i^2 - 2x_i\bar{x} + \bar{x}^2) \right].$$

Then using

$$\sum_{i}^{n} x_i = n\bar{x} \text{ and } \sum_{i}^{n} x_i^2 = n\overline{x^2},$$

gives

$$s^2 = \frac{n}{n-1} \left[\overline{x^2} - 2\bar{x}^2 + \bar{x}^2 \right] = s^2 = \frac{n}{n-1} \left[\overline{x^2} - \bar{x}^2 \right].$$

The definitions of the sample and population variance differ in their external factors, although for large sample sizes, the difference is of little consequence. The reason for the difference is a theoretical one related to the fact that (1.5) contains $\bar{x}$, which has itself been calculated from the data, and we require that for large samples, sample statistics should provide values that on average are close to the corresponding population parameters. In this case we require the sample variance to provide a 'true', or so-called 'unbiased', estimate of the population variance. This will be discussed in Chapter 5, when we consider sampling in more detail, and in Section 5.2 it will be proved that the choice of the external factor in (1.5) ensures this for the sample variance. The role of s in determining how well the sample mean is determined is also discussed in Chapter 5.

Example 1.9

The price of laboratory consumables from 10 randomly selected suppliers showed the following percentage price increases over a period of 1 year.

Supplier	1	2	3	4	5	6	7	8	9	10
Price increase (%)	15	14	20	19	18	13	15	16	22	17

Find the average price increase and the sample standard deviation.
Solution: To find the average percentage price increase, we calculate the sample mean. This is

$$\bar{p} = \frac{1}{10} \sum_{i=1}^{10} p_i = 16.9.$$

This can then be used to find the sample variance from (1.6),

$$s^2 = \frac{1}{9} \sum_{i=1}^{10} (p_i - \bar{p})^2 = 8.10,$$

and hence the sample standard deviation is $s = \sqrt{8.10} = 2.85$. Thus the outcome of the observations could be quoted as a mean price increase of 16.9% with a standard deviation of 2.9%. An empirical interpretation of statements such as this is given in Section 1.5.

It is worth noting that if we define a new data set by $y_i = a + bx_i$, a simple calculation, in an obvious notation, shows that $\bar{y} = a + b\bar{x}$, and $s_y^2 = b^2 s_x^2$. Thus in some cases it may be quicker to subtract a constant from each value before calculating the mean and the variance.

The mean and variance involve the first and second powers of x, but higher powers may also be used to define *moments*. In general, the nth *moment* of a discrete population about a point λ is defined as

$$\mu_n' = \frac{1}{N} \sum_{i=1}^{N} (x_i - \lambda)^n. \tag{1.8}$$

Although λ can be an arbitrary point, it is usual to use $\lambda = \mu$, in which case the moments are called *central* and are written without a prime. For example, $\mu_0 = 1, \mu_1 = 0$, and $\mu_2 = \sigma^2$. Moments can also be defined for samples by formulas analogous to those above and can be used to compare with the predictions of theoretical distributions by using a *moment generating function*. This will be discussed in Chapter 3 after we have discussed probability distributions.

In the case of grouped data taking the frequencies to be those at the midpoints of the intervals is an approximation, and so some error is thereby introduced. This was mentioned previously for the case of the sample mean and sample variance. In these circumstances some statisticians advocate applying the so-called *Sheppard's corrections*, although not all agree with this approach. If μ_n are the true central moments, and $\bar{\mu}_n$ the central moments as calculated from the grouped data with interval width h, then Sheppard's corrections to the first four moments are[10]

$$\mu_1 = \bar{\mu}_1, \quad \mu_2 = \bar{\mu}_2 - \frac{1}{12}h^2, \quad \mu_3 = \bar{\mu}_3, \quad \mu_4 = \bar{\mu}_4 - \frac{1}{2}\bar{\mu}_2\, h^2 + \frac{7}{240}h^4. \tag{1.9}$$

Statistics that measure additional properties can be defined in terms of low-order moments, such as *skewness* (the degree to which a distribution of data is asymmetric) and *kurtosis* (the degree of peaking of a distribution), although more than one definition of such statistics exists. In practice, they are not very useful, because, using the same data, different distributions can be constructed with similar values of these statistics, and they are rarely used in physical science applications.

Finally, the definition (1.6) can be used to prove an important general constraint on how the data points x_i are distributed about the sample mean. If the data are divided into two sets, one denoted S_k, with N_k elements having $|x_i - \bar{x}| < ks$, taking the positive square root of the sample variance so that $s > 0$, and the other containing the rest of the points having $|x_i - \bar{x}| > ks$, then from (1.6)

$$(n-1)s^2 = \sum_{i=1}^{n}(x_i - \bar{x})^2 = \sum_{x_i \in S_k}(x_i - \bar{x})^2 + \sum_{x_i \notin S_k}(x_i - \bar{x})^2 \geq \sum_{x_i \notin S_k}(x_i - \bar{x})^2,$$

where the expression $x_i \in S_k$ ($x_i \notin S_k$) means that the quantity x_i is (is not) in the set S_k, and the inequality follows from the fact that the terms in the summations are all positive. Using the condition $(x_i - \bar{x})^2 \geq k^2 s^2$ for points not in the set S_k, the right-hand side may be replaced so that

$$(n-1)s^2 \geq k^2 s^2 (n - N_k).$$

Finally, dividing both sides by $nk^2 s^2$ gives

$$\frac{n-1}{nk^2} \geq 1 - \frac{N_k}{n}, \tag{1.10a}$$

---

[10] The derivation of the relations (1.9) is given, for example, on pp. 359–361 in Cramér, *Mathematical Methods of Statistics*, listed in the Bibliography.

or, equivalently,

$$\frac{N_k}{n} \geq 1 - \frac{1}{k^2} + \frac{1}{nk^2} \geq 1 - \frac{1}{k^2}. \tag{1.10b}$$

This result is called *Chebyshev's inequality*. Recalling that N_k is the set with elements having $|x_i - \bar{x}| < ks$ $(s > 0)$, it follows from (1.10b) that for any value of $k \geq 1$, at least $100\left(1 - 1/k^2\right)$ percent of the data lie within an interval from $\bar{x} - ks$ to $\bar{x} + ks$. For example, if $k = 2$, then at least 75% of the data lie within $2s$ of the sample mean, and 25% without. For $k = 3$, at least 88.8% of the data lie within $3s$ of the sample mean and 11.2% without. This indicates the extent to which data are grouped around the sample mean, and we should expect the number of data points outside of any interval around the sample mean to decrease as the size of that interval increases. Although this result is qualitatively consistent with observed data sets, in practice, it is only a relatively weak bound, and for the distributions commonly met, the actual percentage of data that lie within the interval $(\bar{x} - ks)$ to $(\bar{x} + ks)$ is considerably larger than that given by (1.10b). This will be illustrated in Section 1.5, where we discuss the form of many common histograms when the number of events becomes large.

There is a stronger version of the result (1.10b), called the *one-sided Chebyshev's inequality*. Without proof, this replaces the result (1.10b) by

$$\frac{N_u}{n} \leq \frac{1}{k^2 + 1}, \tag{1.11}$$

where N_u is the number of elements having $x_i - \bar{x} \geq ks$. For example, if $k = 2$, at most 20% of the data should be more than two standard deviations above the mean, compared to 12.5% from the standard Chebyshev inequality, assuming the elements are symmetrical about the mean.

1.4.3 More than one variable: Correlation

The mean and variance are the most useful quantities that characterize a set of data, but if the data are defined by more than one variable, then other quantities are needed. The most important of these is the *covariance* (abbreviated as *cov*). If the data depend on two variables and consist of pairs of numbers $\{(x_1, y_1), (x_2, y_2), \ldots\}$, their population covariance is defined as

$$\mathrm{cov}(x, y) = \frac{1}{N} \sum_{i=1}^{N} (x_i - \mu_x)\left(y_i - \mu_y\right), \tag{1.12a}$$

where μ_x and μ_y are the population means of the quantities x and y, respectively. The related *sample covariance* is defined as

$$\mathrm{cov}(x, y) = \frac{1}{n - 1} \sum_{i=1}^{n} (x_i - \bar{x})(y_i - \bar{y}) = \frac{n}{n - 1}(\overline{xy} - \bar{x}\bar{y}), \tag{1.12b}$$

1. Statistics, experiments, and data

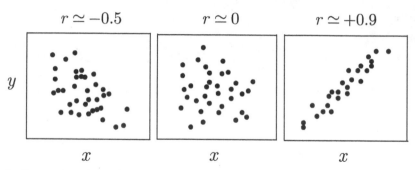

FIGURE 1.4   Scatter plot of two-dimensional data and approximate values of their correlation coefficients r.

where overbars again denote averages. Just as for the variance, the external factor differs in the definitions (1.12a) and (1.12b). The reason is the same: to ensure that for large sample sizes, sample values are unbiased estimates of the equivalent population values. The covariance can be used to test whether quantity x depends on quantity y. If small values of x tend to be associated with small values of y, then both terms in the summation will be negative and the sum itself will be positive. Likewise, if large values of x are associated with small values of y, the sum will be negative. If there is no general tendency for values of x to be associated with particular values of y, the sum will be close to zero.

Because the covariance has dimensions, a more convenient quantity is the *correlation coefficient* (also called *Pearson's correlation coefficient*), defined for a sample by

$$r = \frac{\text{cov}(x, y)}{s_x s_y}. \tag{1.13}$$

In this formula s_x and s_y are the standard deviations of the quantities x and y, respectively, and r is a dimensionless number between -1 and $+1$.[11] An analogous relation to (1.13) with the sample standard deviation replaced by the population value defines the *population correlation coefficient* ρ. A positive value of r, that is, a *positive correlation*, implies that values of x that are larger than the mean $\bar{x}$ tend on average to be associated with values of y that are larger than the mean $\bar{y}$. Likewise, a negative value of r, that is, a *negative correlation*, implies that values of x that are larger than $\bar{x}$ tend on average to be associated with values of y that are smaller than $\bar{y}$. If r is $+1$ or -1, then x and y are *totally correlated*, that is, knowing one completely determines the other. If $r = 0$, then x and y are said to be *uncorrelated*. Examples of scatter plots for data showing various degrees of correlation are shown in Figure 1.4. If there are more than two variables present, correlation coefficients for pairs of variables can be defined and form a matrix.[12]

---

[11]This will be proved in Section 3.3.3 after *expectation values* are discussed in Section 3.2.2.

[12]A brief review of matrices is given in Appendix A.

   Correlation coefficients must be interpreted with caution, because they measure *associa-tion*, which is not necessarily the same as *causation*. This is a common misconception. For example, although the failure rate of a piece of equipment in a particular month may show an association, that is, be correlated, with an increase in the number of users of the equipment, this does not necessarily mean that the latter has caused the former. The failures might have occurred because of other reasons, such as disruptions in the power supply to the equipment during the month in question. Another example is the ownership of mobile phones over the past decade or so. The number of mobile phones in use correlates positively with a wide range of disparate variables, including the increase in the total prison population and the increase in the consumption of 'organic' food. But common sense would say that neither of these increases could possibly have caused the increased number of mobile phones. The answer lies in realizing that each of the former quantities has also increased with time and it is this that has led to the observed correlations. Thus time is acting as a 'hidden variable' and without knowing this, a misleading conclusion may be drawn from the correlation.

Example 1.10

The lengths and electrical resistances (in arbitrary units) of a sample of 10 pieces of copper wire were measured with the results below. Calculate the correlation coefficient for the sample.

Number of pieces	1	2	3	4	5	6	7	8	9	10
Length (L)	15	13	10	11	12	11	9	14	12	13
Resistance (R)	21	18	13	15	16	14	10	16	15	12

   Solution: From these data we can calculate the sample means from (1.3) to be $\bar{L} = 12$ and $\bar{R} = 15$; the sample variances from (1.6) are $s_L^2 = 30/9$ and $s_R^2 = 86/9$; and the covariance from (1.12b) is $\text{cov}(L/R) = 40/9$. So, from (1.13), the correlation coefficient of the sample is 0.79, which indicates a strong positive correlation between the length and resistance of the pieces of wire, as one would expect.

   Just as the mean and variance can be calculated using binned data, so can the correlation coefficient, although the calculations are a little more complicated, as the following example shows.

Example 1.11

A class of 100 students has taken examinations in mathematics and physics. The number of students obtaining marks in various bins is shown in the table below. Use them to calculate the correlation coefficient.

		Mathematics marks					
		40–49	50–59	60–69	70–79	80–89	90–99
Physics marks	40–49	2	5	4			
	50–59	3	7	6	2		
	60–69	2	4	8	5	2	
	70–79	1	1	5	7	8	1
	80–89			2	4	6	5
	90–99				2	4	4

Solution: Here we are working with binned data, but for the whole population, that is, $N = 100$ students. The variances and covariance are easiest to calculate from formulas analogous to (1.6) and (1.12b), but for binned data. For the population, using x for the mathematics marks and y for the physics mark, these are:

$$\sigma_x^2 = \frac{1}{N}\sum_{i=1}^{6} f_i^{(x)} x_i^2 - \frac{1}{N^2}\left(\sum_{i=1}^{6} f_i^{(x)} x_i\right)^2, \quad \sigma_y^2 = \frac{1}{N}\sum_{i=1}^{6} f_i^{(y)} y_i^2 - \frac{1}{N^2}\left(\sum_{i=1}^{6} f_i^{(y)} y_i\right)^2,$$

and

$$\mathrm{cov}(x,y) = \frac{1}{N}\sum_{i,j=1}^{6} h_{ij}\, x_i y_j - \frac{1}{N^2}\left(\sum_{i=1}^{6} f_i^{(x)} x_i\right)\left(\sum_{i=1}^{6} f_i^{(y)} y_i\right),$$

where

$$f_i^{(x)} = \sum_{j=1}^{6} h_{ij} \quad \text{and} \quad f_j^{(y)} = \sum_{i=1}^{6} h_{ij}$$

and h_{ij} is the frequency in the individual bin corresponding to (x_i, y_i); that is, $f_i^{(x)}\left(f_j^{(y)}\right)$ is the total frequency of the bin having a central value x, (y_i). Using the frequencies in the table gives $\sigma_x^2 = 206.51$, $\sigma_y^2 = 220.91$, and $\mathrm{cov}(x,y) = 153.21$. Hence the correlation coefficient is $\rho = \mathrm{cov}(x,y)/(\sigma_x \sigma_y) = 0.72$. As one would expect, ability in maths and physics tend to 'go together'.

1.5 Large samples

If, as is usually done for a bar chart, the height of each bar in a histogram is equal to the frequency of each bin, the total area under the histogram is equal to the total number of entries n multiplied by the bin width $\triangle x$. Thus the histogram may be normalized to unit area by dividing each entry by the product of the bin width (assumed for convenience to be all equal) and the total number of entries, as was done for the data shown in Figure 1.2. As the number of entries increases and the bin widths are reduced, the normalized histogram frequently approximates to a smooth curve, and in the limit that the bin width tends to

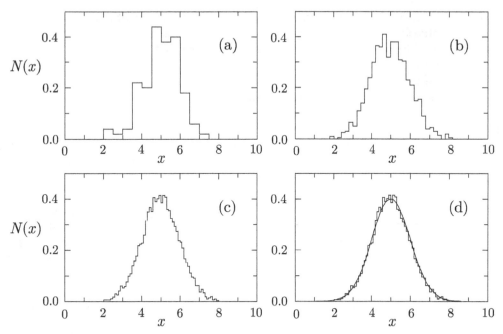

FIGURE 1.5 Normalized histograms $N(x)$ obtained by observations of a random variable x: (a) $n = 100$ observations and bin width $\Delta x = 0.5$; (b) $n = 1000$, $\Delta x = 0.2$; (c) $n = 1000$, $\Delta x = 0.1$; (d) as for (c), but also showing the density function $f(x)$ as a smooth curve.

zero and the number of events increases, the resulting continuous function $f(x)$ is called a *probability density function*, abbreviated to *pdf*, mentioned earlier, or simply a *density function*. This is illustrated in Figure 1.5, which shows the results of repeated measurements of a quantity, represented by the random variable x. The three normalized histograms (a), (b), and (c) show the effect of increasing the number of measurements and at the same time reducing the bin width. Figure 1.5(d) shows the final normalized histogram, together with the associated density function $f(x)$.

The properties of density functions will be discussed in detail in Chapter 3, but one thing worth noting here is that $f(x)$ is very often a symmetrical *normal distribution*, also known as the Gaussian function in physical science, like that shown in Figure 1.5(d), the former name indicating its central importance in statistical analysis.[13] If the function $f(x)$ is of approximately normal form, an empirical rule is that:

(1) approximately 68.3% of observations lie within 1 sample standard deviation of the sample mean;
(2) approximately 95.4% of observations lie within 2 sample standard deviations of the sample mean;
(3) approximately 99.7% of observations lie within 3 sample standard deviations of the sample mean.

[13]The normal distribution was mentioned in Section 1.3 and is discussed in detail in Section 4.1.2.

These results could be used in principle to interpret the results of experiments like that in Example 1.4, although in that case with only 10 events, the distribution of observations is unlikely to closely approximate a normal distribution. Note that the bounds above confirm the remarks made in Section 1.4.2 about Chebyshev's inequality.

The question of to what extent a set of measurements may be assumed *a priori* to be normally distributed is an interesting and important one. It is sometimes remarked that physical scientists make this assumption because they believe that mathematicians have proved it and that mathematicians assume it because they believe that it is experimentally true. In fact, there is evidence both from mathematics and experiments that the approximation is frequently very good, but this is not universally true. In later chapters we will discuss the circumstances in which one can be confident about the assumption, and those where it is provably wrong.

A problem with the standard deviation defined in (1.6) as a measure of spread is that because the terms in the definition of the variance are squared, its value can be strongly influenced by a few points far from the mean. For this reason, another measure sometimes used is related to the form of the pdf. This is the *full width at half maximum height (FWHM)*, often (rather confusingly) called the *half-width*, which is easily found by measuring the width of the distribution at half its maximum height. This quantity depends far less on values a long way from the maximum, that is, data points in the tails of the distribution. For an exact normal distribution, the half-width is 2.35σ. (See Problem 1.10.)

The above discussion is for the repeated measurement of a quantity, but not all histograms lead to an approximately normal distribution (see, e.g., the histogram in Figure 1.2). This may be because large samples are not available, or the phenomenon studied may have an intrinsically different distribution. In these cases it may be useful to have a function that can be used to interpolate the data. This was mentioned briefly in Section 1.3 when discussing a mathematical form to interpolate the data in a histogram. As before, the form of such a function, usually referred to as a *kernel*, should contain features that approximate important features of the data and contain parameters that can be varied to fit the histogram more accurately. Thus, for example, the kernel might have more than one maximum and a parameter, called the *band width*, that plays the role of the bin width in the histogram. Several specific functions have been suggested that might be suitable, depending on the form of the histogram. This approach leads to so-called *kernel density estimates* and is essentially the same as that mentioned in Section 1.3, but there the fitting function was tailored to fit the data. For example, in fitting data containing a resonance signal in the presence of a nonresonance background, it would be usual to fit the data using a combination of a standard resonance formula[14] and a polynomial, or a piece-wise polynomial (the latter also called a spline function).

Example 1.12

The table below shows the lifetimes in days of 30 samples of an electronic component. Calculate the sample mean $\bar{x}$ and the sample standard deviation s. What percentage of the data lie in the range $(\bar{x} \pm 1.5s)$? Compare

---

[14]In physics-related science this is called the Breit-Wigner formula. In statistics it is known as the Cauchy distribution. It will be discussed in Section 4.1.6.

your answer with the prediction of Chebyshev's bound, and the prediction of 87% if the data were distributed as a normal distribution.

346	372	395	355	362	380	386	371	356	349
396	390	370	361	372	365	348	387	382	391
369	367	372	366	379	375	393	374	350	383

Solution: Using Equations 1.2(a) and 1.5, give $\bar{x} = 372.07$ and $s = 14.57$ so that

$$\bar{x} - 1.5s = 350.14 \text{ and } \bar{x} + 1.5s = 393.93.$$

Six data values, corresponding to 20% of the total data, lie outside these bounds. Thus 80% lie within 1.5 standard deviations from the mean. This is close to the prediction from the normal distribution, whereas Chebyshev's bound of at least 56% is weaker.

1.6 Experimental errors

In making inferences from a set of data it is essential that experimenters are able to assess the reliability of their results. Consider the simple case of an experiment to measure a single parameter, the length of a rod. The rod clearly has a true length, although unknown, so the results of the experiment would be expressed as an average value, obtained from a sample of measurements, together with a precise statement about the relationship between it and the true value, that is, a statement about the experimental uncertainty, or 'error'. Without such a statement, the measurement has little value. The closeness of the measured value to the true value defines the *accuracy* of the measurement, and an experiment producing a measured value closer to the true value than that of a second experiment is said to be more accurate.

For the experiments discussed in earlier sections, errors were only briefly mentioned because, in such very simple cases, they are usually irrelevant, but measurement errors, of which there are several types, are a fundamental challenge in conducting many experiments. The first is a simple mistake; for example, a reading of 23 from a measuring device may have been recorded incorrectly as 32. These types of errors usually quickly reveal themselves as gross discrepancies with other measurements, particularly if data are continually recorded and checked during the experiment. By repeating the measurement, they can usually be eliminated. The second is *bias* that we met in Section 1.1 when discussing investigating the skills of different teachers. This is more usually called a *systematic error*, and as mentioned earlier (where it was referred to as a *nonrandom error*), it is one of the aims of a well-designed experiment to eliminate such errors as far as possible. A third type is called a *random (or statistical) error* and is intrinsic to the measuring process and cannot be totally removed but can also sometimes be minimized by careful experimental design.

Statistical errors occur when a measurement is repeated multiple times, but the results may vary slightly because of the limitations of the measuring equipment, even when the quantity being measured is unchanging. As a very simple example, consider measuring the length of a rod by using a meter rule. If the end of the rod does not lie exactly on one of the marks on the rule, the experimenter will have to estimate what fraction of the distance between two adjacent marks is indicated by the position of the end of the rod. If it is equally likely that the experimenter will overestimate, or underestimate, this distance, the resulting error is random. Analogous errors are also present in realistic experiments, such as those that involve electronic counting equipment, where small random fluctuations in electronic detectors may occur. Mathematical statistics is mainly concerned with the analysis of statistical errors. One general result that we have assumed earlier, and that will be made more precise in later chapters, is that statistical errors can be reduced by accumulating larger quantities of data, that is, taking more readings.

The statistical error on a measurement is a measure of its *precision*. Denoting the measurement as y and the statistical error as $\triangle$, the result of an experiment is expressed as $y \pm \triangle$. An experiment with a smaller statistical error is said to be more *precise* than one with a larger statistical error. In statistics 'precision' and 'accuracy' are not the same. This is illustrated in Figure 1.6, which shows a set of measurements made at different values of x of a quantity y that is known to be a linear function of x, as shown by the straight line. The data in Figure 1.6(a) are more precise than those in Figure 1.6(b) because they have smaller errors, as shown by the error bars. (These are the vertical lines of length $2\triangle$ drawn vertically through the data points to show the range of values $y \pm \triangle$.) But the data clearly show a systematic deviation from the straight line that gives the known dependence of y on x. The data in graph (b) have larger error bars and so are less precise, but they are scattered

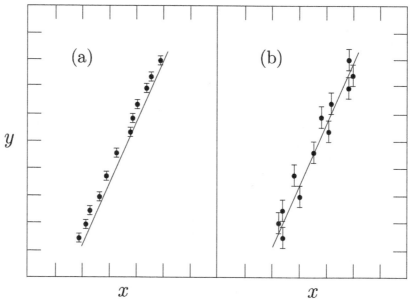

FIGURE 1.6 Illustration of the difference between precision and accuracy. The measurements in (a) are more precise, but less accurate, than those in (b).

about the line and are a better representation of the true relationship between x and y. Thus they are more accurate.

Simple historical examples of systematic and statistical error are found in the early attempts of astronomers to determine the orbits of the planets by measuring the angle the planet made to the horizon, at the same time of night, over many nights. One astronomer could make several measurements on one night and get slightly different results for each, mainly due to slightly different alignment of the eye with the instrument when targeting the horizon, and again when targeting the planet. Two different astronomers using different instruments would get different results, not only because of this statistical alignment error but also because of small differences between their instruments. The small, but unavoidable, errors in position of the markings used to measure the angles also contribute to systematic error in the measurements.

A modern example from the field of elementary particle physics concerns experiments to measure a property of a particle called the *muon*, which is a heavier version of the electron. Like the electron, the muon has an electric charge, and a magnetic moment, analogous to that possessed by a simple iron bar magnet, but vastly smaller in magnitude. Quantum theory predicts the value of the muon's magnetic moment, and it can also be measured experimentally. The calculated theoretical value, and the experimentally measured value obtained in a large experiment, agreed to at least 1 part in 10^6, but because of the extreme smallness of the magnetic moment, the difference was still about 3 standard deviations and, if confirmed, would be very significant,[15] as it would be evidence for new quantum theories beyond the existing ones, and the existence of hitherto unknown particles, thus changing our understanding of the universe. To establish whether this apparent discrepancy was real, a major new effort was made to calculate the theoretical value, which involved several years of work using the largest supercomputers. At the same time, a new experiment was commissioned to improve the determination of the experimental value. This involved taking far more measurements than in the original experiment, to reduce the statistical error, and carefully examining all sources of possible systematic errors that might originate from a variety of sources, such as the exact value and stability of the very strong magnetic fields used in the experiment. In 2021, after several years of work, both groups announced their findings: the result from the new experiment was consistent with the earlier measurement, but there was still a discrepancy with the theoretical calculation. Taking account of the experimental values from both experiments, the discrepancy is now 4.2 standard deviations. Thus the existence of a new force of nature remains a possibility.

Systematic errors are a potentially serious problem, because you can never be sure that all of them have been taken into account. There is no point in producing a very precise measurement by taking more data to reduce the statistical error, if the systematic error is larger. This would only lead to a spurious precision. The deviation of the data in Figure 1.6(a) from the true values is an indication of the presence of a systematic error. There are many possible sources of these, which may, or may not, be known and they are by no means so obvious as Figure 1.6 might suggest. If a meter rule is used to measure the rod, it may have been wrongly calibrated during its manufacture so that each scale unit, for example, 1 mm, is smaller than it should be. In a more realistic case where an experiment counts the number

---

[15]Recall the previous discussion of the normal distribution in Section 1.5.

of particles emitted by a radioactive isotope, the detectors could also have been wrongly calibrated, but in addition, the source might contain other isotopes, or the detectors may be sensitive to particles other than those that identify the decay. In the simple case of measuring the length of a rod, repeating the experiment with another meter rule would reveal the problem and enable it to be eliminated, but in real situations using substantial equipment this may well not be possible. It is therefore important to try to constrain the range of systematic uncertainties, as was done in the muon experiment mentioned above. Several different strategies may be used, and computers are often used to simulate different operating conditions of the experiment and, by generating 'pseudodata' under these conditions, to examine the range of systematic errors generated. A brief discussion of this approach, called *simulation*, is given in Section 5.5.2.

One of the skills of a good experimentalist is to anticipate possible sources of systematic errors and, if possible, design them out at the planning stage of the experiment. Those that cannot be eliminated must be carefully investigated and taken into account in the final estimation of the overall error. In practice, it is better for clarity to quote the statistical (random) and systematic (nonrandom) errors separately, by writing $x \pm \triangle_R \pm \triangle_S$, where the subscripts stand for 'random' and 'systematic', respectively. This also allows a new source of systematic error to be incorporated into the results, should one be revealed later. The number of significant figures quoted for the measurement should be one more than that dictated by the experimental precision to avoid arithmetical errors that might be caused by rounding errors in later calculations that use the result.

Books on statistics usually have little to say about systematic errors, or ignore them altogether, because, in general, a full mathematical treatment of them cannot be made. But in the real world of science we do not have the luxury of ignoring this type of error, and a limited analysis is possible in some circumstances, particularly if the systematic effect is the same for all data points, or its dependence on the measuring process is known, as is often the case. We will return to this point in Section 5.4 and Section 8.1.1, when we discuss how to combine data from different experiments and take account of outliers mentioned previously in Section 1.4.1.

Problems 1

1.1 An experiment consists of tossing a die followed by tossing a coin (both unbiased). The coin is tossed once if the number on the die is odd and twice if it is even. If the order of the outcomes of each toss of the coin is taken into account, list the elements of the sample space.

1.2 Five students, denoted by S_1, S_2, S_3, S_4, and S_5, are divided into pairs for laboratory classes. List the elements of the sample space that define possible pairings.

1.3 The table gives the examination scores out of a maximum of 100 for a sample of 40 students.

22	67	45	76	90	87	27	45	34	36
67	68	97	73	56	59	76	67	63	45
55	59	90	82	74	34	68	56	53	68
28	39	43	66	67	59	38	39	56	61

Cast the data in the form of a frequency histogram with eight equally spaced bins. What is the frequency of each bin and the numbers in the bins of the cumulative distribution?

1.4 A histogram of a sample of a variable x has four bins (0,2), (2,5), (5,7), and (7,9), with heights 0.25, 0.10, 0.075, and 0.025, respectively. Calculate the value of the empirical distribution function at $x = 6$.

1.5 Calculate the median and the percentile $P_{0.67}$ for the unbinned data of Problem 1.3.

1.6 Use the data of Problem 1.3 to compute the sample mean $\bar{x}$ and the sample standard deviation s, both for the binned and unbinned data. How would Shepard's corrections change the results? What percentage of the unbinned data falls within $\bar{x} \pm 2s$? Compare this with the predictions that would follow if the data were approximately normally distributed.

1.7 A sample of six data $x_i (i = 1, 6)$ has a mean value of 3.1 and a variance of 2.0. If the first four data values are $x_1 = 1.0$, $x_2 = 2.1$, $x_3 = 3.2$, and $x_4 = 4.0$, what are the values of x_5 and x_6?

1.8 Verify that the second moment of a population about an arbitrary point λ is given by $\mu_2' = \sigma^2 + d^2$, where $d = \mu - \lambda$, and μ and σ^2 are the mean and variance, respectively, of the population.

1.9 A group of 20 students, 10 males and 10 females, attend a short week-long course on a physics topic. In addition to lectures, they are told that they would be expected to do some additional work on the topic in their own time (homework), and at the end of the course, they will all sit an examination. The table below shows, for each member of the two groups, how much time they spent on homework during the week, and their final examination mark.

Female students	1	2	3	4	5	6	7	8	9	10
Homework hours (h)	2	5	4	3	4	3	6	1	1	3
Exam mark M (%)	40	76	65	40	76	41	82	30	25	38

Male students	1	2	3	4	5	6	7	8	9	10
Homework hours (h)	7	4	4	1	2	3	4	5	3	2
Exam mark M (%)	85	64	63	35	38	38	47	45	55	25

What can be deduced from the data?

1.10 Show that for the normal (Gaussian) density

$$f(x) = \frac{1}{\sigma\sqrt{2\pi}}\exp\left[-\frac{(x-\mu)^2}{2\sigma^2}\right],$$

the half-width is equal to 2.34σ.

1.11 One measure of the skewness of a population is the parameter

$$\gamma = \frac{1}{N\sigma^3}\sum_{i=1}^{N}(x_i - \mu)^3.$$

Show that this may be written

$$\gamma = \frac{1}{\sigma^3}\left(\overline{x^3} - 3\bar{x}\overline{x^2}\right) + 2\bar{x}^3,$$

where the overbars denote averages over the population.

1.12 The table below shows the values of a random sample of 40 numbers drawn from an unknown distribution. Is there evidence that would support the view that the data are distributed as a normal distribution?

32	38	40	41	44	46	47	51	53	53	54	55	56	57	58	61	62	63	63	64
65	65	65	66	67	68	71	72	73	73	75	75	77	82	84	85	86	90	93	96

CHAPTER

2

Probability

Overview

This short chapter starts by listing the axioms of probability and is followed by a section showing how these are used in the calculus of probabilities to calculate probabilities, conditional probabilities, and marginal probabilities. The binomial and multinomial formulae are discussed in the context of using permutations to calculate discrete probabilities. The meaning of probability is problematic and there are different views on this, so there is a section discussing the two main interpretations used in physical science: The frequency and subjective (or Bayesian) approaches.

Statistics is closely connected to the branch of mathematics called *probability theory*, and so before the subject of statistics can be meaningfully discussed and used to analyze data, we must say something about probabilities. This chapter therefore starts by defining the notations we will use to describe sample spaces and events. This is followed by a brief review of the axioms of probabilities, and the rules for their application. The meaning of probability is subject to multiple interpretations, but it is important to note that whatever interpretation is adopted, providing it satisfies the axioms, the same rules will apply. The final section examines two interpretations that are widely used in physical sciences.

2.1 Sample spaces and events

Let S denote a sample space consisting of a set of events $E_i (i = 1, 2, ..., n)$,[1] where for specific events, subscripts will be avoided by using the notation A, B, C, etc. If we have two events in S, denoted by A and B, then:

1. The event that includes all the outcomes that are in both A and B is called the intersection of A and B and is denoted by $A \cap B$, or equivalently $B \cap A$. If $A \cap B = \varnothing$, where the symbol $\varnothing$ denotes a sample space with no events (called a null space), the events are said to be mutually exclusive (also called disjoint or distinct).

---

[1] In Chapter 1 a sample space was defined in terms of events, and likewise probabilities are invariably defined in terms of events. For simple events, these are the same as outcomes, but a complex event will involve subsets of outcomes, the probabilities of which are used to find the probabilities of events.

Probability and Statistics for Physical Sciences, Second Edition
https://doi.org/10.1016/B978-0-443-18969-2.00002-7

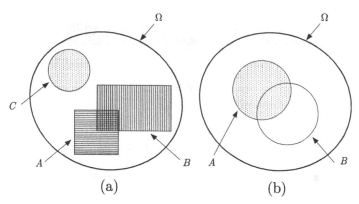

FIGURE 2.1   A Venn diagram. (a) illustration of union and intersection (b) illustration of de Morgan's law.

2. The event that includes the outcomes of A or B, or both, is called the union of A and B and is denoted $A \cup B$. An outcome that is common to both A and B is included in the union just once.

3. If $\overline{A}$ denotes the event 'not A', called the complement of A, then $\overline{A} = S - A$, although other notations are also used for the complement.[2]

These relations are illustrated geometrically in the *Venn diagram* shown in Figure 2.1(a). The sample space S consists of all events within the boundary Ω. A, B, and C are three such events. The doubly shaded area is $A \cap B$; the sum of both the shaded regions of A and B is $A \cup B$, with any common outcomes appearing only once; and the area outside the region occupied by A and B is $(A \cup B)^c$ and includes the area occupied by event C. The latter is disjoint from both A and B, and so $C \cap A = C \cap B = \varnothing$; that is, A and C, and B and C are pairwise mutually exclusive. The following example illustrates these relations.

Example 2.1

A sample space S consists of all the numbers from 1 to 8 inclusive. Within S there are four events:

$$A = (2,4,7,8), \ B = (1,3,5,7), \ C = (2,3,4,5), \text{ and } D = (1,7,8)$$

Construct the content of the events:

$$\overline{A} \cup C, \ B \cap \overline{C}, \ \overline{S \cap \overline{B}}, \ (\overline{C} \cap D) \cup B, (B \cap \overline{C}) \cup A, \text{ and } (A \cap C \cap \overline{D}).$$

Solution: The complements of events A, B, C, and D are

$$\overline{A} = (1,3,5,6), \ \overline{B} = (2,4,6,8), \ \overline{C} = (1,6,7,8), \ \overline{D} = (2,3,4,5,6),$$

and thus the structures of the required events are

$$\overline{A} \cup C = (1,2,3,4,5,6), \ B \cap \overline{C} = (1,7), \ \overline{S \cap \overline{B}} = (1,3,5,7),$$

---

[2]The notations in probability theory are not unique. For example, the complement of A is often written A^c, and other differences in notation, particularly for the union, are common.

and

$$(\overline{C} \cap D) \cup B = (1,3,5,7,8), \quad (B \cap \overline{C}) \cup A = (1,2,4,7,8), \quad A \cap C \cap \overline{D} = (2,4).$$

In Example 2.1 we have used in places a double bar notation, which is equivalent to a double negative, as in the statement: 'It is untrue to say that neither the rain nor the wind caused the damage'. This is equivalent to saying, 'Either the wind, or the rain, or both, caused the damage'. If we denote by W the event 'wind caused the damage' and R the event 'rain caused the damage', then formally $\overline{(\overline{W} \cap \overline{R})} = (W \cup R)$. This is an example of the general results for any two events A and B:

$$\overline{(A \cup B)} = (\overline{A} \cap \overline{B}) \quad \text{and} \quad \overline{(A \cap B)} = (\overline{A} \cup \overline{B}).$$

These are known as De Morgan's laws and may be easily understood by examining Figure 2.1(b).

The discussion so far is relevant to a single experiment. For example, in a single throw of a coin the outcomes are just two, head (H) or tail (T), and the sample space is $\{H, T\}$. But if the 'experiment' is to toss the coin twice, the sample space is the product (mentioned in Example 1.2),

$$\{H,T\} \times \{H,T\} = \{(HH),(HT),(TH),(TT)\}$$

and contains four events.

2.2 Axioms and calculus of probability

Although, as mentioned in Chapter 1, the concept of probability has been around for hundreds of years, it was not until the 20th century that the mathematical axioms of probability were formulated.[3] These may be stated as follows:

1. Every event E_i in a sample space S can be assigned a *number* $P[E_i]$, called the *probability* of E_i, that is, the probability that E_i will occur, and

$$0 \leq P[E_i] \leq 1 . \tag{2.1}$$

2. The sum of the probabilities $P[E_i]$ over all events E_i is unity, that is,

$$P[S] = 1. \tag{2.2}$$

[3]This was done in 1933 in a classic book by Kolmogorov.

3. The probability of the union of any set of mutually exclusive events E_1, E_2, ..., is the sum of the probabilities of those events; that is

$$P[E_1 \cup E_2 \cup E_3 \cdots] = \sum_i P[E_i] = P[E_1] + P[E_2] = P[E_3] \cdots. \tag{2.3}$$

This is called the *additive rule*. If each event is simple, and so coincides with an outcome, this result (2.3) is obviously true, but for complex events, some explanation is needed, because such events will contain combinations of outcomes. In general, events A and B are not disjoint, and in this case

$$P[A \cup B] = P[A] + P[B] - P[A \cap B], \tag{2.4}$$

which can be seen geometrically from the Venn diagram Figure 2.1(b), where it is clear that the sum of the probabilities of A and B includes twice the probability of the intersection of A and B. The algebraic proof of (2.4) is straightforward and is given in Example 2.2. Two results that follow from the axioms are:

1. If $A \subset B$, that is, A is a subset of B, then $P[A] \leq P[B]$ and $P[B - A] = P[B] - P[A]$.
2. $P[\overline{A}] = 1 - P[A]$, then for any two events A and B, A may be written in the form

$$A = (A \cap B) + (A \cap \overline{B}).$$

Example 2.2

Derive Equation (2.4).
Solution:
Using (2.3), we can write the probability of the event A in the form

$$P[A] = P[A \cap B] + P[A \cap \overline{B}]. \tag{2.5a}$$

If this procedure is repeated for the event $(A \cup B)$, we get

$$P[A \cup B] = P[(A \cup B) \cap B] + P[(A \cup B) \cap \overline{B}],$$

but

$$(A \cup B) \cap B = B \text{ and } (A \cup B) \cap \overline{B} = (A \cap \overline{B}),$$

so

$$P[A \cup B] = P[B] + P(A \cap \overline{B}). \tag{2.5b}$$

Finally, eliminating $P[A \cap \overline{B}]$ from (2.5a) and (2.5b) gives the probability of the union of A and B as

$$P[A \cup B] = P[A] + P[B] - P[A \cap B]. \tag{2.6}$$

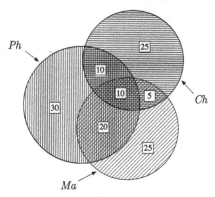

FIGURE 2.2 Venn diagram for Example 2.3.

Example 2.3

A class of 100 students are studying physical sciences, and each must make a choice of courses from physics (Ph), chemistry (Ch), and mathematics (Ma). As a result, 30 choose to take physics and mathematics, 20 take physics and chemistry, 15 take chemistry and mathematics, and 10 take all three courses. If the total numbers of students taking each subject are 70 (Ph), 60 (Ma), and 50 (Ch), find the probabilities that a student chosen at random from the group will be found to be taking the specific combinations of courses: (1) physics but not mathematics, (2) chemistry and mathematics, but not physics, and (3) neither physics nor chemistry.

Solution: The probabilities may be found using the formulas given above, which give:

$$\text{(a) } P\left[Ph \cap \overline{Ma}\right] = P[Ph] - P[Ph \cap Ma] = 0.70 - 0.30 = 0.40,$$

$$\text{(b) } P\left[Ch \cap Ma \cap \overline{Ph}\right] = P[Ch \cap Ma] - P[Ch \cap Ma \cap Ph] = 0.15 - 0.10 = 0.05,$$

$$\text{(c) } P\left[\overline{Ph} \cap \overline{Ch}\right] = P[Ma] - P[Ch \cap Ma] - P[Ph \cap Ma] + P[Ph \cap Ch \cap Ma] = 0.06 - 0.15 - 0.30 + 0.10 = 0.25.$$

They can also be found by constructing the Venn diagram shown in Figure 2.2.

2.3 Conditional and marginal probabilities

In this section some relations involving multiple events are introduced. Firstly, if the sample space contains two subsets, A and B, then provided $P[B] \neq 0$, the probability of the occurrence of A, *given that B has occurred*, is called the *conditional probability of A*, written $P[A|B]$, and is given by

$$P[A|B] = \frac{P[A \cap B]}{P[B]}, \quad P[B] \neq 0. \tag{2.7a}$$

Some authors take (2.7a) to be the definition of conditional probability, but it also follows from the intuitive interpretation of probability that most scientists use in practice. This is called the *frequency interpretation*,[4] which we have used in earlier examples. Thus if an experiment is repeated many times, the proportion of repetitions in which the event B occurs is interpreted as approximately $P[B]$, and the proportion in which both A and B occur is $P[A \cap B]$. Therefore, among the repetitions in which B occurs, the proportion of repetitions where A also occurs is given approximately by (2.7a).

If the occurrence of A does not depend on the fact that B has occurred, that is,

$$P[A|B] = P[A], \tag{2.7b}$$

then the event A is said to be *independent* of the event B. (Note that independence is not the same as being distinct.) An important result that follows in a simple way from the previous relations is the *multiplicative rule*, which follows from rewriting (2.7a) as

$$P[A \cap B] = P[B]P[A|B] = P[A]P[B|A], \tag{2.8a}$$

and reduces to

$$P[A \cap B] = P[A]P[B], \tag{2.8b}$$

if A and B are independent. This may be generalized in a straightforward way. For example, if A, B, and C are three events, then

$$P[A \cap B \cap C] = P[A]P[B|A]P[C|A \cap B]. \tag{2.8c}$$

Finally, if the event A must result in one of the mutually exclusive events B, C, ..., then

$$P[A] = P[B]P[A|B] + P[C]P[A|C] + ... \tag{2.9}$$

Use of these various results is illustrated in the following two examples.

Example 2.4

A student takes a multiple-choice exam where each question has $n = 4$ choices from which to select an answer. If $p = 0.5$ is the probability that the student knows the answer, what is the probability that a correct answer indicates that the student really did know the answer and that it was not just a 'lucky guess'?

Solution: Let Y be the event where the student answers correctly and let $+$ and $-$ be the events where the student knows and does not know the answer, respectively. Then we need to find $P[+|Y]$, which from (2.7a) is given by

$$P[+|Y] = \frac{P[+ \cap Y]}{P[Y]}.$$

[4]Interpretations of probability are discussed in more detail in Section 2.5.

From (2.7a) the numerator is

$$P[+ \cap Y] = P[+]P[Y|+] = p \times 1 = p,$$

and from (2.8a), the denominator is

$$P[Y] = P[+]P[Y|+] + P[-]P[Y|-] = p + (1-p) \times \frac{1}{n}.$$

So, finally,

$$P[+|Y] = \frac{p}{p + (1-p)/n} = \frac{np}{1 + (n-1)p}.$$

Thus for $n = 4$ and $p = 0.5$, the probability is 0.8 that the student gave the correct solution because they really did know the answer to the question.

Example 2.5

What is the probability that four people chosen at random were born in different months of the year?

Solution: Whatever the birthday month of the first person, to satisfy the requirement the second person cannot be born in the same month. So, suppose M_2 is this event:

$$P[M_2] = \left(1 - \frac{1}{12}\right).$$

If we now define A_3 as the event where the third person has a birthday month that is not the same as those of the first two people, then using the multiplicative rule, we can write

$$P[M_3] = P[A_3 \cap M_2] = P[A_3|M_2]\, P[M_2],$$

where

$$P[A_3|M_2] = \left(1 - \frac{2}{12}\right),$$

and so

$$P[M_3] = \left(1 - \frac{2}{12}\right)\left(1 - \frac{1}{12}\right).$$

Continuing this sequence yields

$$P[M_4] = \left(1 - \frac{3}{12}\right)\left(1 - \frac{2}{12}\right)\left(1 - \frac{1}{12}\right) = 0.573,$$

that is, approximately 57%.

These ideas can be generalized to the situation where an event can be classified under multiple criteria. Consider, for example, the case of three classifications. If the classifications under the criteria are

$$A_1, A_2, ..., A_r; B_1, B_2, ..., B_s; \text{and } C_1, C_2, ..., C_t;$$

with

$$\sum_{i=1}^{r} P[A_i] = \sum_{i=1}^{s} P[B_i] = \sum_{i=1}^{t} P[C_i] = 1,$$

then a table of the possible values of the three random variables, together with their associated probabilities, defines the *joint probability* of A, B, and C. The *marginal probability* of A_i and C_k is then defined as

$$P[A_i \cap C_K] \equiv \sum_{j=1}^{s} P[A_i \cap B_j \cap C_k] \tag{2.10a}$$

and likewise, the marginal probability of C_k is

$$P[C_k] \equiv \sum_{i=1}^{r} \sum_{j=1}^{s} P[A_i \cap B_j \cap C_k] = \sum_{i=1}^{r} P[A_i \cap C_k]$$

$$= \sum_{j=1}^{s} P[B_j \cap C_k] = \sum_{j=1}^{s} P[B_j | C_k] P[B_j], \tag{2.10b}$$

where (2.7a) has been used in the last expression. This result is known as the *law of total probability* and is a generalization of (2.9).

2.4 Permutations and combinations

Calculation of probabilities often involves determining the number of ways elements can be selected from a set or the number of possible arrangements, or ordering, of elements in a sequence. Each distinct arrangement is called a *permutation* and the basic theorem of permutations gives us an easy way to calculate numbers of possible selections and arrangements. The number of permutations of m objects selected from n distinct objects is

$$_nP_m \equiv \frac{n!}{(n-m)!} \tag{2.11}$$

where $n!$, called *n factorial*, is defined as $n! \equiv n(n-1)(n-2)\cdots 1$, with $0! \equiv 1$. The proof of (2.11) is quite simple. There are n ways to select the first object, $n-1$ ways to select the second, and so on. The number of objects remaining available from which the mth object can be selected is obviously $n - m + 1$. The total number of ways to select the m objects is just the product of the numbers of ways of selecting each object, $n(n-1)(n-2)\cdots(n-m+1)$, which is the same as the right-hand side of (2.11).

Often the order in which the m objects are selected is not important. To determine the number of ways of selecting m objects from n objects, without regard to the order of selection, we must divide $_nP_m$ by the number of permutations possible for m specific objects. Since there are m choices for which of the m objects is selected first, $m - 1$ choices for which is selected second, and so on, with only one choice for which of the m objects is selected last, the number of permutations is $m!$. The total number of *combinations* of m objects taken from a set of n objects, without regard to arrangement is then

$$_nC_m = \binom{n}{m} = \frac{_nP_m}{m!} = \frac{n!}{m!(n-m)!},$$

(2.12)

that is, the coefficient of x^m in the binomial expansion of $(1+x)^n$.

The coefficients of the multinomial expansion of $\left(\sum_{i=1}^{k} x_i \right)^n$, for k variables $x_1, x_2, \ldots, x_k$, also play an important role in calculating combinations of objects and events. Suppose we have a set of n objects, and each object can exhibit one of k distinct and mutually exclusive characteristics. On measuring the characteristics, we find n_i of the objects have the ith characteristic, with $n = n_1 + n_2 + \ldots + n_k$. The number of ways the objects can take on the characteristics with the result being n_1 having the first characteristic, n_2 having the second, and so on, without regard to the order in which the objects are measured, is easily determined by using (2.12). There are $_nC_{n1}$ ways to select the first n_1 objects and $_{(n-n_1)}C_{n_2}$ ways to select the set of n_2 objects. Continuing in this way, and multiplying all the numbers of combinations together, we find

$$_nP_{n_1,n_2,\ldots n_k} = \frac{n!}{n_1!n_2!\ldots n_k!}.$$

(2.13)

Example 2.6

Nine students each donate a textbook to a traveling library. The library needs physics, chemistry, and mathematics books and will not accept other subjects. The students donate in total four physics books, three chemistry books, and two mathematics books. What is the total number of possible ways the students could have chosen their donations to achieve this result?

Solution: The number of different arrangements N can be found by applying (2.13) with $n_p = 4, n_c = 3, n_m = 2$, and $n = n_p + n_c + n_m = 9$. Thus

$$N = {}_nP_{n_p, n_c, n_m} = \frac{9!}{4!3!2!} = 1260.$$

Example 2.7

A committee of four people is to be selected at random from a group of six physicists and nine engineers. What is the probability that the committee consists of two physicists and two engineers?

Solution: There is a total of $_{15}C_4$ possible choices of two people for the committee and each choice has an equal probability of being chosen. But there are $_6C_2$ possible choices of two physicists from the six in the group, and $_9C_2$ possible choices of two engineers from the nine in the group. So the required probability is

$$\frac{_6C_2 \times \,_9C_2}{_{15}C_4} = \frac{540}{1365} = 0.396.$$

2.5 The meaning of probability

The axioms and definitions discussed so far specify the rules that probabilities satisfy and can be used to calculate the probability of complex events from the probabilities of simple events, but they do not tell us how to assign specific probabilities to actual events. Mathematical statistics proceeds by assigning a *prior* probability to an event on the basis of a given mathematical model (specified by known parameters) about the possible outcomes of the experiment. In physical situations, even if the mathematical form is known, its parameters rarely are, and one of the prime objectives of statistical analysis is to obtain values for them when there is access to only incomplete information. Without complete knowledge, we cannot make absolutely precise statements about the correct mathematical form and its parameters, but we can make less precise statements in terms of probabilities. So we now turn to examine in more detail what is meant by the word 'probability'.

2.5.1 Frequency interpretation

We all use the word 'probability' intuitively in everyday language. We say that an unbiased coin when tossed has an equal probability of coming down 'heads' or 'tails'. What we mean by this is that if we were to repeatedly toss such a coin, we would expect the average number of heads and tails to be very close to, but not necessarily exactly equal to, 50%. Thus we are adopting a view of probability operationally defined as *the limit of the relative frequency of occurrence*. While this is a common-sense approach to probability, it does have an element of circularity. What do we mean by an unbiased coin? It is presumably one that when tossed many times tends to give an equal number of heads and tails! We have already used the word 'random' in this context in Chapter 1 when discussing statistical errors, and again in using the words 'equally likely' when discussing the example of an experiment to test lecturers' teaching capabilities.

The *frequency definition of probability* may be stated formally as follows. In a sequence of n trials of an experiment in which the event E of a given type occurs n_E times, the ratio $R[E] = n_E/n$ is called the relative frequency of the event E, and the probability $P[E]$ of the event E is the limit approached by $R[E]$ as n increases indefinitely, with it being assumed that this limit exists.

Of course, it is impossible to repeat every experiment indefinitely, and usually impractical to repeat any experiment a very large number of times. Usually, an assumption is made about the probabilities of outcomes and calculation proceeds as if that assumption was the result of many experiments. Quite often, the probabilities are assumed to take the Gaussian distribution mentioned in Chapter 1. When no information is available against it, we often assume that each simple event (outcome) in a sample space has the same probability of occurring.

For example, if we return to the sample space of Example (1.2), and denote by X the number of heads obtained in a single throw, then

$$X(E_1) = 2; \quad X(E_2) = 1; \quad X(E_3) = 1; \quad X(E_4) = 0,$$

and we can calculate $P[X]$ using the frequency approach as follows, assuming each simple event $X(E_i)$ has equal probability:

$$P[X = 2] = \frac{1}{4}; \quad P[X = 1] = \frac{1}{2}; \quad P[X = 0] = \frac{1}{4}.$$

It is worth noting that this approach allows us to avoid the problem of calculating the mathematical limit for the frequentist definition: That for some arbitrary small quantity ε, there exists a large n, say n_L, such that $|R[E] - P[E]| < \varepsilon$ for all $n > n_L$. The frequency definition has an element of uncertainty in it, because in practice only a finite number of trials can ever be made. The frequency way of approaching probability is essentially experimental. A probability, referred to as the *posterior* probability, is assigned to an event on the basis of experimental observation. A typical situation that occurs in practice is when a model of nature is constructed and from the model certain *prior* probabilities concerning the outcomes of an experiment are computed. These computations require assumptions in the model about the probabilities of events. The experiment is then performed, and from the results obtained *posterior* probabilities are calculated for the same events using the frequency approach.[5] The correctness of the model is judged by the agreement of these two sets of probabilities and on this basis, modifications may be made to the model. These ideas will be put on a more quantitative basis in later chapters when we discuss estimation and the testing of hypotheses.

Most physical scientists would claim that they use the frequency definition of probability, and this is what has been used in the previous sections and examples, but it is not without its difficulties, as we have seen. It also must be used in context and the results interpreted with care. A much-quoted example that illustrates this is where an insurance company analyzes the death rates of its insured men and finds there is about a 1% probability of them dying at age 40. This does *not* mean that a particular insured man has a 1% probability of dying at this age. For example, he may be a regular participant in a hazardous sport, or having a dangerous occupation. So had an analysis been made of members of those groups, the probability of his death at age 40 may well have been much greater. Another example is where canvassers questioning people on a busy street claim to deduce the 'average' view of the

---

[5]The *prior* and *posterior* probabilities were formerly, and sometimes still are, called the *a priori* and *a posteriori* probabilities.

population about a specific topic. Even if the sample of subjects approached really is random (a dubious assumption), the outcome would only be representative of the people who frequent that particular street at that particular time and day and may well not represent the views of people using other streets, or in other towns, at different times of the day.

Crucially, the frequency approach ignores prior information. Consider, for example, a situation where several experiments have measured a certain quantity. As expected, all of them find slightly different values, but all are consistent with each other within the errors of each experiment. A new experiment is then built with the aim of measuring the quantity more accurately. In the analysis of the data it would be reasonable to incorporate in some way the results of the earlier experiments, and we will briefly return to this in Section 8.4.3, where we discuss Bayes' estimators.

The frequency approach also assumes the repeatability of experiments, under identical conditions and with the possibility of different outcomes, for example, tossing a coin many times. So what are we to make of an everyday statement such as 'It will probably rain tomorrow', when there is only one tomorrow? Critics also argue that quoting the result of the measurement of a physical quantity, such as the mass of a body, for example, as (10 ± 1) kg, together with a statement about the probability that the quantity lies within the range specified by the uncertainty of 1 kg, is incompatible with the frequency definition. This is because the quantity measured presumably does have a true value and so either it lies within the error bars or it does not; that is, the probability is either 1 or 0. These various objections are addressed in the next section.

2.5.2 Subjective interpretation

The calculus of probabilities as outlined above proceeds from the definition of probabilities for simple events to the probabilities of more complex events. In practice, what is required in physical applications is the inverse; that is, given certain experimental observations, we would like to deduce something about the parent population and the generating mechanism by which the events were produced. This, in general, is the problem of *statistical inference* alluded to in Chapter 1.

To illustrate how this leads to an alternative interpretation of probability, we return to the definition of conditional probability, which can be written using (2.8a) as

$$P[B \cap A] = P[A]P[B|A]. \tag{2.14a}$$

Since $A \cap B$ is the same as $B \cap A$, we also have

$$P[B \cap A] = P[B]P[A|B], \tag{2.14b}$$

and by equating these two quantities, we deduce that

$$P[B|A] = \frac{P[B]P[A|B]}{P[A]}, \tag{2.15}$$

provided $P[A] \neq 0$. Finally, we can generalize to the case of multiple criteria and use the law of total probability (2.10b) to write

$$P[A] = \sum_{j}^{n} P[A \cap B_j] = \sum_{j}^{n} P[B_j]P[A|B_j],$$

where $B_j(j = 1, n)$ is a set of n mutually exclusive and exhaustive hypotheses (i.e., all possible hypotheses are included in the set) so that

$$P[B_i|A] = \frac{P[B_i]P[A|B_i]}{\sum_{j}^{n} P[B_j]P[A|B_j]}. \tag{2.16}$$

This result was first published in the 18th century by an English Clergyman, Thomas Bayes, and is known as Bayes' theorem. It differs from the frequency approach to probability by introducing an element of subjectivity into the definition — hence its name: *Subjective probability*.

In this approach the sample space is interpreted as the set B_i of n mutually exclusive and exhaustive hypotheses. Suppose an event A can be explained by the mutually exclusive hypotheses represented by $B_1, B_2, ..., B_n$. These hypotheses have certain *prior* probabilities $P[B_i]$ of being true. Each of them can give rise to the occurrence of the event A, but with distinct probabilities $P[A|B_i]$, which are the probabilities of observing A *when B_i is known to be satisfied*. The interpretation of (2.16) is then: The probability of the hypothesis B_i, given the observation of A, is the *prior* probability that B_i is true, multiplied by the probability of the observation assuming that B_i is true, divided by the sum of the same product of probabilities for all allowed hypotheses. In this approach the *prior* probabilities are statements of one's belief that a particular hypothesis is true. Thus in the subjective approach quoting the measurement of a mass as (10 ± 1) kg is a valid statement that expresses one's current belief about the true value of that quantity. The following examples will illustrate the use of Bayes' theorem.

Example 2.8

A football team is in a knockout tournament and will play either team A or team B next depending on their performance in earlier rounds. The manager assesses their prior probabilities of winning against A or B as $P[A] = 3/10$ and $P[B] = 5/10$; and their probabilities of winning given that they know their opponents as $P[W|A] = 5/10$ and $P[W|B] = 7/10$. If the team win their next game, what are the probabilities that their opponents were either A or B?

Solution: If the team wins their next game, we can calculate from (2.16) the probabilities that their opponents were either A or B as $P[A|W] = 3/10$ and $P[B|W] = 7/10$. So the odds favor the hypothesis that the opponents were team B.

Example 2.9

A device D consists of three components $C_i (i = 1, 2, 3)$, each of which functions independently with probabilities $1/2, 1/3$, and $1/5$, respectively, of not being defective. The device works if at least one of the components is not defective. If the device functions correctly, what is the probability that component 2 is functioning?

Solution: Let $P[C_i]$ be the probability that component C_i is functioning and $P[D]$ be the probability that the device is working. The required probability is then $P[C_2|D]$, which from (2.15) is

$$P[C_2|D] = \frac{P[C_2]\, P[D|C_2]}{P[D]}.$$

The probability that the device is working is the probability that not all three components are defective at the same time, which is given by

$$P[D] = \{1 - P[C_1]\}\{1 - P[C_2]\}\{1 - P[C_3]\} = 1 - 0.733$$

Then, using $P[D|C_2] = 1$, and $P[C_2] = 0.333$, gives $P[C_2|D] = 0.4545$.

Examples using the subjective approach are often in agreement with the frequency approach, but the following examples illustrate that Bayes' theorem can sometimes lead to results that at first sight are somewhat surprising.

Example 2.10

The process of producing microchips at a particular factory is known to result in 0.2% that do not satisfy their specification, that is, are faulty. A test is developed that has a 99% probability of detecting these chips if they are faulty. There is also a 3% probability that the test will give a false positive, that is, a positive result even though the chip is not faulty. What is the probability that a chip really is faulty if the test gives a positive result?

Solution: If we denote the presence of a fault by f and its absence by $\bar{f}$, then $P[f] = 0.002$ and $P[\bar{f}] = 0.998$. The test has a 99% probability of detecting a fault if present, so it follows that the test yields a 'false-negative' result in 1% of tests, that is, the probability is 0.01 that the test will be negative even though the chip tested does have a fault. So if we denote a positive test by + and a negative one by −, then $P[+|f] = 0.99$ and $P[-|f] = 0.01$. There is also a 3% probability of the test giving a false positive, that is, a positive result even though the chip does not have a fault, so $P[+|\bar{f}] = 0.03$ and $P[-|\bar{f}] = 0.97$. Then from Bayes' theorem,

$$P[f|+] = \frac{P[+|f]P[f]}{P[+|f]P[f] + P[+|\bar{f}]P[\bar{f}]}$$

$$= \frac{0.99 \times 0.002}{(0.99 \times 0.002) + (0.03 \times 0.998)} = 0.062$$

So the probability of a chip having a fault given a positive test result is only 6.2%.[6]

---

[6]The same reasoning applied to medical test for rare conditions shows that a positive test result often means only a low probability for having the condition.

Example 2.11

An experiment is set up to detect particles of a particular type A in a beam, using a detector that has a 95% efficiency for their detection. However, the beam also contains 15% of particles of a second type B and the detector has a 10% probability of mistakenly recording these as particles of type A. What is the probability that the signal is due to a particle of type B?

Solution: If the observation of a signal is denoted by S, we have

$$P[A] = 0.85, P[B] = 0.15 \text{ and } P[S|A] = 0.95, P[S|B] = 0.10.$$

Then from Bayes' theorem,

$$P[B|S] = \frac{P[B]P[S|B]}{P[A]P[S|A] + P[B]P[S|B]} = \frac{0.15 \times 0.10}{(0.85 \times 0.095) + (0.15 \times 0.10)} = 0.018.$$

Thus there is a probability of only 1.8% that the signal is due to a particle of type B, even though 15% of the particles in the beam are of type B.

If we had to choose a hypothesis from the set B_i, we would choose the one with the greatest posterior probability. But (2.16) shows that this requires knowledge of the prior probabilities $P[B_i]$ and these are, in general, unknown. *Bayes' postulate* is the hypothesis that, in the absence of any other knowledge, the prior probabilities should all be taken as equal. We have seen the assumption of equal probability for all the outcomes in a sample space used in the frequentist approach in the previous section. Bayes' postulate is more general and applies to hypotheses rather than events. The next example illustrates the use of this postulate.

Example 2.12

A container has four balls, which could be either all white (hypothesis 1), or two white and two black (hypothesis 2). If n balls are withdrawn, one at a time, replacing them after each drawing, what are the probabilities of obtaining an event E with n white balls under the two hypotheses? Comment on your answer.

Solution: If n balls are withdrawn, one at a time, replacing them after each drawing, the probabilities of obtaining an event E with n white balls under the two hypotheses are

$$P[E|H_1] = 1 \quad \text{and} \quad P[E|H_2] = 2^{-n}.$$

Now from Bayes' postulate,

$$P[H_1] = P[H_2] = \frac{1}{2}$$

and so from (2.16),

$$P[H_1|E] = \frac{2^n}{1 + 2^n} \quad \text{and} \quad P[H_2|E] = \frac{1}{1 + 2^n}.$$

Provided no black ball appears, the first hypothesis should be accepted because it has the greater posterior probability.

While Bayes' postulate might seem reasonable, it is the subject of controversy and can lead to erroneous conclusions. In the frequency theory of probability it would imply that events corresponding to the various B_i are distributed with equal frequency in some population from which the actual B_i has arisen. Many statisticians reject this as unreasonable.[7]

Later in this book we will examine some of the many other suggested alternatives to Bayes' postulate, including the principle of least squares and minimum chi-squared. That discussion will be anticipated here by briefly mentioning one principle of general application, that of *maximum likelihood*. From (2.16) we see that

$$P[B_i|A] \propto P[B_i]L, \tag{2.17}$$

where $L = P[A|B_i]$ is called the *likelihood*. Note that L is defined in terms of hypotheses, not events, and is not a pdf. *The Principle of Maximum Likelihood* states that when confronted with a set of hypotheses B_i, we choose the one that maximizes L, if one exists, that is, the one that gives the greatest probability to the observed event. Note that this is *not* the same as choosing the hypothesis with the greatest probability. It is not at all self-evident why one should adopt this choice as a principle of statistical inference, and we will return to this point in Chapter 7. For the simple case above, the maximum likelihood method clearly gives the same result as Bayes' postulate.

There are other ways of defining probabilities and statisticians do not agree among themselves on the 'best' definition. We will not dwell too much on the differences between them in this book,[8] except to note that the frequency definition is usually used in physical sciences, although the subjective approach will be important when discussing some aspects of interval estimation and hypothesis testing.[9]

Problems 2

2.1 In the sample space $S = \{1, 2, 3, \dots 9\}$, the events A, B, C, and D are defined by $A = (2, 4, 8)$, $B = (2, 3, 5, 9)$, $C = (1, 2, 4)$, and $D = (6, 8, 9)$. List the structure of the events

$$\text{(a) } \overline{A} \cap D, \text{ (b)} (B \cap \overline{C}) \cup A, \text{ (c)} (A \cap \overline{B} \cap C), \text{ (d) } A \cap (\overline{B} \cup \overline{D}).$$

2.2 The diagram in Figure 2.3 shows an electrical circuit with four switches S, labeled 1, 2, 3, and 4, that when closed allow a current to flow. If the switches act independently and have a probability p for being closed, what is the probability for a current

---

[7]Bayes himself may have had some doubts about it, as it was not published until after his death.

[8]To quote a remark attributed to the eminent statistician Sir Maurice Kendall: 'In statistics it is a mark of immaturity to argue over much about the fundamentals of probability theory'.

[9]For example, Bayesian probability has become popular with researchers in some fields of physics research, including particle physics and cosmology.

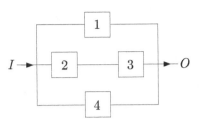

FIGURE 2.3 Circuit diagram.

to flow from I to O? Check your calculation by calculating the probability for no current to flow.

2.3 Over several seasons, two teams A and B have met 10 times. Team A has won five times, team B has won three times, and two matches have been drawn. What is the probability that in their next two encounters (1) team B will win both games and (2) team A will win at least one game?

2.4 A technician is 70% convinced that a circuit is failing due to a fault in a particular component. Later it emerges that although the quality assurance test does not fail any good components, there is a 15% chance that a faulty component will also pass. How should this evidence alter the technician's view?

2.5 Five physics books, four maths books, and three chemistry books are to be placed on a shelf so that all books on a given subject are together. How many arrangements are possible?

2.6 (a) One card is drawn at random from a standard deck of 52 cards. What is the probability that the card is a nine (9) or a club (C)?
(b) If four cards are drawn, what is the probability that at least three will be of the same suite?

2.7 Box A contains four red and two blue balls, and box B contains two red and six blue balls. One of the boxes is selected by tossing a die and then a ball is chosen at random from the selected box. If the ball selected is red, what is the probability that it came from box A?

2.8 A lie detector test is used to detect people who have committed a crime. It is known to be 95% reliable when testing people who have actually committed a crime and 99% reliable when they are innocent. The test is given to a suspect chosen at random from a population of which 3% have been convicted of a crime, and the result is positive, that is, indicates guilt. What is the probability that the suspect has not actually committed a crime?

2.9 Three balls are drawn at random successively from a bag containing 12 balls, of which 3 are red, 4 are white, and 5 are blue. In case (1) each ball is not replaced after it has been drawn, and in case (2) they are replaced. In both cases all three balls are found to be in the following order: red, white, and blue. What are the probabilities for this in the two cases?

2.10 Referring to Figure 2.4, a man enters a park at the In gate (marked I) and leaves at the Out gate. (marked O). At junctions with more than one choice of path, he chooses a direction at random and never retraces his steps. What is the probability that he

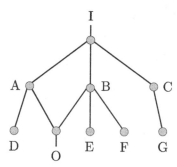

FIGURE 2.4 Possible pathways from I to O.

goes from I to O in just two stages, that Is, walking along only two of the links between the points marked I, O, and A–G?

2.11 A beam of particles is scattered from a nuclear target and the outgoing particles are recorded on a photographic film. This is scanned manually twice to identify a special type of event. The scans are done independently, and all events on the film are equally likely to be found. Suppose the two scans find, respectively, $(C+D_1)$ and $(C+D_2)$ examples of the event, where $D_{1,2}$ denotes the numbers of events found only in the first and second scan, respectively, and C denotes the number of events found in both scans. Determine: (1) the overall scanning efficiency and (2) the total number of events on the film.

2.12 A factory produces identical components using three machines $M_1, M_2,$ and M_3. The contributions of each to the total output are 20%, 30%, and 40%, respectively, and each machine produces defective components at rates of 3%, 2%, and 1%, respectively. One component is selected at random from the entire output of all three machines and is found to be defective. What is the probability that the component was made by machine M_2?

Probability distributions I
Basic concepts

Overview

This chapter is an introduction to the main theoretical properties of probability distributions. It starts by defining the concept of a random variable and then discusses the properties of the probability distribution for the case of a single variable. Definitions are given for expectation values, moment-generating and characteristic functions, key quantities that are used throughout the rest of the book. The discussion is then generalized to the case of several variables, which introduces additional quantities such as marginal and conditional distributions. Finally, there is a short discussion of how distributions are constructed for functions of random variables.

3.1 Random variables

The events discussed in Chapters 1 and 2 could be arbitrary quantities, like heads and tails, or numerical values, but it is useful to associate a set of real numbers with the outcomes of an experiment even if the basic data are nonnumeric. This association can be expressed by a real-valued function that transforms the points in the space S to points on the x axis. The function is called a *random variable*,[1] introduced in Chapter 1, and if we denote it by X, then formally $X(O)$ is the function that maps every outcome O_i in the sample space S to the space of real numbers $\mathbb{R}$, where each O_i is mapped to only one value in $\mathbb{R}$, but multiple outcomes can be mapped to the same number.

Returning to the coin-tossing experiment of Example 1.2, the four events were

$$E_1 = (H, H); \quad E_2 = (H, T); \quad E_3 = (T, H); \quad E_4 = (T, T).$$

---

[1]To distinguish between random variables and other variables, some authors call the former *variates*. Other authors use the word 'variate' to mean any variable, random or otherwise, that can take on a numerical value. In practice, it is usually clear what is meant, so in this book, the two terms will be interchangeable.

So, if we let X be a random variable[2] that can assume the values given by the number of 'heads', then

$$X(E_1) = 2; \quad X(E_2) = 1; \quad X(E_3) = 1; \quad X(E_4) = 0,$$

and, using the frequency definition of probability,

$$P[X = 0] = P[E_4] = 1/4; \quad P[X = 1] = P[E_2 \cup E_3] = 1/2; \quad P[X = 2] = P[E_1] = 1/4,$$

where we have assumed an unbiased coin so that throwing a 'head' or a 'tail' is equally likely. From this example we see that a random variable can assume an ensemble of numerical values in accordance with the underlying probability distribution. These definitions can be extended to continuous variates and to situations involving multivariates, as we will see below. In general, it is the quantities $P[X]$ that are the objects of interest, and it is to these that we now turn.

3.2 Single variable

In this section we will examine the case of a single random variable. The ideas discussed here will be extended to the multivariate case in Section 3.3.

3.2.1 Probability distributions

Firstly, we will need some definitions that extend those given in Chapter 2 in the discussion of the axioms of probability, starting with the case of a single discrete random variable. If X is a discrete random variable that can take on the values $x_k (k = 1, 2, \ldots)$ with probabilities $P[X]$, then we can define a probability distribution $f[x]$ by

$$P[X = x] = f(x). \tag{3.1a}$$

Thus

$$P[X = x_k] = f(x_k) \text{ for } X = x_k, \text{otherwise} f(x) = 0. \tag{3.1b}$$

To distinguish between the cases of discrete and continuous variables, we will use the term *probability density function*,[3] abbreviated to *pdf* (or sometimes simply *density function*), for a distribution of continuous variables and the term *probability function* for discrete variables. Some authors call the latter the *probability mass function* (or simply the *mass function*), sometimes abbreviated to *pmf*.

---

[2] A convention that is often used is to denote random variables by upper-case letters and the values they can take by the corresponding lower-case letters. This convention is used in this chapter to formally define the relation between random variables and their values, but as there is usually no ambiguity, a less formal notation of using lower-case letters for both will be used in later chapters.

[3] This term was introduced briefly in Section 1.4 in the context of the common behavior of large samples of data.

A probability function satisfies the following two conditions:

1. $f(x)$ is a single-valued nonnegative real number for all real values of x, that is, $f(x) \geq 0$;
2. $f(x)$ summed over all values of x is unity:

$$\sum_x f(x) = 1. \tag{3.1c}$$

We saw in Chapter 1 that we are also interested in the probability that x is less than or equal to a given value. This was called the *cumulative distribution function* (or simply the *distribution function*), sometimes abbreviated to *cdf*, and is given by

$$F(x) = \sum_{x_k \leq x} f(x_k). \tag{3.2a}$$

So, if X takes on the values $x_k (k = 1, 2, ..., n)$, the cumulative distribution function is

$$F(x) = \begin{cases} 0 & -\infty < x < x_1 \\ f(x_1) & x_1 \leq x < x_2 \\ f(x_1) + f(x_2) & x_2 \leq x < x_3 \\ \vdots & \vdots \\ f(x_1) + \cdots + f(x_n) & x_n \leq x < \infty \end{cases} \tag{3.2b}$$

The quantile x_α of order α, defined in Chapter 1, is thus the value of x such that $F(x_\alpha) = \alpha$, with $0 \leq \alpha \leq 1$, and so $x_\alpha = F^{-1}(\alpha)$, where F^{-1} is the inverse function of F. For example, the median is $x_{0.5}$. $F(x)$ is a nondecreasing function with limits 0 and 1 as $x \to -\infty$ and $x \to +\infty$, respectively.

If the random variable X can take on continuously varying values, the empirical distribution function (defined in Chapter 1) is similar to a discrete *cdf*, in that the experimental values of X are limited to a finite number of values. We take the distribution function as uniform on each interval $x_i < x < x_{i+1}$, so the empirical distribution function $F_n(x)$ for a set of n data has the form of a staircase with $F_n(x) = 0$ for all values of x smaller than the minimum of the data set, and $F_n(x) = 1$ for all values of x greater than the maximum of the data set. Between the minimum and maximum of the range, we assume that each $x_i < x < x_{i+1}$ has the same probability of occurring, so F increases by an amount $1/n$ at each element of the data set and is a constant between successive elements. An example is shown in Figure 3.1b below.

As sample sizes become larger, frequency plots tend to approximate smooth curves, and if the area of the histogram is normalized to unity, as in Figure 1.5(d), the resulting function $f(x)$ is a pdf introduced in Chapter 1. The definitions above for the cumulative distribution may be extended to continuous random variables with the appropriate changes. Thus, for a continuous random variable X with a *pdf* $f(x)$, (3.2a) becomes

$$F(x) = \int_{-\infty}^{x} f(x')dx', \qquad (-\infty < x < \infty). \qquad (3.3)$$

It follows from (3.3) that if a member of a population is chosen at *random*, that is, by a method that makes it equally likely that each member will be chosen, then $F(x)$ is the probability that the member will have a value $\leq x$. While all this is clearly consistent with earlier definitions, once again, we should note the element of circularity in the concept of randomness defined in terms of probability. In mathematical statistics it is usual to start from the cumulative distribution and define the density function as its derivative. For the mathematically well-behaved distributions usually met in physical science, the two approaches are equivalent.

The density function $f(x)$ has the following properties analogous to those for discrete variables:

(i) $f(x)$ is a single-valued nonnegative real number for all real values of x.

In the frequency interpretation of probability $f(x)dx$ is the probability of observing the random variable X in the range $(x, x + dx)$. Thus the second condition is:

(ii) $f(x)$ is normalized to unity:

$$\int_{-\infty}^{+\infty} f(x)dx = 1.$$

It follows from property (ii) that the probability of X lying between any two real values a and b, for which $a < b$, is given by

$$P[a \leq X \leq b] = \int_{a}^{b} f(x)dx, \qquad (3.4)$$

and so, unlike a discrete random variable, the probability of a continuous random variable assuming *exactly* any of its values is 0. This result may seem rather paradoxical at first until you consider that between any two values a and b, there is an infinite number of other values and so the probability of selecting an exact value from this infinity of possibilities must be 0. The density function cannot therefore be given in a tabular form like the pmf of a discrete random variable.

Example 3.1

A family has five children. Assuming that the birth of a girl (g) or boy (b) is equally likely, construct a frequency table of possible outcomes and plot the resulting probability function $f(g)$ and the associated cumulative distribution function $F(g)$.

Solution: The probability of a sequence of births containing g girls (and hence $b = 5 - g$ boys) is $(1/2)^g(1/2)^b = (1/2)^5$. But there are $_5C_g$ such sequences, and so the probability of having g girls is $P[g] = {_5C_g}/32$. The probability function $f(g)$ is thus as given in the table below.

g	0	1	2	3	4	5
$f(g)$	1/32	5/32	10/32	10/32	5/32	1/32

From this table we can find the cumulative distribution function using (3.2b). $f(g)$ and $F(g)$ are plotted in Figures 3.1(a) and (b), respectively.

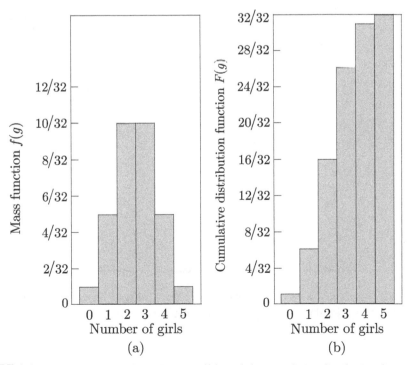

FIGURE 3.1 Plots of the probability function $f(g)$, and the cumulative distribution function $F(g)$.

Example 3.2

Find the value of N in the continuous density function

$$f(x) = \begin{cases} Ne^{-x}x^2/2 & x \geq 0 \\ 0 & x < 0 \end{cases},$$

and its associated distribution function $F(x)$. Plot $f(x)$ and $F(x)$.

Solution: Since $f(x)$ must be correctly normalized, to find N, we evaluate the integral

$$\frac{N}{2} \int_0^\infty e^{-x}x^2 \mathrm{d}x = 1.$$

Integrating by parts gives

$$\frac{1}{N} = \frac{1}{2} \int_0^\infty e^{-x}x^2 \mathrm{d}x = -\frac{1}{2}\left[e^{-x}(x^2 + 2x + 2)\right]_0^\infty = 1$$

so that $N = 1$. The resulting density function is plotted in Figure 3.2(a). The associated distribution function is

$$F(x) = \frac{1}{2} \int_0^x e^{-u}u^2 \mathrm{d}u = -\frac{1}{2}\left[e^{-u}(u^2 + 2u + 2)\right]_0^x = -\frac{1}{2}e^{-x}(x^2 + 2x + 2) + 1$$

and is shown in Figure 3.2(b).

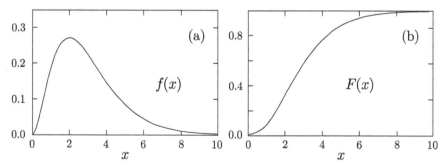

FIGURE 3.2 (a) Probability density function $f(x) = e^{-x}x^2/2 (x \geq 0)$ and (b) the corresponding distribution function $F(x)$.

Some of the earlier definitions of Chapter 1 may now be rewritten in terms of these formal definitions. Thus the general moments about an arbitrary point λ are, for a continuous variate,

$$\mu_n' = \int_{-\infty}^\infty f(x)(x - \lambda)^n \mathrm{d}x. \tag{3.5}$$

so that the mean and variance, also with respect to the point λ, are

$$\mu_\lambda = \int_{-\infty}^{+\infty} f(x)(x-\lambda)\mathrm{d}x \quad \text{and} \quad \sigma^2 = \int_{-\infty}^{+\infty} f(x)(x-\mu_\lambda)^2\mathrm{d}x, \tag{3.6}$$

respectively. The integrals in (3.5) may not converge for all n, and some distributions possess only the trivial zero-order moment. For convenience, usually $\lambda = 0$ will be used in what follows.

Example 3.3

Find the mean of the continuous distribution of Example 3.2.
Solution: Using (3.6), with $\lambda = 0$, the mean is

$$\mu = \frac{1}{2}\int_0^\infty x^3 e^{-x}\mathrm{d}x,$$

and integrating by parts gives

$$\mu = -\left[\frac{e^{-x}}{2}(x^3+3x^2+6x+6)\right]_0^\infty = 3.$$

3.2.2 Expectation values

The *expectation value*, also called the *expected value*, of a random variable is an important concept in statistics and has a wide range of uses. It is defined here and used in applications in later sections. It is obtained by finding the average value of the variable over all its possible values weighted by the probability of their occurrence. Thus, if X is a discrete random variable with the possible values $x_1, x_2, ..., x_n$, then the expectation value of X is given by

$$E[X] \equiv \sum_{i=1}^{n} x_i P[X = x_i] = \sum_x x f(x), \tag{3.7a}$$

where the second sum is over all relevant values of x and $f(x)$ is their probability mass distribution. The analogous quantity for a continuous variate with density function $f(x)$ is

$$E[X] = \int_{-\infty}^{\infty} x f(x) \ \mathrm{d}x. \tag{3.7b}$$

We can see from this definition that the nth moment of a distribution about any point λ is

$$\mu_n' = E[(X-\lambda)^n]. \tag{3.8a}$$

In particular, the nth *central moment* is

$$\mu_n = E[(X - E[X])^n] = \int_{-\infty}^{\infty} (x - \mu)^n f(x) \mathrm{d}x \tag{3.8b}$$

and for $\lambda = 0$, the nth *algebraic moment* is

$$\mu_n' = E[x^n] = \int_{-\infty}^{\infty} x^n f(x) \mathrm{d}x. \tag{3.8c}$$

Thus the mean is the first algebraic moment, and the variance is the second central moment. It also follows from (3.7) and property (ii) that if a and b are constants, then

$$E[aX + b] = aE[X] + b. \tag{3.9a}$$

In addition, the expectation of a sum of random variables is equal to the sum of the expectation values of each random variable, and in general,

$$E[a_1 X_1 + a_2 X_2 + \cdots + a_n X_n + b] = a_1 E[X_1] + a_2 E[X_2] + \cdots + a_n E[X_n] + b. \tag{3.9b}$$

Example 3.4

Three 'fair' dice are thrown and yield face values A, B, and C. What is the expectation value for the sum of their faces?
Solution: From (3.7a),

$$E[A] = \sum_1^6 i(1/6) = 7/2,$$

and since

$$E[A] = E[B] = E[C], \text{then from (3.9b)}, E[A + B + C] = 21/2.$$

Unlike the sum of random variables, the expectation of a product of random variables is not equal to the product of their expectation values, except in the special case where the variables are independent. The proof of this is discussed in Section 3.3.

3.2.3 Moment-generating and characteristic functions

The usefulness of moments partly stems from the fact that knowledge of them usually determines the form of the density function. Formally, if the moments μ_n' of a random variable X exist and the series

$$\sum_{n=1}^{\infty} \frac{\mu_n'}{n!} r^n \tag{3.10}$$

converges absolutely for some $r > 0$, then the set of moments μ'_n uniquely determines the density function. There are exceptions to this statement, but fortunately, it is true for all the distributions commonly met in physical science. In practice, knowledge of the first few moments essentially determines the general characteristics of the distribution and so it is worthwhile to construct a method that gives a representation of all the moments. Such a function is called a *moment generating function (mgf)* and is defined by

$$M_X(t) \equiv E\left[e^{Xt}\right],$$ (3.11)

For a discrete random variable X, this is

$$M_X(t) = \sum e^{xt} f(x),$$ (3.12a)

and for a continuous variable,

$$M_X(t) = \int\limits_{-\infty}^{+\infty} e^{xt} f(x) \mathrm{d}x.$$ (3.12b)

The moments about $\lambda = 0$ may be generated from (3.11) by first expanding the exponential,

$$M_X(t) = E\left[1 + xt + \frac{1}{2!}(xt)^2 + \cdots\right] = \sum_{n=0}^{\infty} \frac{1}{n!} \mu'_n t^n,$$

then differentiating n times and setting $t = 0$, that is,

$$\mu'_n = \left.\frac{\partial^n M_X(t)}{\partial t^n}\right|_{t=0}.$$ (3.13)

For example, setting $n = 0$ and $n = 1$ gives $\mu'_0 = 1$ and $\mu'_1 = \mu$. Also, since the *mgf* about any point λ is

$$M_\lambda(t) = E[\exp\{(x - \lambda)t\}] = e^{-\lambda t} M_X(t)$$

then if $\lambda = \mu$,

$$M_\mu(t) = e^{-\mu t} M_X(t).$$ (3.14)

An important use of the *mgf* is to compare the density functions $f(x)$ and $g(y)$ of two random variables X and Y. If x and y are finite and possess *mgfs* that are identical for all values of t in any interval symmetric about the point $t = 0$, then $f(x)$ and $g(x)$ are identical density functions. It is also straightforward to show that the *mgf* of a sum of independent random variables is equal to the product of their individual *mgfs*.

It is sometimes convenient to consider, instead of the *mgf*, its logarithm. The Taylor expansion[4] for this quantity is

$$\ln M_X(t) = \kappa_1 t + \kappa_2 \frac{t^2}{2} + \cdots,$$

where k_n is called the *cumulant* of order n, and

$$\kappa_n = \left. \frac{\partial^n \ln M_X(t)}{\partial t^n} \right|_{t=0}.$$

Cumulants are simply related to the central moments of the distribution, the first few relations being

$$k_i = \mu_i \ (i = 1, 2, 3), k_4 = \mu_4 - 3\mu_2^2.$$

For some distributions the integral defining the *mgf* may not exist, and in these circumstances the Fourier transform of the density function, defined as

$$\phi_X(t) \equiv E\left[e^{itx}\right] = \int_{-\infty}^{+\infty} e^{itx} f(x) \mathrm{d}x = M_X(it), \tag{3.15}$$

may be used. In statistics $\phi_X(t)$ is called the *characteristic function (cf)*. The density function is then obtainable using the Fourier transform theorem (known in this context as the *inversion theorem*):

$$f(x) = \frac{1}{2\pi} \int_{-\infty}^{+\infty} e^{-itx} \phi_X(t) \mathrm{d}t. \tag{3.16}$$

The *cf* obeys analogous theorems to those obeyed by the *mgf*, that is, (a) if two random variables possess *cfs* of all orders that are equal for some interval symmetric about the origin, then they have identical density functions; and (b) the *cf* of a sum of independent random variables is equal to the product of their individual *cfs*. The converse of (b) is, however, untrue.

To prove (b), consider the case of two independently distributed random variables, X and Y. The *cf* of their sum is

$$\phi_{X+Y}(t) = E\left[e^{it(X+Y)}\right] = E\left[e^{itX} e^{itY}\right] = E\left[e^{itX}\right] E\left[e^{itT}\right] = \phi_X(t)\phi_Y(t),$$

which may easily be generalized to the case of many variables.

---

[4]Some essential mathematics is reviewed briefly in Appendix A.

Example 3.5

Find the moment generating function of the density function used in Example 3.2 and calculate the three moments $\mu'_1, \mu'_2,$ and μ'_3.

Solution: Using definition (3.12b),

$$M_X(t) = \int_0^\infty e^{xt} f(x) dx = \frac{1}{2} \int_0^\infty e^{xt} x^2 e^{-x} dx = \frac{1}{2} \int_0^\infty e^{-x(1-t)} x^2 dx,$$

which, integrating by parts, gives

$$M_X(t) = \left\{ -\frac{e^{-x(1-t)}}{2(1-t)^3} \left[(1-t)^2 x^2 + 2(1-t)x + 2\right] \right\}_0^\infty = \frac{1}{(1-t)^3}.$$

Then, using (3.13), the first three moments of the distribution are found to be

$$\mu'_1 = 3, \mu'_2 = 12, \mu'_3 = 60.$$

Example 3.6

(a) Find the characteristic function of the density function

$$F(x) = \begin{cases} 2x/a^2 & 0 < x < a \\ 0 & \text{otherwise} \end{cases}$$

and (b) the density function corresponding to a characteristic function $e^{-|t|}$.

Solution: (a) From (3.15),

$$\phi_X(t) = E[e^{itx}] = \frac{2}{a^2} \int_0^a e^{itx} x \, dx.$$

Again, integration by parts gives

$$\phi_X(t) = \frac{2}{a^2} \left[\frac{e^{itx}}{(it)^2} (itx - 1) \right]_0^a = -\frac{2}{a^2 t^2} \left[e^{ita}(ita - 1) + 1 \right].$$

(b) From the inversion theorem,

$$f(x) = \frac{1}{2\pi} \int_{-\infty}^{+\infty} e^{-|t|} e^{-itx} dt = \frac{1}{\pi} \int_0^{+\infty} e^{-t} \cos(tx) dt,$$

where the symmetry of the circular functions has been used. The second integral may be evaluated by parts to give

$$\pi f(x) = \left[-e^{-t}\cos(tx) \right]_0^\infty - x \int_0^\infty e^{-t}\sin(tx)\,dt$$

$$= 1 - x\left\{ \left[-e^{-t}\sin(tx) \right]_0^\infty + x \int_0^\infty e^{-t}\cos(tx)\,dt \right\} = 1 - \pi x^2 f(x)$$

Thus

$$f(x) = \frac{1}{\pi(1 + x^2)}, \quad -\infty \le x \le \infty.$$

This is the density function of the Cauchy distribution and will be discussed in Section 4.1.6.

3.3 Several variables

All the results of the previous sections may be extended to multivariate distributions. We will concentrate on continuous variates, but the formulas may be transcribed in a straightforward way to describe discrete variates.

3.3.1 Joint probability distributions

The *multivariate joint density function* $f(x_1, x_2, \cdots, x_n)$ of the n continuous random variables $X_1, X_2, \cdots, X_n$ is a single-valued nonnegative real number for all real values of $x_1, x_2, \cdots, x_n$, normalized so that

$$\int_{-\infty}^{+\infty} \cdots \int_{-\infty}^{+\infty} f(x_1, x_2, ..., x_n) \prod_{i=1}^n dx_i = 1, \tag{3.17}$$

and the probability that x_1 falls between any two numbers a_1 and b_1, and x_2 falls between any two numbers a_2 and b_2, $\cdots$, and x_n falls between any two numbers a_n and b_n, *simultaneously*, is defined by

$$P[a_1 \le X_1 \le b_1; ...; a_n \le X_n \le b_n] \equiv \int_{a_n}^{b_n} \cdots \int_{a_1}^{b_1} f(x_1, x_2, ..., x_n) \prod_{i=1}^n dx_i. \tag{3.18}$$

Similarly, the *multivariate joint distribution function* $F(x_1, x_2, \cdots, x_n)$ of the n random variables $X_1, X_2, \cdots, X_n$ is

$$F(x_1, x_2, ..., x_n) \equiv \int_{-\infty}^{x_n} \cdots \int_{-\infty}^{x_1} f(t_1, t_2, ..., t_n) \prod_{i=1}^n dt_i. \tag{3.19}$$

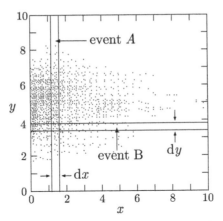

FIGURE 3.3 A scatter plot of 1000 events that are functions of two random variables X and Y showing two infinitesimal bands dx and dy. The area of intersection of the bands is $dx\,dy$ and $f(x,y)dxdy$ is the probability of finding X in the interval $(x, x + dx)$ and Y in the interval $(y, y + dy)$.

For simplicity, consider the case of just two random variables X and Y. These could correspond to the energy and angle of emission of a particle emitted in a nuclear scattering reaction. If an event A corresponds to the variable X being observed in the range $(x, x + dx)$ and the event B corresponds to the variable Y being observed in the range $(y, y + dy)$, then

$$P[A \cap B] = \text{probability of } x \text{ being in } (x, x + dx) \text{ and } y \text{ being in } (y, y + dy)$$
$$= f(x, y)dx\,dy.$$

As noted in Chapter 1, the joint density function corresponds to the density of points on a scatter plot of X and Y, in the limit of an infinite number of points. This is illustrated in Figure 3.3, using the data shown on the scatter plot of Figure 1.3(b).

3.3.2 Marginal and conditional distributions

We may also be interested in the density function of a subset of variables. This is called the *marginal density function* f^M, and in general, for a subset $X_i(i = 1, 2, ..., m < n)$ of the variables, is given by integrating the joint density function over all the variables other than $X_1, X_2, ..., X_m$. Thus

$$f^M(x_1, x_2, ..., x_m) = \int_{-\infty}^{+\infty} \cdots \int_{-\infty}^{+\infty} f(x_1, x_2, ..., x_m, x_{m+1}, ..., x_n) \prod_{i=m+1}^{n} dx_i. \tag{3.20a}$$

In the case of two variables we may be interested in the density function of X regardless of the value of Y, or the density function of Y regardless of X. For example, the failure rate of a resistor may be a function of its operating temperature and the voltage across it, but in

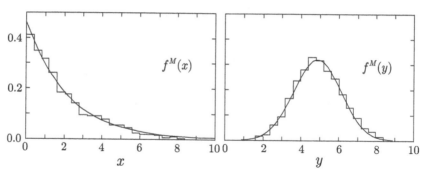

FIGURE 3.4 Normalized histograms obtained by projecting the data of Figure 1.3(b) onto the x and y axes, together with the corresponding marginal probability density functions $f^M(x)$ and $f^M(y)$.

some circumstances we might be interested in just the dependence on one of these variables. In these cases (3.20a) becomes

$$f^M(x) = \int_{-\infty}^{+\infty} f(x,y)\mathrm{d}y \quad \text{and} \quad f^M(y) = \int_{-\infty}^{+\infty} f(x,y)\mathrm{d}x. \tag{3.20b}$$

These density functions correspond to the normalized histograms obtained by projecting a scatter plot of X and Y onto one of the axes. This is illustrated in Figure 3.4, again using the data of Figure 1.3(b).

We can also define the *multivariate conditional density function* of the random variables $x_i (i = 1, 2, ..., m < n)$ by

$$f^C(x_1, x_2, ..., x_m | x_{m+1}, x_{m+2}, ..., x_n) \equiv \frac{f(x_1, x_2, ..., x_n)}{f(x_{m+1}, x_{m+2}, ..., x_n)}. \tag{3.21}$$

Again, if we consider the case of two random variables X and Y, the probability for Y to be in the interval $(y, y + \mathrm{d}y)$ with any X (event B), given that X is in the interval $(x, x + \mathrm{d}x)$ with any y (event A), is

$$P[B|A] = \frac{P[A \cap B]}{P[A]} = \frac{f(x,y)\mathrm{d}x\mathrm{d}y}{f^M(x)\mathrm{d}x},$$

where $f^M(x)$ is the marginal density function for X. The conditional density function for Y, given X, is thus

$$f^C(y|x) = \frac{f(x,y)}{f^M(x)} = \frac{f(x,y)}{\int f(x,y')\mathrm{d}y'}. \tag{3.22a}$$

This is the density function of the single random variable Y where x is treated as a constant. It corresponds to projecting the events in a band dx centered on some value x onto the y axis and renormalizing the resulting density so that it is unity when integrated over y. The form of $f^C(y|x)$ will therefore vary as different values of x are chosen.

The conditional density function for X given Y is obtained from (3.22a) by interchanging x and y so that

$$f^C(x|y) = \frac{f(x,y)}{f^M(y)} = \frac{f(x,y)}{\int f(x',y)dx'}, \tag{3.22b}$$

and combining these two equations gives

$$f^C(x|y) = \frac{f^C(y|x)f^M(x)}{f^M(y)}, \tag{3.22c}$$

which is Bayes' theorem for continuous variables.

We can use these definitions to generalize the law of total probability (2.6b) to the case of continuous variables. Using conditional and marginal density functions, we have

$$f(x,y) = f^C(y|x)f^M(x) = f^C(x|y)f^M(y), \tag{3.23}$$

so the marginal density functions may be written as

$$f^M(y) = \int_{-\infty}^{+\infty} f^C(y|x)f^M(x)dx$$

and

$$f^M(x) = \int_{-\infty}^{+\infty} f^C(x|y)f^M(y)dy.$$

With more than one random variable, we must also consider the question of statistical independence (by analogy with the work of Chapter 2). If the random variables can be split into groups such that their joint density function is expressible as a product of marginal density functions of the form

$$f(x_1, x_2, ..., x_n) = f_1^M(x_1, x_2, ..., x_i)f_2^M(x_{i+1}, x_{i+2}, ..., x_k)... f_n^M(x_{l+1}, x_{l+2}, ..., x_n),$$

then the sets of variables

$$(X_1, X_2, ..., X_i); (X_{i+1}, X_{i+2}, ..., X_k); ...; (X_{l+1}, X_{l+2}, ..., X_n)$$

are said to be *statistically independent,* or *independently distributed.* Thus two random variables X and Y are independently distributed if

$$f(x,y) = f^M(x) f^M(y). \tag{3.24}$$

It follows from (3.22) that in this case the conditional density function of one variate does not depend on knowledge about the other variate.

Example 3.7

The joint probability function for two discrete random variables X and Y is given by

$$f(x,y) = \begin{cases} k(2x+3y) & 0 \le x \le 3, \ 0 \le y \le 2 \\ 0 & otherwise \end{cases},$$

where k is a constant. Find: (a) the value of k, (b) $P[X \ge 2, Y \le 1]$, and (c) the marginal density of X.
Solution: The probability function is tabulated below.

x	y 0	1	2	Total
0	0	$3k$	$6k$	$9k$
1	$2k$	$5k$	$8k$	$15k$
2	$3k$	$7k$	$10k$	$21k$
3	$6k$	$9k$	$12k$	$27k$
Total	$12k$	$24k$	$36k$	$72k$

(a) The normalization condition is $\sum_{x,y} f(x,y) = 1$, so $k = 1/72$.

(b)
$$P[X \ge 2, Y \le 1] = P[X = 2, Y = 1] + P[X = 2, Y = 0] + P[X = 3, Y = 1] + P[X = 3, Y = 0]$$
$$= 7k + 4k + 9k + 6k = 26k = 13/36.$$

(c) The marginal probability of x is

$$P[X = x] = \sum_y P[X = x, Y = y] = \begin{cases} 9k = 3/24 & x = 0 \\ 15k = 5/24 & x = 1 \\ 21k = 7/24 & x = 2 \\ 27k = 9/24 & x = 3 \end{cases}$$

Example 3.8

If $f(x,y)$ is the joint density function of two continuous random variables X and Y, defined by

$$f(x,y) = \begin{cases} e^{-(x+y)} & x,y \geq 0 \\ 0 & \text{otherwise} \end{cases},$$

find their conditional distribution.

Solution: From (3.22b),

$$f^C(x|y) = \frac{f(x,y)}{f^M(y)},$$

where the marginal density of Y is given from (3.20b) as

$$f^M(y) = \int_0^\infty f(x,y)dx = e^{-y}[-e^{-x}]_0^\infty = e^{-y}.$$

Thus

$$f^C(x|y) = \frac{e^{-(x+y)}}{e^{-y}} = e^{-x}.$$

3.3.3 Expectation values of sums and products

We can now return to the discussion of expectation values of sums and products of random variables started in Section 3.2.2. We will consider just the case of continuous variables, but the discrete case follows in a similar way. For simplicity, we consider the case of two random variables X_1 and X_2 with a continuous joint pdf $f(x_1,x_2)$. Then,

$$E[X_1 + X_2] = \int_{-\infty}^{\infty} \int_{-\infty}^{\infty} (x_1 + x_2)f(x_1,x_2) \, dx_1dx_2$$

$$= \int_{-\infty}^{\infty} \int_{-\infty}^{\infty} x_1f(x_1,x_2) \, dx_1dx_2 + \int_{-\infty}^{\infty} \int_{-\infty}^{\infty} x_2f(x_1,x_2) \, dx_1dx_2$$

$$= \int_{-\infty}^{\infty} x_1 f_1(x_1)dx_1 + \int_{-\infty}^{\infty} x_2f_2(x_1)dx_2 = E[X_1] + E[X_2],$$

where f_1 and f_2 are the marginal probability densities of X_1 and X_2, respectively. The result in (3.9a) then follows. Note that the result does not depend on the form of the joint probability distribution; only the marginal distributions appear in the final result.

The derivation of the formula for the product of expectation values is done in an analogous way and is given in Problem 3.11.

3.3.4 Moments and expectation values

The definition of moments and expectation values can be generalized to the multivariable case. Thus the rth algebraic moment of the random variable X_i is given by

$$E[X_i^r] = \int\limits_{-\infty}^{\infty} \cdots \int\limits_{-\infty}^{\infty} x_i^r f(x_1, x_2 \ldots, x_n) \prod_{j=1}^{n} \mathrm{d}x_j, \tag{3.25}$$

from which we obtain the results

$$\mu_i = \int\limits_{-\infty}^{+\infty} \cdots \int\limits_{-\infty}^{+\infty} x_i f(x_1, x_2, \ldots, x_n) \prod_{j=1}^{n} \mathrm{d}x_j, \tag{3.26a}$$

and

$$\sigma_i^2 = \int\limits_{-\infty}^{+\infty} \cdots \int\limits_{-\infty}^{+\infty} (x_i - \mu_i)^2 f(x_1, x_2, \ldots, x_n) \prod_{j=1}^{n} \mathrm{d}x_j, \tag{3.26b}$$

for the mean and variance. The latter can also be written as

$$\mathrm{var}(X) = E\big[(X - \mu)^2\big],$$

from which a more useful form can be derived for calculating the variance, given as follows:

$$\begin{aligned} \mathrm{var}(X) &= E\big[(X - \mu)^2\big] = E\big[X^2 - 2\mu X + \mu^2\big] = E\big[X^2\big] - E\big[2\mu X\big] + E\big[\mu^2\big] \\ &= E\big[X^2\big] - 2\mu E\big[X\big] + \mu^2 = E\big[X^2\big] - \mu^2 = E\big[X^2\big] - (E[X])^2. \end{aligned}$$

In addition to the individual moments of (3.25), we can also define *joint moments*. In general, these are given by

$$E\big[X_a^i X_b^j \ldots X_c^k\big] \equiv \int\limits_{-\infty}^{+\infty} \cdots \int\limits_{-\infty}^{+\infty} \Big(x_a^i x_b^j \ldots x_c^k\Big) f(x_1, x_2, \ldots, x_n) \prod_{m=1}^{n} \mathrm{d}x_m. \tag{3.27}$$

The most important of these is the *covariance*, introduced in Chapter 1 for two random variables, and now defined more generally for any pair of variates as

$$\text{cov}(X_i, X_j) \equiv \sigma_{ij} = \int\limits_{-\infty}^{+\infty} \cdots \int\limits_{-\infty}^{+\infty} (x_i - \mu_i)(x_j - \mu_j) f(x_1, x_2, \ldots, x_n) \prod_{m=1}^{n} dx_m, \qquad (3.28)$$

where the means are given by (3.26a). In Chapter 1 the *correlation coefficient* $\rho(X_1, X_2)$ was defined as

$$\rho(X_i, X_j) \equiv \frac{\text{cov}(X_i, X_j)}{\sigma(X_i)\sigma(X_j)}. \qquad (3.29)$$

This is a number lying between -1 and $+1$, which may be proved by first considering the variance of a sum of two random variables X/σ_X and Y/σ_Y. The variance is

$$\text{var}\left(\frac{X}{\sigma_X} + \frac{Y}{\sigma_Y}\right) = \frac{\text{var}(X)}{\sigma_X^2} + \frac{\text{var}(Y)}{\sigma_Y^2} + 2\frac{\text{cov}(X, Y)}{\sigma_X \sigma_Y} \geq 0 = 1 + 1 + 2\rho(X, Y) \geq 0,$$

and hence $-1 \leq \rho$. Repeating the above for the difference $(X/\sigma_X - Y/\sigma_Y)$ leads directly to $\rho \leq 1$ and hence $-1 \leq \rho \leq 1$. It is a necessary condition for statistical independence that $\rho(X_i, X_j) = 0$. However, this is *not* a sufficient condition and $\rho(X_i, X_j) = 0$ does *not* always imply that X_i and X_j are independently distributed. (See Example 3.9 below.)

Example 3.9

If the random variable X has a density function that is antisymmetric about the mean, find the covariance of the two random variables $X_1 = X$ and $X_2 = Y = X^2$. Comment on your answer.
Solution: The covariance is

$$\text{cov}(X, Y) = E[XY] - E[X]E[Y] = E[X^3] - E[X]E[X^2].$$

But since x has a density function that is antisymmetric about the mean, all the odd-order moments vanish and, in particular,

$$E[X] = E[X^3] = 0.$$

Thus $\text{cov}(X, Y) = 0$, and hence $\rho(X, Y) = 0$, even though x and y are not independent. Thus $\text{cov}(X, Y) = 0$ is a necessary, but not sufficient, condition for statistical independence.

3.4 Functions of a random variable

In practice, it is common to consider a function of a random variable, for example Y(X). This is also a random variable, and one question that arises is: what is the density function of Y, given that we know the density function of X? The cumulative distribution of X must equal that of Y(X). If Y is monotonic (strictly increasing or decreasing) then differentiating the distributions:

$$f(y(x)) = f(x(y)) \left| \frac{dx}{dy} \right|, \tag{3.30}$$

the absolute value being necessary to ensure that probabilities are always non-negative. If, instead, Y has a continuous nonzero derivative at all but a finite number of points, the range must be split into a finite number of sections in each of which $y(x)$ is a strictly monotonic increasing or decreasing function of x with a continuous derivative, and then (3.30) applied to each section separately. Thus, at all points where (i) $dy/dx \neq 0$ *and* (ii) $y = y(x)$ has a real finite solution for $x = x(y)$, the required density function is

$$g(y(x)) = f(x(y)) \left| \frac{dx}{dy} \right|. \tag{3.31}$$

If the above conditions are violated, then $g(y(x)) = 0$ at that point.

The method may be extended to multivariate distributions. Consider n random variables $X_i(i = 1, 2, ..., n)$ with a joint probability density $f(x_1, x_2, ...x_n)$, and suppose we wish to find the joint probability density $g(y_1, y_2, ...y_n)$ of a new set of variates Y_i, which are themselves a function of the n variables $X_i(i = 1, 2, ..., n)$, defined by $Y_i = Y_i(X_1, X_2, ..., X_n)$. To do this, we impose the probability condition

$$|f(x_1, x_2, ..., x_n) dx_1 dx_2 \cdots dx_n| = |g(y_1, y_2 \cdots y_n) dy_1 dy_2 \cdots dy_n|.$$

It follows that

$$g(y_1, y_2, ..., y_n) = f(x_1, x_2, ..., x_n)|J|, \tag{3.32a}$$

where $|J|$ is the modulus of the determinant of the Jacobian[5] of x_i with respect to y_i, that is, the matrix

$$J = \frac{\partial(x_1, x_2, ..., x_n)}{\partial(y_1, y_2, ..., y_n)} = \begin{pmatrix} \dfrac{\partial x_1}{\partial y_1} & \cdots & \dfrac{\partial x_n}{\partial y_1} \\ \vdots & \ddots & \vdots \\ \dfrac{\partial x_1}{\partial y_n} & \cdots & \dfrac{\partial x_n}{\partial y_n} \end{pmatrix}, \tag{3.32b}$$

where again the absolute value is necessary to ensure that probabilities are always nonnegative. If the partial derivatives are not continuous, or there is not a unique solution for x_i in terms of the y_i, then the range of the variables can always be split into sections as for the single variable case and (3.31) applied to each section. The marginal density of one of the random variables can then be found by integrating the joint density over all the other variables. For several random variables it is usually too difficult in practice to carry out the above program analytically and numerical methods are used.

---

[5] The Jacobian J is defined in Appendix A.

Example 3.10

A random variable X has a density function

$$f(x) = \frac{1}{\sqrt{2\pi}} \exp\left(\frac{-x^2}{2}\right).$$

What is the density function of $Y = X^2$?

Solution: We have

$$x = \pm\sqrt{y} \quad \text{and} \quad \frac{dy}{dx} = 2x = \pm 2\sqrt{y}.$$

Thus for $y < 0$, x is not real and so $g(y\{x\}) = 0$. For $y = 0$, $dx/dy = 0$, so again $g(y\{x\}) = 0$
Finally, for $y > 0$, we may split the range into two parts, $x > 0$ and $x < 0$. Then, applying (3.30) gives

$$g(y) = \frac{1}{2\sqrt{y}}[f(x = -\sqrt{y}) + f(x = +\sqrt{y})] = \frac{1}{\sqrt{(2\pi y)}} \exp\left(\frac{-y}{2}\right).$$

Example 3.11

The single-variable density function of Example 3.10 may be generalized to two variables, that is,

$$f(x_1, x_2) = \frac{1}{2\pi} \exp\left\{ -\frac{1}{2}(x_1^2 + x_2^2) \right\}.$$

Find the joint density function $g(y_1, y_2)$ of the variables $Y_1 = X_1/X_2$ and $Y_2 = X_1$.

Solution: The Jacobian of the transformation is

$$J = \det\begin{pmatrix} x_2 & -x_2^2/x_1 \\ 1 & 0 \end{pmatrix} = \frac{x_2^2}{x_1} = \frac{y_2}{y_1^2}.$$

Thus applying (3.32) (*provided* $y_1 \neq 0$) gives

$$g(y_1, y_2) = \frac{1}{2\pi} \frac{|y_2|}{y_1^2} \exp\left\{ -\frac{1}{2}\left(y_2^2 + \frac{y_2^2}{y_1^2} \right) \right\} \quad (y_1 \neq 0).$$

A particular example of interest is the probability density of the sum of two random variables X and Y. If we set $U = X + Y$ and $V = X$, where the second choice is arbitrary, the modulus of the Jacobian of the transformation is

$$J = \begin{vmatrix} \dfrac{\partial x}{\partial u} & \dfrac{\partial y}{\partial u} \\ \dfrac{\partial x}{\partial v} & \dfrac{\partial y}{\partial v} \end{vmatrix} = \begin{vmatrix} 1 & 1 \\ 1 & 0 \end{vmatrix} = -1.$$

Thus the joint density of u and v is

$$g(u,v) = f(x,y) = f(v, u-v).$$

The density function of u, denoted $h(u)$, is then

$$h(u) = g^M(u) = \int_{-\infty}^{\infty} f(x, u-x) \ dx, \qquad\qquad (3.33a)$$

and in the special case where X and Y are independent, so that $f(x,y) = f_1(x)f_2(y)$, (3.33a) reduces to

$$h(u) = \int_{-\infty}^{\infty} f_1(x)f_2(u-x)dx, \qquad\qquad (3.33b)$$

which is called the *convolution* of f_1 and f_2 and is denoted by $f_1 * f_2$.

Convolutions obey the (a) commutative, (b) associative, and (c) distributive laws of algebra, that is,

$$(a)\, f_1 * f_2 = f_2 * f_1, (b)\, f_1 * (f_2 * f_3) = (f_1 * f_2) * f_3, \quad \text{and} \quad (c)\, f_1 * (f_2 + f_3) = (f_1 * f_2) + (f_1 * f_3).$$

They occur frequently in physical science applications. An example is the problem of determining a physical quantity represented by a random variable X with a density $f_1(x)$, from measurements having experimental errors Y distributed like the normal distribution of Problem 1.6 with zero mean and variance σ^2. The measurements yield values of the sum $U = X + Y$. Then, using the form of the normal distribution given in Problem 1.6, Equation (3.33b) gives

$$f(u) = \frac{1}{\sqrt{2\pi\sigma}} \int_{-\infty}^{\infty} f_1(x) \exp\left[-\frac{(u-x)^2}{2\sigma^2}\right] dx,$$

and we wish to find $f_1(x)$ from the experimental values of $f(u)$ and σ^2. In general, this is a difficult problem unless $f(u)$ turns out to be particularly simple. More usually, the form of $f_1(x)$ is assumed but is allowed to depend on one or more parameters. The integral is then evaluated and compared with the experimental values of $f(u)$ and the parameters varied until a match is obtained. Even so, exact evaluation of the integral is rarely possible and numerical methods must be used. One of these is the so-called Monte Carlo method that we will meet briefly in Chapter 5.

Example 3.12

If two random variables X and Y have probability densities of the form

$$f(x) = \frac{1}{\sigma_x \sqrt{2\pi}} \exp\left[-\frac{x^2}{2\sigma_x^2} \right],$$

and similarly for Y, find the density function $h(u)$ of the random variable $U = X + Y$.

Solution: From (3.33b), the density $h(u)$ is given by

$$h(u) = \int_{-\infty}^{\infty} f_1(x) f_2(u-x) dx = \frac{1}{2\pi\sigma_x\sigma_y} \int_{-\infty}^{\infty} \exp\left[-\frac{x^2}{2\sigma_x^2} - \frac{(u-x)^2}{2\sigma_y^2} \right] dx.$$

Completing the square for the exponent gives

$$-\frac{\sigma^2}{2\sigma_x^2\sigma_y^2}\left[\left(x - \frac{\sigma_x^2}{\sigma^2} u \right)^2 - \frac{\sigma_x^4}{\sigma^4} u^2 + \frac{\sigma_x^2}{\sigma^2} u^2 \right],$$

where $\sigma^2 = \sigma_x^2 + \sigma_y^2$. Then changing variables to $v = (\sigma/\sigma_x\sigma_y)(x - \sigma_x^2 u/\sigma^2)$ in the integral and simplifying yields

$$h(u) = \frac{1}{2\pi\sigma} \exp\left[\left\{ \frac{(\sigma_x^2 - \sigma^2)\sigma_x^2}{2\sigma^2\sigma_x^2\sigma_y^2} \right\} u^2 \right] \int_{-\infty}^{\infty} \exp\left(-\frac{v^2}{2} \right) dv = \frac{1}{\sigma\sqrt{2\pi}} \exp\left(-\frac{u^2}{2\sigma^2} \right).$$

Expectation values may also be found for functions of X. If $h(X)$ is a function of X, then its expectation value is

$$E[h(X)] = \sum_{i=1}^{n} h(x_i) P[x_i] = \sum_{x} h(x) f(x), \tag{3.34a}$$

if the variate is discrete, and

$$E[h(X)] = \int_{-\infty}^{+\infty} h(x) f(x) dx, \tag{3.34b}$$

if it is continuous. Note that, in general, $E[h(X)] \neq h(E[X])$.

Problems 3

3.1 Use the method of characteristic functions to find the first two moments of the distribution whose pdf is

$$f(x) = \frac{a^\gamma}{\Gamma(\gamma)} e^{-ax} x^{\gamma-1}, \quad 0 \le x \le \infty; \ a > 0, \gamma > 0,$$

where $\Gamma(\gamma)$ is the gamma function, defined by

$$\Gamma(\gamma) \equiv \int_0^\infty e^{-x} x^{\gamma-1} dx, \quad 0 < \gamma < \infty$$

3.2 A disgruntled employee types n letters and n envelopes but assigns the letters randomly to the envelopes. What is the expected number of letters that will arrive at their correct destination?

3.3 An incompetent purchasing clerk repeatedly forgets to specify the magnitude of capacitors when ordering them from a manufacturer. If the manufacturer makes capacitors in 10 different sizes and sends 1 at random, what is the expected number of different capacitor values received after 5 orders are placed?

3.4 (a) Find the characteristic function of the exponential random variable defined by the following pdf:

$$f(x) \equiv f(x; \lambda) = \begin{cases} \lambda e^{-\lambda x} & \lambda > 0, x \geq 0 \\ 0 & \text{otherwise.} \end{cases}$$

(b) Find the characteristic function of a sum of two independent identical exponential random variables and verify that it is the square of the characteristic function of a single exponential random variable.

3.5 Two random variables X and Y have a joint density function

$$f(x, y) = \begin{cases} 3e^{-x} e^{-3y} & 0 < x, y < \infty \\ 0 & \text{otherwise.} \end{cases}$$

Find (a) $P[X < Y]$ and (b) $P[X > 1, Y < 2]$.

3.6 Two random variables X and Y have a joint density function

$$f(x, y) = \begin{cases} cx^2(1 + x - y) & 0 < x, y < 1 \\ 0 & \text{otherwise,} \end{cases}$$

where c is a constant. Find the conditional density of X given Y.

3.7 If $H = aX + bY$, where X and Y are random variables and a and b are constants, find the variance of H in terms of the variances of X and Y and their covariance.

3.8 Find the probability density of the random variable $U = X + Y$, where X and Y are two independent random variables distributed with densities of the form

$$f_x(x) = \begin{cases} 1 & 0 \leq x < 1 \\ 0 & \text{otherwise} \end{cases},$$

and similarly for $f_y(y)$.

3.9 The table shows the joint probability function of two discrete random variables X and Y defined in the ranges $1 \leq x \leq 3$ and $1 \leq y \leq 4$, respectively. (Note that the probabilities are correctly normalized.)

y	1	2	3	Row totals
1	3/100	4/25	1/20	6/25
2	3/25	7/50	1/20	31/100
3	1/10	9/100	3/50	1/4
4	1/20	1/20	1/10	1/5
Column totals	3/10	11/25	13/50	1

Construct the following marginal probabilities:

(a) $P[X \leq 2, Y = 1]$, (b) $P[X > 2, Y \leq 2]$, and (c) $P[X + Y = 5]$.

3.10 Show that

$$E\left(\prod_{i=1}^{n} X_i \right) = \prod_{i=1}^{n} E[X_i],$$

provided the random variables $X_i (i = 1, n)$ are independently distributed.

3.11 The three independent random variables $X_i(i = 1, 2, 3)$ have expectation values $E[X_i] = i$ and $E[X_i^2] = i^2$, for $i = 1, 2, 3$. Calculate $E[X_1^2(X_2 - 2X_3)^2]$.

3.12 Find the means, variances, and covariance for the density of Example 3.8.

CHAPTER

4

Probability distributions II: Examples

Overview

This chapter builds on the work of Chapter 3 and applies the concepts and definitions given there to several practical distributions frequently met in physical science. These are as follows: The uniform, exponential, Cauchy, binomial, multinomial, Poisson, and normal (Gaussian) distributions. Because of the great importance of the normal distribution, this is discussed in some detail for both the single variate and the multivariate cases, with a separate section on the bivariate case. The basic features and properties of these various distributions, and any relations between them, are derived.

4.1 Continuous variables

We will start with distributions of continuous random variables, because in the case of large samples many distributions often approximate that of the all-important and ubiquitous distribution of continuous random variables called the normal distribution, which we met briefly in earlier chapters.

4.1.1 Uniform distribution

The *uniform distribution*, the simplest of all distributions for a continuous random variable x,[1] has a density function

$$f(x) \equiv u(x; a, b) = \begin{cases} \dfrac{1}{b - a} & a \leq x \leq b \\ 0 & \text{otherwise} \end{cases}, \tag{4.1}$$

where a and b are constants[2]. The distribution function from (4.1) is

---

[1] As mentioned in Chapter 3, in the present chapter and all the chapters that follow, where there is no ambiguity, we revert to the simpler notation of using lower-case letters for both random and other variables.

[2] Here, and in the distributions that follow, we use the convention of separating the random variable from any constants by a semicolon.

Probability and Statistics for Physical Sciences, Second Edition
https://doi.org/10.1016/B978-0-443-18969-2.00004-0

$$F(x) = \begin{cases} 0 & x < a \\ \dfrac{x-a}{b-a} & a \leq x < b \\ 1 & x \geq b \end{cases} \qquad (4.2)$$

The uniform distribution describes situations where events occur with equal probability, for example when rounding the result of numerical calculations to the nearest integer. An example for a discrete variable is rolling an unbiased six-sided die, where each face has the same probability of being uppermost. Using (3.5) and (3.6) we can easily show that the mean and variance are given by

$$\mu = \frac{a+b}{2}; \qquad \sigma^2 = \frac{(b-a)^2}{12}. \qquad (4.3)$$

The value of a random variable uniformly distributed in the interval (0,1) is called a *uniform random number*. These random numbers are useful because they enable various probabilities and expectation values to be evaluated empirically. The theoretical importance of the uniform distribution is enhanced by the fact that any density function $f(x)$ of a continuous random variable x may be transformed into the uniform density function

$$g(u) = 1, \qquad 0 \leq u \leq 1,$$

by the transformation $u = F(x)$, where $F(x)$ is the distribution function of x. That is, we may select a value x from the random variable with distribution $F(x)$ by selecting a value u from the uniform random variable on the interval (0,1) and then calculate $x = F^{-1}(u)$. This follows from the fact that

$$\frac{du}{dx} = \frac{d}{dx} \int_{-\infty}^{x} f(x')dx' = f(x),$$

and hence by changing variables,

$$g(u) = f(x)\left|\frac{du}{dx}\right|^{-1} = 1, \qquad 0 \leq u \leq 1.$$

This property is useful in generating random numbers from an arbitrary distribution by transforming a set of uniformly distributed random numbers. It enables many properties of continuous distributions to be exhibited, by proving them for the particular case of the uniform distribution. It also follows that there is at least one transformation that transforms any continuous distribution to any other; it is simply the product of the transformation that takes one distribution into the uniform distribution, with the inverse of the transformation that takes the other distribution into the uniform distribution.

Example 4.1

Trains to a given destination depart on the hour and at 30 minutes past the hour. A passenger arrives at the station at a time t that is uniformly distributed in the interval from one hour to the next. What is the probability that they will have to wait at least 10 minutes for a train?

Solution: Let t denote the time in minutes past the hour that the passenger arrives at the station. Since t is a random variable uniformly distributed in the interval $(0, 60)$, it follows that the passenger will have to wait at least 10 minutes if they arrive up to 20 minutes past the hour, or between 30 and 50 minutes past the hour. Thus the required probability is

$$P[0 < t < 20] + P[30 < t < 50] = \frac{20}{60} + \frac{20}{60} = \frac{2}{3}.$$

4.1.2 Univariate normal (Gaussian) distribution

This distribution is by far the most important in statistics because many distributions encountered in practice are found to be of approximately this form, particularly when dealing with large samples, a point that was mentioned in Section 1.4 and that will be discussed in more detail in Chapter 5. Examples are measurements of the blood pressure, heights, and other characteristics of groups of people, provided the size of the group is large. The name 'normal' is perhaps unfortunate, because it might imply that all other distributions are somehow 'abnormal', which of course they are not. In physical sciences the normal distribution is more usually known as a *Gaussian distribution*, although several people in addition to Gauss have claimed to have studied this function. In this book the name used in statistics has been adopted. We start with the case of a single variate.

The *normal density function* for a single continuous random variable x is defined as

$$f(x) \equiv n(x; \mu, \sigma) = \frac{1}{\sqrt{2\pi}\sigma} \exp\left[-\frac{1}{2}\left(\frac{x-\mu}{\sigma}\right)^2\right], \quad (\sigma > 0) \tag{4.4}$$

and its distribution function is

$$F(x) \equiv N(x; \mu, \sigma) = \frac{1}{\sqrt{2\pi}\sigma} \int_{-\infty}^{x} \exp\left[-\frac{1}{2}\left(\frac{t-\mu}{\sigma}\right)^2\right] dt. \tag{4.5}$$

Graphs of $f(x)$ and $F(x)$ are shown in Figure 4.1 for $\mu = 0$, and $\sigma = 0.5, 1.0$, and 2.0. Keeping the value of σ fixed but changing the value of the parameter μ simply moves the curves along the x axis.

As this is the first nontrivial distribution we have encountered, it will be useful to implement some of our previous definitions. Firstly, it is clear from (4.4) that $f(x)$ is a single-valued nonnegative real number for all values of x. Furthermore, by the substitution

$$t^2 = \frac{1}{2}\left(\frac{x-\mu}{\sigma}\right)^2,$$

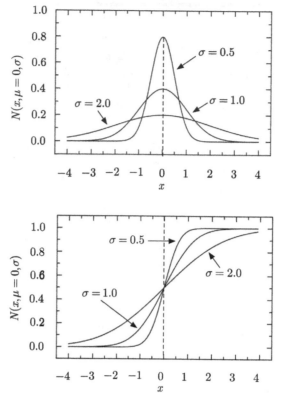

FIGURE 4.1 Normal (Gaussian) density function (upper graphs) and distribution function (lower graphs) for $\mu = 0$ and $\sigma = 0.5, 1.0$ and 2.0.

we can write

$$\int\limits_{-\infty}^{+\infty} f(x)\,\mathrm{d}x = \frac{1}{\sqrt{\pi}} \int\limits_{-\infty}^{+\infty} e^{-t^2}\,\mathrm{d}t.$$

Since the integral on the right-hand side has the well-known value of $\sqrt{\pi}$, we see that $f(x)$ is normalized to unity and is thus a valid density function.

To find the moments of the normal distribution, we first find the *mgf*. From Equation (3.11),

$$M_x(t) = E[\exp(tx)] = \exp(t\mu)E[\exp\{t(x - \mu)\}]$$

$$= \frac{\exp(t\mu)\exp(\sigma^2 t^2/2)}{(2\pi)^{1/2}\sigma} \int\limits_{-\infty}^{\infty} \exp\left[\frac{-(x - \mu - \sigma^2 t)^2}{2\sigma^2}\right]\mathrm{d}x.$$

The integral is related to the area under a normal curve with mean $(\mu + \sigma^2 t)$ and variance σ^2. Thus

$$M_x(t) = \exp(t\mu + \sigma^2 t^2/2). \tag{4.6}$$

On differentiating (4.6) twice and setting $t = 0$, we have

$$\mu_1' = \mu, \qquad \mu_2' = \sigma^2 + \mu^2,$$

and

$$\mathrm{var}(x) = \mu_2' - \left(\mu_1'\right)^2 = \sigma^2.$$

The mean and variance of the normal distribution are therefore μ and σ^2, respectively. The same technique for moments about the mean gives

$$\mu_{2n} = \frac{(2n)!}{n!2^n}\sigma^{2n} \quad \text{and} \quad \mu_{2n+1} = 0, \qquad n \geq 1. \tag{4.7}$$

The odd order moments are zero by virtue of the symmetry of the distribution. Using (4.7) we can calculate quantities that are sometimes used to measure skewness and the degree of peaking in a distribution. These are denoted β_1 and β_2, respectively (β_2 is also called the *kurtosis*), and are

$$\beta_1 \equiv \mu_3^2/\mu_2^3 = 0 \quad \text{and} \quad \beta_2 \equiv \mu_4/\mu_2^2 = 3. \tag{4.8}$$

This value of β_2 is taken as a standard against which the kurtosis of other distributions may be compared.

Using essentially the same technique that was used to derive the *mgf* we can show that the *cf* of the normal distribution is

$$\phi(t) = \exp\left[it\mu - t^2\sigma^2/2\right], \tag{4.9}$$

which agrees with (3.15) and which may be confirmed by applying the result of the inversion theorem.

Any normal distribution may be transformed to a normal distribution in z with $\mu = 0$ and $\sigma^2 = 1$ by setting $z = (x - \mu)/\sigma$. Then, from (4.4) and (4.5),

$$n(z; 0, 1) = \frac{1}{(2\pi)^{1/2}}\exp\left(-z^2/2\right) \tag{4.10}$$

and

$$N(z; 0, 1) = \frac{1}{(2\pi)^{1/2}} \int_{-\infty}^{z} \exp(-u^2/2)\,du. \tag{4.11}$$

These forms are called the *standard normal density function* and the *standard normal distribution function*, respectively, and will usually be denoted by $n(z)$ and $N(z)$, omitting the constants. Values of $N(z)$ are given in Appendix C, Table C.1. If these functions are required for negative values of z, they may be found from the relations

$$n(-z) = n(z), \tag{4.12}$$

and

$$N(-z) = 1 - N(z), \tag{4.13}$$

that follow from the symmetry of the distribution. Another useful relation that follows from (4.13) is

$$2\int_{0}^{z} n(u)\,du = \int_{-z}^{z} n(u)\,du = 2N(z) - 1. \tag{4.14}$$

Using (4.12)–(4.14) and Table C.1, the following results may be deduced:

i. The proportion of standard normal variates contained within 1, 2, and 3 standard deviations from the mean are 68.3%, 95.4%, and 99.7%, respectively.

ii. If t_α denotes that value of the standard normal distribution for which

$$\int_{t_n}^{\infty} n(t; 0, 1)\,dt = \alpha, \tag{4.15}$$

then $(\mu \pm t_\alpha \sigma)$ defines a $100(1 - 2\alpha)\%$ symmetric interval centered on μ.

The first of these results was mentioned in Section 1.4, when discussing the behavior of experimental frequency plots for cases where the sample size becomes large. These are stronger conditions than the constraints implied by Chebyshev's inequality (1.11). The reason why such plots tend to the normal distribution is embedded in the so-called Law of Large Numbers and will be discussed in Chapter 5. The usefulness of the second result will be evident when confidence intervals are discussed in Chapter 9.

We will see in Chapter 5 when we consider sampling in more detail that one can often assume that the measurement errors on a quantity x are distributed according to a normal distribution $n(x)$ with mean 0. This means that the probability of obtaining a value between x and $x + dx$ is $n(x)dx$. The dispersion σ of the distribution is called the *standard error* and

(4.15) tells us the probabilities for the true value being within an interval of plus or minus one, two, or three standard errors about the measured value, given a single measurement of x.

One final useful result is that the distribution of a linear sum

$$T = \sum_i a_i x_i$$

of n independent random variables x_i, having normal distributions $N(x_i; \mu_i \sigma_i^2)$, is distributed as $N(T; \mu, \sigma^2)$, where

$$\mu = \sum_{i=1}^{n} a_i \mu_i \quad \text{and} \quad \sigma^2 = \sum_{i=1}^{n} a_i^2 \sigma_i^2,$$

that is, it is also normally distributed. To show this, we can use the characteristic function. Because the x_i are independent, this may be written

$$\phi_T(t) = \prod_{i=1}^{n} E[\exp(it a_i x_i)] = \prod_{i=1}^{n} \phi_i(t),$$

where $\phi_i(t)$ is the *cf* of the random variable $a_i x_i$, and we have used the result proved in Section 3.2.3, that relates the *cf* of a sum of independent random variables to the product of the individual *cfs*. We have previously shown in (4.9) that

$$\phi_i(t) = \exp\left[it a_i \mu_i - t^2 \sigma_i^2 a_i^2 / 2\right],$$

and so

$$\phi_T(t) = \exp\left\{ \sum_{i=1}^{n} (it a_i \mu_i - t^2 a_i^2 \sigma_i^2 / 2) \right\} = \exp(it\mu - t^2 \sigma / 2),$$

where

$$\mu = \sum_{i=1}^{n} a_i \mu_i \quad \text{and} \quad \sigma^2 = \sum_{i=1}^{n} a_i^2 \sigma_i^2.$$

But this is the *cf* of a normal variate whose mean is μ and whose variance is σ^2. Thus, by the inversion theorem, T is distributed as $N(T; \mu, \sigma^2)$.

Example 4.2

Resistors are manufactured with a mean value of $R = 50$ ohms, and values less than 48.0 ohms, or greater than 51.5 ohms, are considered defective. If the values of R are assumed to be normally distributed with a standard deviation of 1 ohm, what percentage of resistors would be expected to be defective?

Solution: In standard measure $R = 48.0 \Rightarrow z = -2.0$ and $R = 51.5 \Rightarrow z = 1.5$. So the probability that a resistor is defective is

$$P = P[z < -2.0] + P[x > 1.5] = [1 - N(2.0; 0, 1)] + [1 - N(1.5; 0, 1)] = 0.0896, \text{ about } 9\%.$$

Example 4.3

If x is a random variable normally distributed with mean μ and variance σ^2, show that for any constants a and b, with $b \neq 0$, $y = a + bx$ is a normally distributed random variable with mean $(a + b\mu)$ and variance $b^2\sigma^2$.

Solution: If $F_x(x)$ and $F_u(y)$ are the distribution functions of x and y, respectively, then for $b > 0$,

$$F_y(y) = P[(a + bx) \leq y] = P[x \leq (y - a)/b] = F_x(\{y - a\}/b).$$

Similarly, for $b < 0$,

$$F_y(y) = P[(a + bx) \leq y] = P[x \geq (y - a)/b] = 1 - F_x(\{y - a\}/b).$$

The associated density functions $F_x(x)$ and $F_u(y)$ are obtained by differentiating the distribution functions. Thus

$$f_y(y) = \begin{cases} \dfrac{1}{b} f_x\left[\dfrac{y - a}{b}\right] & b > 0 \\[2ex] -\dfrac{1}{b} f_x\left[\dfrac{y - a}{b}\right] & b < 0 \end{cases},$$

which when combined is

$$f_y(y) = \frac{1}{|b|} f_x\left(\frac{y - a}{b}\right) = \frac{1}{\sigma|b|\sqrt{2\pi}} \exp\left[-\frac{1}{2\sigma^2}\left(\frac{y - a}{b} - \mu\right)^2\right]$$

$$= \frac{1}{\sigma|b|\sqrt{2\pi}} \exp\left[-\frac{1}{2}\left(\frac{y - a - b\mu}{b\sigma}\right)^2\right].$$

This is a normal distribution with mean $(a + b\mu)$ and variance $b^2\sigma^2$.

Example 4.4

A sample of three random variables $x_i (i = 1, 2, 3)$ is taken from a normal distribution with mean $\mu \neq 0$ and variance $\sigma^2 = 1/20$. What values of the constants a and b would ensure that the function

$$w = ax_1 + bx_2 - x_3$$

is distributed as a standard normal variable.

Solution: For a standard normal distribution, $\mu = 0$ and $\sigma^2 = 1$, so

$$E[w] = E[ax_1 + bx_2 - x_3] = (a + b - 1)\mu = 0,$$

and hence

$$a + b - 1 = 0.$$

The variance of w is

$$\text{var}(w) = (a^2 + b^2 + 1)\sigma^2 = 1,$$

and hence

$$a^2 + b^2 + 1 = 20.$$

Solving these two equations for a and b yields $a = 3.54$ and $b = -2.54$ (a and b are interchangeable).

4.1.3 Multivariate normal distribution

If $(x_1, x_2, ..., x_n \equiv \mathbf{x})$ are n random variables, then the *multivariate normal density function*, of order n, also called the *n-variate normal*, is

$$f(x) = \frac{1}{(2\pi)^{n/2}|\mathbf{V}|^{1/2}} \exp\left[-\frac{1}{2}(\mathbf{x} - \boldsymbol{\mu})^T \mathbf{V}^{-1}(\mathbf{x} - \boldsymbol{\mu}) \right], \tag{4.16}$$

where the constant vector $\boldsymbol{\mu}$ is the mean of the distribution, and $\mathbf{V}$ is an $n \times n$ symmetric positive-definite matrix, which is the variance matrix of the vector $\mathbf{x}$. The quantity

$$Q = (\mathbf{x} - \boldsymbol{\mu})^T \mathbf{V}^{-1}(\mathbf{x} - \boldsymbol{\mu}), \tag{4.17}$$

is called the *quadratic form* of the multivariate normal distribution. The distribution possesses a number of important properties, and three are discussed below.

The first concerns the form of the joint marginal distribution of a subset of the n variables. If the n random variables $(x_1, x_2, ..., x_n)$ are distributed as an n-variate normal distribution, then the joint marginal distribution of any set $x_i (i = 1, 2, ..., m < n)$ is the m-variate normal. This result can be proved in a straightforward way by constructing the joint marginal distribution from (4.16) using the definition (3.20a). It follows that the distribution of any single random variable in the set x_i (this is the case $m = 1$) is distributed as the univariate normal.

This result can be used to derive the second property: The necessary condition under which the variables of the distribution are independent. If we set $\text{cov}(x_i, x_j) = 0$ for $i \neq j$, then this implies that $\mathbf{V}$ is diagonal, so the quadratic form becomes

$$(\mathbf{x} - \boldsymbol{\mu})^T \mathbf{V}^{-1}(\mathbf{x} - \boldsymbol{\mu}) = \sum_{i=1}^{n} (x_i - \mu_i)^2 V_{ii}^{-1},$$

and the density function may be written as

$$f(\mathbf{x}) = \prod_{i=1}^{n} f_i(x_i),$$

where

$$f_i(x_i) = \frac{1}{(2\pi)^{1/2}} \frac{1}{V_{ii}^{1/2}} \exp\left[-\frac{(x_i - \mu_i)^2}{2V_{ii}} \right].$$ (4.18)

Equation (4.18) is the density function for a univariate normal distribution and so, by virtue of the earlier result on the marginal distribution, and the definition of statistical independence, (3.24), the variables x_i are independently distributed. Thus the necessary condition for the components of $\mathbf{x}$ to be jointly independent is $\text{cov}(x_i, x_j) = 0$ for all $i \neq j$. In the case of the multivariate normal distribution this is also a sufficient condition. It is straightforward, by an analogous argument, to establish the inverse, that is, that if x_i are jointly independent, then $\mathbf{V}$ is diagonal.

The third, and final, property concerns the distribution of linear combinations

$$S = \sum_{i=1}^{n} a_i x_i = \mathbf{x}^T \mathbf{A},$$ (4.19)

of random variables $\mathbf{x} = x_i (i = 1, 2, ..., n)$, each of which has a univariate normal distribution, where a_i are constants and $\mathbf{A} = a_i (i = 1, 2, ..., n)$. The moment generating function of S is

$$M_S(t) = E\left[\exp(St)\right] = E\left[\exp(\mathbf{x}^T \mathbf{A} t)\right] = \exp(\boldsymbol{\mu}^T \mathbf{A} t) E\left[\exp\{(\mathbf{x} - \boldsymbol{\mu})^T \mathbf{A} t\}\right].$$

Now, if $\mathbf{x}$ has a multivariate normal distribution with mean $\boldsymbol{\mu}$ and variance matrix $\mathbf{V}$, then

$$E\left[\exp\left[(x - \mu)^T \mathbf{A} t\right]\right] = \exp\left[(\mathbf{A}^T \mathbf{V} \mathbf{A}) t^2 / 2\right],$$

and thus

$$M_s(t) = \exp\left[(\boldsymbol{\mu}^T \mathbf{A}) t + (\mathbf{A}^T \mathbf{V} \mathbf{A}) t^2 / 2\right].$$

But from (4.6), this is the *mgf* of a normal variate with mean

$$\mu = \boldsymbol{\mu}^T \mathbf{A} = \sum_{i=1}^{n} a_i \mu_i$$

and variance

$$\sigma^2 = \mathbf{A}^T \mathbf{V} \mathbf{A} = \sum_{i=1}^{n} \sum_{j=1}^{n} a_i a_j V_{ij},$$

and so S is distributed as $N(S; \mu, \sigma^2)$. This result is a generalization of the result obtained at the end of Section 4.1.2 for the sum of two random variables.

4.1.4 Bivariate normal distribution

An important example of a multivariate normal distribution is the bivariate case, which occurs frequently in practice. Its density function is

$$n\left(x, y; \mu_x, \mu_y, \sigma_x, \sigma_y \rho\right) \equiv n(x, y) = \frac{1}{2\pi\sigma_x\sigma_y(1 - \rho^2)^{1/2}} \exp\left[\frac{-R}{2(1 - \rho^2)}\right], \qquad (4.20)$$

where

$$R \equiv \left(\frac{x - \mu_x}{\sigma_x}\right)^2 - 2\rho\left(\frac{x - \mu_x}{\sigma_x}\right)\left(\frac{y - \mu_y}{\sigma_y}\right) + \left(\frac{y - \mu_y}{\sigma_y}\right)^2, \qquad (4.21)$$

and ρ is the correlation coefficient, defined in (3.29). If the exponent in (4.20) is a constant $(-K)$, that is,

$$R = 2\left(1 - \rho^2\right)K,$$

then the points (x, y) lie on an ellipse with center $\left(\mu_x, \mu_y\right)$. The density function (4.20) is a bell-shaped surface, and any plane parallel to the xy plane that cuts this surface will intersect it in an elliptical curve. Any plane perpendicular to the xy plane will cut the surface in a curve of the normal form.

Just as for the univariate normal distribution, we can define a *standard bivariate normal density function*

$$n(u, v) = \frac{1}{2\pi(1 - \rho^2)} \exp\left[-\frac{(u^2 - 2\rho uv + v^2)}{2(1 - \rho^2)}\right], \qquad (4.22)$$

where

$$u = \frac{x - \mu_x}{\sigma_x}; \qquad v = \frac{y - \mu_y}{\sigma_y}.$$

A feature of this distribution is that for $\rho = 0$

$$n(u, v) = n(u)n(v), \qquad (4.23)$$

which implies that u and v are independently distributed, a result that is *not* generally true for all bivariate distributions.

Finally, the joint moment generating function may be obtained from the definition

$$M_{xy}(t_1, t_2) = E[\exp(t_1 x + t_2 y)] = \int\limits_{-\infty}^{\infty} \int\limits_{-\infty}^{\infty} \exp(t_1 x + t_2 y) f(x, y) dx dy. \qquad (4.24)$$

After changing variables to u and v, this becomes

$$M_{xy}(t_1, t_2) = \frac{\exp\left(t_1 \mu_x + t_2 \mu_y\right)}{2\pi(1 - \rho^2)^{1/2}} \iint e^{(t_1 \sigma_x u + t_2 \sigma_y v)} \exp\left[\frac{u^2 - 2\rho u v + v^2}{-2(1 - \rho^2)}\right] du dv,$$

which, after some algebra, gives

$$M_{xy}(t_1, t_2) = \exp\left[t_1 \mu_x + t_2 \mu_y + \frac{1}{2}\left(t_1^2 \sigma_x^2 + 2\rho t_1 t_2 \sigma_x \sigma_y + t_2^2 \sigma_y^2\right)\right]. \qquad (4.25)$$

The moments may be obtained in the usual way by evaluating the derivatives of (4.25) at $t_1 = t_2 = 0$. For example,

$$E[x^2] = \left.\frac{\partial^2 M_{xy}(t_1, t_2)}{\partial t_1^2}\right|_{t_1 = t_2 = 0} = \sigma_x^2 + \mu_x^2.$$

4.1.5 Exponential and gamma distributions

The *exponential* density function for a continuous random variable x with parameter λ is

$$f(x) \equiv f(x; \lambda) = \begin{cases} \lambda e^{-\lambda x} & \lambda > 0, x \geq 0 \\ 0 & \text{otherwise} \end{cases}. \qquad (4.26)$$

Its *mgf* $M_x(t) = E[e^{tx}]$ may be found from (4.26) and is

$$M_x(t) = \lambda \int\limits_0^{\infty} e^{tx} e^{-\lambda x} dx = \frac{\lambda}{\lambda - t}, \qquad t < \lambda. \qquad (4.27)$$

Differentiating, as usual, gives the mean μ and variance σ^2 as

$$\mu = 1/\lambda \quad \text{and} \quad \sigma^2 = 1/\lambda^2. \qquad (4.28)$$

The exponential density is used to model probabilities where there is an interval of time before an event occurs. Examples are the lifetimes of electronic components. In this context the parameter λ is the *(failure) rate*, or *inverse lifetime*, of the component. An interesting property of exponential random variables is that they are *memoryless*, that is, for example, the probability that a component will function for at least an interval $(t + s)$, having already operated for at least an interval t, is the same as the probability that a new component would operate for at least the interval s, if it were activated at time 0.

The proof of this key property, which is unique to exponentially distributed random variables, follows from the definition of conditional probability:

$$P[x > t + s | x > t] = \frac{P[x > t + s]}{P[x > t]}. \tag{4.29}$$

The probabilities on the right-hand side may be calculated from (4.26):

$$P[x > t + s] = \lambda \int_{t+s}^{\infty} e^{-\lambda x'} dx' = e^{-\lambda(t+s)},$$

and

$$P[x > t] = \lambda \int_{t}^{\infty} e^{-\lambda x'} dx' = e^{-\lambda t}.$$

$$P[x > t + s | x > t] = e^{-\lambda s} = P[x > s].$$

Example 4.5

A system that has been operating for 1000 hours has a critical component whose average lifetime is exponentially distributed with a mean value of 2000 hours. What is the probability that this component will continue to operate for more than another 1500 hours?

Solution: From the memoryless property of the exponential distribution, the distribution of the remaining lifetime of the component is independent of the time the system has already been operating and is exponential with parameter $\lambda = 1/2000$. Then,

$$P[\text{remaining lifetime} > 1500] = 1 - F[1500] = e^{-3/4} \approx 0.47.$$

If there are several independent random variables $x_1, x_2, ..., x_n$, each exponentially distributed with parameters $\lambda_1, \lambda_2, ..., \lambda_n$, respectively, then since the smallest value of a set of numbers is greater than some value x if, and only if, all values are greater than x,

$$P[\min(x_1, x_2, ..., x_n) > x] = P[x_1 > x, ... x_n > x].$$

But, because the variables are independently distributed,

$$P[\min(x_1, x_2, ..., x_n) > x] = \prod_{i=1}^{n} P[x_i > x] = \exp\left[-\sum_{i=1}^{n} \lambda_i x \right]. \tag{4.30}$$

This result may be used to model the lifetime of a complex system of several independent components, all of which must be working for the system to function. Thus, for example, if a device consists of n such components, each independently distributed as an exponential density with parameter λ_i, the probability that the lifetime T of the device will exceed a given time t is given by

$$P[T > t] = \exp\left[-\sum_{i=1}^{n} \lambda_i t\right].$$

In the exponential distribution the quantity λ is a constant, but there are many situations where it is more appropriate to assume that λ is not constant. An example is when calculating the failure rate with time of aging components that are subject to wear. In this case we could assume that $\lambda(t) = \alpha\beta t^{\beta-1}(t > 0)$, where α and β are positive constants, so that $\lambda(t)$ increases or decreases when $\beta > 1$ or $\beta < 1$, respectively. It was for precisely this situation, where the components were light bulbs, that the *Weibull distribution* was devised, with a density function

$$f(x; \alpha, \beta) = \alpha\beta x^{\beta-1} \exp\left(-\alpha x^{\beta}\right), \qquad x > 0, \tag{4.31}$$

which reduces to the exponential distribution when $\beta = 1$. It is a useful distribution for representing a situation where a probability rises from small values of x to a maximum and then falls again at large values of x.

At the end of Section 4.1.2 we gave a derivation of the probability density function for the sum of independent random variables, independently distributed as normal distributions. The same technique can be used to derive the *pdf* of a sum z of n independent random variables independently distributed as exponential distributions with the same value of λ. Problem 3.4 asks for the result for $n = 2$. Repeated application of the convolution formula gives for a sum of n exponential distributions

$$f(z) = \begin{cases} \dfrac{1}{(n-1)!}\lambda(\lambda z)^{n-1}e^{-\lambda z} & z \geq 0 \\ 0 & z < 0 \end{cases},$$

where n is an integer and λ is positive. Generalizing this to the case where n and λ are real and positive, we arrive at the *pdf* of the *gamma distribution*,

$$f(x; \alpha, \lambda) = \begin{cases} \dfrac{\lambda^{\alpha}x^{\alpha-1}e^{-\lambda x}}{\Gamma(\alpha)} & x \geq 0 \\ 0 & x < 0 \end{cases}, \tag{4.32}$$

where the gamma function Γ, introduced in Problem 3.1, is defined by

$$\Gamma(\alpha) = \int_0^\infty e^{-x}x^{\alpha-1}dx, \qquad 0 < \alpha < \infty,$$

and we have relabeled z by x. Graphs of the gamma *pdf* are shown in Figure 4.2 for $\lambda = 1$ and a range of values of α. Note that for large values of α, the *pdf* approaches that of a normal distribution. This is a consequence of the central limit theorem that will be discussed in Section 5.3. Specific values of $\Gamma(\alpha)$ can be found by integrating the defining integral by parts to give

$$\int_0^\infty e^{-x}x^{\alpha-1}dx = (\alpha-1)\int_0^\infty e^{-x}x^{\alpha-2}dx,$$

and in general, by repeated integrations for integer $\alpha = n$, and noting that

$$\Gamma(1) = \int_0^\infty e^{-x}dx = 1,$$

we have the recurrence relation $\Gamma(n) = (n-1)!$. Thus the exponential function is an example of the general class of *gamma distributions* with $\alpha = 1$.

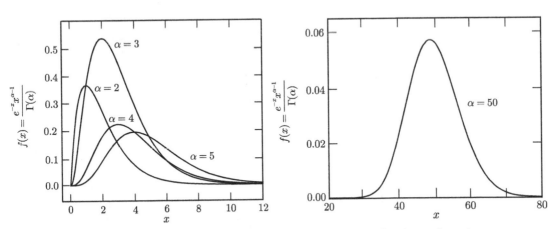

FIGURE 4.2   Graphs of the gamma distribution for $\lambda = 1$ and various values of α.

4.1.6 Cauchy distribution

The density function of the *Cauchy* distribution is

$$f(x,\theta) = \frac{1}{\pi}\cdot\frac{1}{1+(x-\theta)^2}, \qquad -\infty < x < \infty.$$

The parameter θ can be interpreted as the mean μ of the distribution only if the integral over $xf(x)$ is taken to be the principal value, defined by

$$\mu = \lim_{N \to \infty} \int_{-N}^{N} f(x; \theta)x \, dx.$$

In what follows, we will set $\theta = 0$. Then the distribution function becomes

$$F(x) = \frac{1}{2} + \frac{1}{\pi}\arctan(x).$$

The moment about the mean of order $2n$ is then

$$\mu_{2n} = \frac{1}{\pi} \int_{-\infty}^{\infty} \frac{x^{2n}}{1+x^2} \, dx, \tag{4.33}$$

but the integral converges only for $n = 0$, so only the trivial moment $\mu_0 = 1$ exists. Likewise, the *mgf* does not exist, although the *cf* does and is given by (see Example 3.6b)

$$\phi(t) = e^{|t|}.$$

It can be shown that the ratio of two normal variates has a Cauchy density function (see Example 4.6 below), which is one reason why it is encountered in practice. The Cauchy distribution is also met frequently in physical science because it describes the line shape seen in the decay of an excited quantum state, for example, an excited state of an atom. In this context it is usually called the *Lorentz distribution* or *Breit-Wigner formula* and is written as

$$f(E; E_0, \Gamma) = \frac{1}{\pi} \frac{\Gamma/2}{(E - E_0)^2 + \Gamma^2/4},$$

where E is the energy of the system, the parameters Γ and E_0 are interpreted as the 'half-width' of the state, that is, the full width at half maximum height of the line shape (mentioned in Section 1.4), and its energy, respectively. The different factors in the numerators here and in the form given earlier are to ensure that both density functions are correctly normalized. The Cauchy distribution must be treated with care because of the nonconvergence of the moment integrals, which is due to the long tails of the Cauchy density compared to those of the normal density. In these regions the distribution is not necessarily a good approximation to the physical system.

Example 4.6

Two random variables x and y are each distributed with standardized normal density functions. Show that the ratio x/y has a Cauchy probability density.

Solution: Define new variables $r = x/y$ and $s = y$. Then if $h(r,s)$ is the joint probability density of r and s, this may be found from probability conservation; that is, the realization of x and y being in a particular region of (x,y) space is the same as r and s being in the equivalent transformed region of (r,s) space. This implies the relation

$$|h(r,s)drds| = |n(x)n(y)dxdy|,$$

where n is the standard normal density. Changing variables on the right-hand side to r and s gives

$$h(r,s)drds = n(rs)n(s)Jdrds,$$

where

$$J = \begin{vmatrix} \dfrac{\partial x}{\partial r} & \dfrac{\partial y}{\partial r} \\ \dfrac{\partial x}{\partial s} & \dfrac{\partial y}{\partial s} \end{vmatrix} = \begin{vmatrix} y & 0 \\ r & 1 \end{vmatrix} = y = s$$

is the Jacobian of the transformation, as discussed in Section 3.4. Using the symmetry of the normal density about 0, the probability density of r is given by

$$f(r) = 2 \int_0^\infty n(rs)n(s)s \, ds = \frac{1}{\pi} \int_0^\infty \exp\left[-\frac{1}{2}s^2(1+r^2) \right] s \, ds$$

$$= \frac{1}{\pi} \left[-\frac{\exp\left[-\frac{1}{2}s^2(1+r^2) \right]}{(1+r^2)} \right]_0^\infty = \frac{1}{\pi}\frac{1}{1+r^2},$$

which is a Cauchy density.

4.2 Discrete variables

In this section we discuss three important, and frequently met, distributions that for large samples approximate to a normal distribution.

4.2.1 Binomial distribution

The binomial distribution concerns a population of members, each of which either possesses a certain attribute P or does not possess this attribute. We will denote 'not possessing the attribute' by Q. If the proportion of members possessing P is p and that possessing Q is q, then clearly $(p+q) = 1$. An experiment involving such a population is called a *Bernoulli trial*, that is, one with only two possible outcomes. A simple example is tossing a coin, where the two outcomes are 'heads' and 'tails', with $p+q = 0.5$ if the coin is unbiased and thin so that the probability of landing on its edge can be neglected. Suppose we wish to choose sets from the population, each of which contains n members. From the work of Section 2.2, the proportion of cases containing P r times and Q $(n - r)$ times is

$$_nC_r p^r q^{n-r} = \binom{n}{r} p^r q^{n-r}, \tag{4.34}$$

that is, the rth term of the binomial expansion of $f(p, q) = (q + p)^n$, hence the name of the distribution. Expressed in another way, if p is the chance of an event happening in a single trial, then for n independent trials the terms in the expansion

$$f(p, q) = q^n + nq^{n-1}p + \cdots + p^n$$

give the chances of $0, 1, 2, \ldots, n$ events happening. Thus we are led to the following definition. The probability function of the *binomial distribution* is defined as

$$f(r; p, n) = \binom{n}{r} p^r q^{n-r}, \tag{4.35}$$

and gives the probability of obtaining $r = 0, 1, 2, \ldots, n$ successes, that is, events having the attribute P, in an experiment consisting of n Bernoulli trials. Note that f is not a probability *density* but gives the actual probability. Tables of the cumulative binomial distribution are given in Appendix C Table C.2, and plots of the probability function for some values of its parameters are shown in Figure 4.3.

Example 4.7

If a machine making components has a failure rate of 2%, that is, 2% are rejected as being defective, what is the probability that less than three components will be defective in a random sample of size 100?

Solution: Using the binomial distribution, the probability that less than three components will be found to be defective is, with $p = 0.02$

$$P[r < 3] = p[r = 0] + P[r = 1] + P[r = 2]$$

$$= \sum_{r=0}^{2} \binom{100}{r} (0.02)^r (0.98)^{(100-r)} = 0.6767.$$

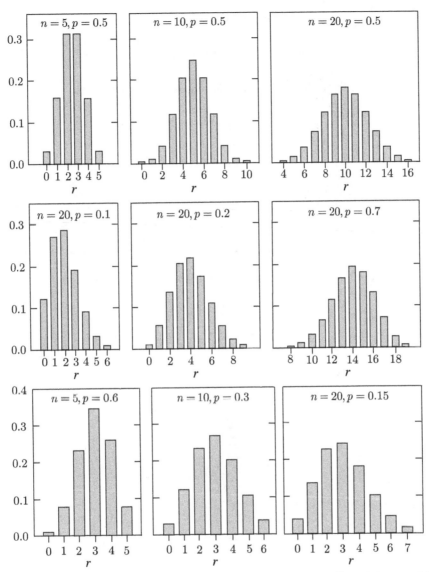

FIGURE 4.3 Plots of the binomial probability function. The top row shows $f(r; p, n)$ for $p = 0.5$ and various values of n; the middle row shows values for $n = 20$ and various values of p; and the lower row shows $f(r; p, n)$ for a fixed value of $np = 3$.

Example 4.8

A device consists of n components, each of which will function independently with a probability p and operates successfully if at least 50% of its components are fully functioning. A researcher can buy a device with four components with p = 0.5 or, for the same price, a device with three components but with a higher value of p. For what higher value of p would it be better to use a three-component system rather than one with four components?

Solution: The probability that a four-component system will fully function is

$$P_4 \binom{4}{2} p^2 (1-p)^2 + \binom{4}{3} p^3 (1-p) + \binom{4}{4} p^4 (1-p)^0,$$

which for $p = 0.5$ is $11/16$. The probability that a three-component system will function is

$$P_3 = \binom{3}{2} p^2 (1-p) + \binom{3}{3} p^3 (1-p)^0 = 3p^2 - 2p^3.$$

A three-component system is more likely to function than a four-component one if $P_3 > P_4$, that is, if

$$3p^2 - 2p^3 - \frac{11}{16} > 0,$$

which is true for (approximately) $p > 0.63$.

The binomial moment generating function may be found directly from (4.34) and the definition (3.12a) and is

$$M_r(t) = \sum_{r=0}^{n} f(r; p, n) e^{tr} = \sum_{r=0}^{n} \binom{n}{r} p^r q^{n-r} e^{tr} = (pe^t + q)^n, \qquad (4.36)$$

from which

$$\mu_1' = \mu = np, \qquad \mu_2' = np + n(n-1)p^2, \qquad (4.37)$$

and

$$\sigma^2 = \mu_2' - (\mu_1')^2 = npq. \qquad (4.38)$$

The *mgf* for moments about the mean is

$$M_\mu(t) = e^{-\mu t} M(t) \qquad (4.39)$$

and gives

$$\mu_3 = npq(q-p), \qquad \mu_4 = npq[1 + 3(n-2)pq]. \qquad (4.40)$$

So using the definitions given in (4.8), we have

$$\beta_1 = (q-p)^2/(npq) \quad \text{and} \quad \beta_2 = 3 + (1-6pq)/npq,$$

which tend to the values for a normal distribution as $n \to \infty$. The plots shown in Figure 4.3 suggest that the limiting form of the binomial distribution is indeed the normal.

The formal proof of this result uses the characteristic function, although it requires several stages. From the relation (3.15) and the form of the *mgf* (4.36), the *cf* is

$$\phi_r(t) = \left(q + pe^{it}\right)^n, \tag{4.41}$$

and the binomial distribution may be expressed in standard measure (i.e., with $\mu = 0$ and $\sigma^2 = 1$) by the transformation

$$x = (r - \mu)/\sigma. \tag{4.42}$$

This can be considered the sum of two independent random variables r/σ and $-\mu/\sigma$, even though the second term is actually a constant. From the work of Chapter 3, the characteristic function of x is the product of the characteristic functions of these two variates. Setting $\mu = np$ and using (4.41), we have

$$\phi_x(t) = \exp\left[\frac{-itnp}{\sigma}\right]\left\{q + p \, \exp\left[\frac{it}{\sigma}\right]\right\}^n.$$

Taking logarithms and using $p = 1 - q$ gives

$$\ln \phi_x(t) = \frac{-itnp}{\sigma} + n \, \ln\left\{1 + p\left[\exp\left(\frac{it}{\sigma}\right) - 1\right]\right\},$$

and since $t/\sigma = t/\sqrt{npq} \to 0$ as $n \to \infty$, the exponential may be expanded, giving

$$\ln \phi_x(t) = \frac{-itnp}{\sigma} + n \, \ln\left\{1 + p\left[\frac{it}{\sigma} - \frac{1}{2}\left(\frac{t}{\sigma}\right)^2 + \cdots\right]\right\}.$$

Next, we expand the logarithm on the right-hand side using

$$\ln(1 + \varepsilon) = \varepsilon - \varepsilon^2/2 + \varepsilon^3/3 - \cdots,$$

with the result

$$\ln \phi_x(t) = \frac{-itnp}{\sigma} + n\left\{\frac{itp}{\sigma} - \frac{1}{2}\left(\frac{t}{\sigma}\right)^2 (p - p^2) + \cdots\right\}.$$

Finally, letting $n \to \infty$ and keeping t finite gives

$$\ln \phi_x(t) = -t^2/2,$$

where we have used $\sigma^2 = npq = np(1 - p)$. So, for any finite t,

$$\phi(t) \to \exp\left(-t^2/2\right).$$

This is the form of the *cf* of a standardized normal distribution, and so by the inversion theorem, the associated density function is

$$f(x) = \frac{1}{(2\pi)^{1/2}} \exp\left(-\frac{x^2}{2}\right),$$ (4.43)

which is the standard form of the normal distribution.

The normal approximation to the binomial is excellent for large values of n and is still good for small values provided p is reasonably close to $1/2$. A working criterion is that the approximation is good if np and nq are both greater than 5. This is confirmed by the plots in Figure 4.3.

Example 4.9

If the probability of a success in a single Bernoulli trial is $p = 0.4$, compare the exact probability of obtaining $r = 5$ successes in 20 trials with the normal approximation.

Solution: The binomial probability is

$$P_B = \binom{20}{5}(0.4)^5(0.6)^{15} = 0.075$$

to three significant figures. In the normal approximation this corresponds to the area under a normal curve in standard form between the points corresponding to $r_1 = 4.5$ and $r_2 = 5.5$. Using $\mu = np = 8$ and $\sigma = \sqrt{npq} = 2.19$, the corresponding standardized variables are $z_1 = -1.60$ and $z_2 = -1.14$. Thus we need to find

$$P_N = P[z < -1.14] - P[z < -1.60] = F(-1.14) - F(-1.60),$$

where F is the standard normal distribution function. Using $F(-z) = 1 - F(z)$ and Table C.1 gives $P_N = 0.072$, so the approximation is accurate to about 4%.

4.2.2 Multinomial distribution

The multinomial distribution is the generalization of the binomial distribution to the case of n repeated trials where there are more than two possible outcomes to each. It is defined as follows. If an event may occur with k possible outcomes, each with a probability $p_i(i = 1, 2, ..., k)$, with

$$\sum_{i=1}^{k} P_i = 1,$$ (4.44)

and if r_i is the number of times the outcome associated with p_i occurs, then the random variables $r_i(i = 1, 2, ..., k-1)$ have a *multinomial probability function* defined as

$$f(r_1, r_2, ..., r_{k-1}) \equiv n! \prod_{i=1}^{k} p_i^{r_i} / \prod_{i=1}^{k} r_i!, \quad r_i = 0, 1, 2, ...n.$$ (4.45)

Note that the factor $n! \Big/ \prod_{i=1}^{k} r_i!$ is the multinomial coefficient given in (2.13) and is the number of ways the combination of outcomes represented by $r_1, r_2, \ldots, r_k$, could occur. Also, each of the r_i may range from 0 to n inclusive, and only $(k-1)$ variables are involved because of the linear constraint

$$\sum_{i=1}^{k} r_i = n.$$

Just as the binomial distribution tends to the univariate normal, so the multinomial distribution tends in the limit to the multivariate normal distribution.

With suitable generalizations the results of Section 4.6 may be extended to the multinomial. For example, the mean and variance of the random variables r_i are np_i and $np_i(1 - np_i)$, respectively. Multiple variables mean that we also have a covariance matrix, given by

$$V_{ij} = E\left[\{r_i - E[r_i]\}\{r_j - E[r_j]\}\right].$$

It is straightforward to show that

$$V_{ij} = \begin{cases} np_i(1 - p_i) & i = j \\ -np_i p_j & \text{otherwise} \end{cases}.$$

An example of a multinomial distribution is if we were to construct a histogram of k bins from n independent observations on a random variable, with r_i entries in bin i. The negative sign in the off-diagonal elements of the covariance matrix shows that if bin i contains a greater than average number of events, then the probability is increased that a different bin j will contain a smaller-than-average number, as expected.

Example 4.10

A bag contains 14 balls: 5 white, 4 red, 3 blue, and 2 yellow. A ball is drawn at random from the bag and then replaced. If 10 balls are drawn and replaced, what is the probability of obtaining 3 white, 3 red, 2 blue, and 2 yellow balls?

Solution: The probability of obtaining a given number of balls of a specified color after n drawings is given by the multinomial probability. We know that in a single drawing

$$P[w] = \frac{5}{14}, P[r] = \frac{4}{14}, P[b] = \frac{3}{14} \text{ and } P[y] = \frac{2}{14}.$$

Thus if 10 balls are drawn and replaced, the required probability is, using (4.45),

$$\frac{10!}{3!3!2!2!}\left(\frac{5}{14}\right)^3\left(\frac{4}{14}\right)^3\left(\frac{3}{14}\right)^2\left(\frac{2}{14}\right)^2 = 0.0251.$$

4.2.3 Poisson distribution

The Poisson distribution is an important distribution occurring frequently in physical science (and elsewhere) where the probability of an event is very small, but the total number of trials is large. An example is the decay of a radioactive material. Here the probability that any particular atom in a sample will decay is very small, but a typical sample will contain a vast number of atoms. This is discussed in more detail at the end of this section, along with other examples.

The Poisson distribution is derived from the binomial distribution by a special limiting process. Consider the binomial distribution for the case when p, the probability of achieving the outcome P, is very small, but n, the number of members of a given sample, is large such that

$$\lim_{p \to 0}(np) = \lambda \tag{4.46}$$

where λ is a finite positive constant, that is, where $n \gg np \gg p$. The kth term in the binomial distribution then becomes

$$\left[\binom{n}{k}p^k q^{n-k}\right] = \frac{n!}{k!(n-k)!}\left(\frac{\lambda}{n}\right)^k \frac{(1-\lambda/n)^n}{(1-\lambda/n)^k}$$

$$= \frac{\lambda^k}{k!}\left[\frac{n(n-1)(n-2)\cdots(n-k+1)}{n^k}\right]\frac{(1-\lambda/n)^n}{(1-\lambda/n)^k}$$

$$= \frac{\lambda^k}{k!}\left(1-\frac{\lambda}{n}\right)^n\left[\frac{(1-1/n)(1-2/n)\cdots(1-(k-1)/n)}{(1-\lambda/n)^k}\right].$$

Now as $n \to \infty$,

$$\lim_{n \to \infty}\left(1-\frac{\lambda}{n}\right)^n = e^{-\lambda},$$

which may be seen by applying the binomial theorem to $(1-\lambda/n)^n$ and then taking the limit as $n \to \infty$. Then all the factors in the last square bracket tend to unity, so in the limit that $n \to \infty$ and $p \to 0$ but $np \to \lambda$,

$$\lim_{np \to \lambda}\left[\binom{n}{k}p^k q^{n-k}\right] = f(k;\lambda) = \frac{\lambda^k}{k!}\exp(-\lambda), \quad \lambda > 0, k = 0,1\dots. \tag{4.47}$$

This is the probability function of the *Poisson distribution*. Although in principle k can take on any integer value, $f(k,\lambda)$ is vanishingly small for large values of k and in practice $f(k,\lambda)$ can

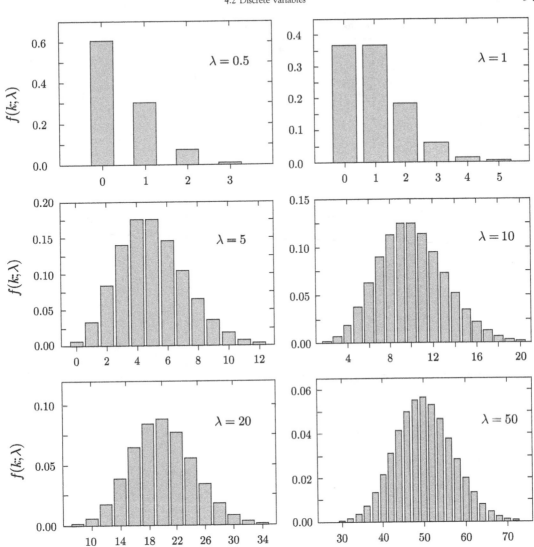

FIGURE 4.4 Plots of the Poisson probability function $f(k, \lambda) = \lambda^k \exp(-\lambda)/k!$ for various values of λ.

be taken as zero for all values of k modestly greater than λ. Some examples of the Poisson probability function are shown in Figure 4.4 and tables of the cumulative distribution are given in Appendix C, Table C.3.

Although in deriving the Poisson distribution we have taken the limit as $n \to \infty$, the approximation works well for modest values of n, provided p is small. This is illustrated in Table 4.1, which shows probability values of the binomial distribution for various values of n and p such that $np = 3$ (see also the plots in Figure 4.2) compared to the probabilities of the Poisson distribution for $\lambda = 3$.

4. Probability distributions II: Examples

TABLE 4.1 Comparison of the binomial and Poisson probability functions for $np = \lambda = 3$.

k bo	Binomial						Poisson $\lambda = 3$
	$p = 0.5$ $n = 6$	$p = 0.2$ $n = 15$	$p = 0.1$ $n = 30$	$p = 0.05$ $n = 60$	$p = 0.02$ $n = 150$	$p = 0.01$ $n = 300$	
0	0.0156	0.0352	0.0424	0.0461	0.0483	0.0490	0.0490
1	0.0937	0.1319	0.1413	0.1455	0.1478	0.1486	0.1486
2	0.2344	0.2309	0.2276	0.2259	0.2248	0.2244	0.2244
3	0.3125	0.2501	0.2361	0.2298	0.2263	0.2252	0.2252
4	0.2344	0.1876	0.1771	0.1724	0.1697	0.1689	0.1689
5	0.0937	0.1032	0.1023	0.1016	0.1011	0.1011	0.1010
6	0.0156	0.0430	0.0474	0.0490	0.0499	0.0501	0.0501
7	0.0000	0.0138	0.0180	0.0199	0.0209	0.0213	0.0213
8		0.0035	0.0058	0.0069	0.0076	0.0079	0.0079
9		0.0007	0.0016	0.0021	0.0025	0.0026	0.0026
10		0.0001	0.0004	0.0006	0.0007	0.0008	0.0008
11		0.0000	0.0001	0.0001	0.0002	0.0002	0.0002
12			0.0000	0.0000	0.0000	0.0000	0.0000

The moment generating function for the Poisson distribution is

$$M_k(t) = E\left[e^{kt}\right] = e^{-\lambda} \sum_{k=0}^{\infty} \frac{(\lambda e^t)^k}{k!} = e^{-\lambda} \exp(\lambda e^t). \tag{4.48}$$

Differentiating (4.48) and setting $t = 0$ gives

$$\begin{aligned} \mu_1' &= \lambda, & \mu_2' &= \lambda(\lambda + 1), \\ \mu_3' &= \lambda\left[(\lambda + 1)^2 + \lambda\right], & \mu_4' &= \lambda\left[\lambda^3 + 6\lambda^2 + 7\lambda + 1\right], \end{aligned} \tag{4.49}$$

and it is easily shown from the definition of moments (1.8) that

$$\mu_2 = \lambda, \quad \mu_3 = \lambda, \quad \mu_4 = \lambda(3\lambda + 1). \tag{4.50}$$

Thus

$$\mu = \sigma^2 = \lambda, \tag{4.51}$$

a simple result which is very useful in practice. Also from (4.50) and (4.8), we have

$$\beta_1 = \frac{1}{\lambda}, \qquad \beta_2 = 3 + \frac{1}{\lambda}. \qquad (4.52)$$

From these results, and the fact that the Poisson distribution is derived from the binomial, one might suspect that as $\lambda \to \infty$ the Poisson distribution tends to the standard form of the normal, and indeed this is the case. It can be proved by again using the characteristic function. Using (3.15) and (4.48), the characteristic function is

$$\phi_k(t) = e^{-\lambda} \exp(\lambda e^{it}).$$

Transforming the distribution to standard measure by the relation

$$z = (k - \mu)/\sigma$$

gives

$$\phi_z(t) = \int\limits_{-\infty}^{\infty} f(z) \exp[it(\sigma z + \mu)] dz = e^{it\sigma} \phi_z(\sigma t).$$

But from (4.51), $\mu = \sigma^2 = \lambda$ and so

$$\phi_z(t) = \exp(-it\lambda^{1/2}) e^{-\lambda} \exp\left(\lambda e^{it\lambda^{-1/2}}\right)$$

and

$$\ln \phi_z(t) = -it\lambda^{1/2} - \lambda + \lambda \exp(it\lambda^{-1/2}).$$

Expanding the exponential letting $\lambda \to \infty$ and keeping t finite gives

$$\begin{aligned}
\ln \phi_z(t) &= -it\lambda^{1/2} - \lambda + \lambda\left[1 + (it\lambda^{-1/2}) + (it\lambda^{-1/2})^2/2! + \cdots\right] \\
&= \lambda(it\lambda^{-1/2})^2/2 = -t^2/2,
\end{aligned}$$

and hence

$$\ln \phi_z(t) = -t^2/2 + O(\lambda^{-1/2}).$$

Thus, for any finite t,

$$\phi(t) \to \exp(-t^2/2),$$

which is the form of the *cf* of a standardized normal distribution and so, by the inversion theorem, the associated density function is the standardized form of the normal distribution. The rate of convergence to normality is the same as for the binomial distribution and so, in particular, the normal approximations to the Poisson distribution are quite adequate for values of $\lambda \geq 10$ and some authors suggest even lower values.

Finally, it can be shown that if each of a Poisson number of events having mean λ is independently classified as being of one of the types $1, 2, \ldots, r$, with probabilities $p_1, p_2, \ldots, p_r$, respectively, where $\sum\limits^{r} p_r = 1$, then the numbers of events of types $1, 2, \ldots, r$ are independent Poisson random variables with means $\lambda p_1, \lambda p_2, \ldots, \lambda p_r$, respectively. As for the normal distribution, the characteristic function may also be used in a straightforward way to show that the sum of quantities independently distributed as Poisson variates is itself a Poisson variate.

We return now to the example of the decay of a radioactive material. A macroscopic amount of the material contains a vast number of atoms, each of which could in principle decay, but the probability of any individual atom decaying in a given time interval is a random event with a very small probability. In this case the quantity $1/\lambda$ is the lifetime of the unstable atom or nucleus. If decays occur randomly in time, with an average of λ events per unit time, then from the Poisson distribution, the probability of N events occurring in an interval t is

$$P[N] = \frac{1}{N!}(\lambda t)^N e^{-\lambda t},$$

and so the probability of no events occurring in time t is an exponential distribution and the probability that the time interval t between events (e.g., the time interval between the detection of the decay particles in a detector) is greater than a specified value x is

$$P[t > x] = e^{-\lambda x}.$$

The memoryless property of the exponential distribution implies that if no events have occurred up to a time y, the probability of no events occurring in a subsequent period x is independent of y.

Example 4.11

If the probability of an adverse reaction to a single exposure to a very low dosage of radiation is 0.1% and 10,000 people are exposed in an accident, use the Poisson distribution to find the probability that less than 3 will be adversely affected. Why is the use of this distribution justified?

Solution: The probability of an adverse reaction is an example of a Bernoulli trial, because there is either a reaction or no reaction. However, if the radiation dose is very low, the probability of an adverse reaction is very small, so in practice, the Poisson distribution may be used to predict how many people will suffer an adverse reaction in a large sample. Then,

$$P[k < 3] = P[k = 0] + P[k = 1] + P[k = 2].$$

Using the Poisson distribution with $\lambda = np = 10000 \times 0.001 = 10$, this is

$$P[k<3] = e^{-10}\left(\frac{10^0}{0!} + \frac{10^1}{1!} + \frac{10^2}{2!}\right) = 61e^{-10} = 0.0028.$$

Example 4.12

Use the data of Example 4.11 to investigate the normal approximation to the Poisson for calculating the probability that exactly five people will have an adverse reaction.

Solution: The Poisson probability that exactly five people will have an adverse reaction is

$$P[k = 5] = \frac{e^{-10}10^5}{5!} = 0.0378.$$

In the normal approximation this corresponds to the area under the normal density curve between the points 4.5 and 5.5, and in standard form these points are, using $\mu = \lambda = 10$ and $\sigma = \sqrt{\lambda} = 3.16$, $z_1 = -1.74$ and $z_2 = -1.42$. Then using Table C.1, as in Example 4.8, we find a probability of 0.0384. So the normal approximation is good.

Problems 4

4.1 A company makes electrical components with a mean life of 800 days and a standard deviation of 20 days. If the distribution of lifetimes is normal, (a) what is the probability that a component chosen at random will last between 780 and 850 days? (b) And what is the minimum lifetime of the longest-lived 12% of the components?

4.2 A supply voltage V is assumed to be a normal random variable with a mean of 100 volts and a variance of 25 volts. It is applied to a resistor with resistance $R = 50$ ohms and the power $W = RV^2$ measured. What is the probability that $W > 6 \times 10^5$ watts?

4.3 A device consists of three sections of different lengths $x_i(i = 1, 2, 3)$. The sections are made by different machines and their lengths are independently distributed as a normal distribution $n(x_i; \mu_i, \sigma_i)$, where the means μ_i and variances σ_i^2 are as follows:

i	1	2	3
μ_i	12	16	18
σ_i^2	0.03	0.04	0.05

If S is the sum of the lengths, calculate $P[45.5 \le S \le 46.5]$.

4.4 The probability of recovering from a certain illness without medical intervention is 50%. A new drug is developed and tested on 20 people with the illness. Fourteen rapidly recover. Is the drug effective?

4.5 In a system designed to destroy incoming missiles, defensive weapons are arranged in layers, each having an efficiency of 95%. To be sure of totally destroying a missile, 'hits' from weapons from at least two defensive layers are required. How many layers would be needed to ensure a probability of at least 99.9% of destroying an incoming missile?

4.6 Find the coefficient of the term $x^6y^4z^6$ in the expansion of $\left(2x^2 - 3xy^2 + z^3\right)^6$.

4.7 A biased coin has a probability of 0.48 to fall 'heads' and 0.49 to fall 'tails'. If the coin is thick such that there is also a probability of it landing on its edge, what is the probability of obtaining 4 heads and 4 tails if it is tossed 10 times?

4.8 A supplier is contracted to produce components and has the option of using one of two methods. Method A produces the components with a mean lifetime of 250 hours at a basic cost of C. Method B produces the components with a mean lifetime of 350 hours at a basic cost of kC, where k > 1. However, if a component's actual lifetime is less than 400 hours, the supplier has to pay a penalty P. Find the condition on k, in terms of C and P, such that the supplier should choose Method A.

4.9 Figure 4.5 shows a system of three units, each of which by itself can ensure continuity of output. When Unit 1 functions, Units 2 and 3 are on standby, but when Unit 1 fails, a switch S activates the standby Unit 2, and likewise, when the latter fails, S activates Unit 3. If the three units have lifetimes $t_i (i = 1, 2, 3)$, independently distributed with densities

$$f(t_i) = \frac{1}{100}\exp(-t/100), \ t \geq 0$$

what is the probability that the system will operate for at least T hours?

FIGURE 4.5   A device consisting of three units.

4.10 A beam of particles is incident on a target with sufficient energy to penetrate it. The particles are mostly absorbed, but there is a small probability p of 5% that this is accompanied by the emission of a new particle from the target. If 100 particles per second are incident on the target, what is the probability that at least 5 particles per second are emitted? Compare your result using the Poisson distribution.

4.11 The average number of car accidents at a dangerous road junction is five per month. What is the probability that there will be more than three accidents next month?

4.12 A nuclear physics experiment uses 80 detectors. They are checked between data runs and any that have failed are replaced. It is found that the detectors have a 1% probability of failing between checks. If a run can be successfully completed, provided no more than three detectors fail during the run, find the probability that a data run will be spoiled because of detector failure.

Sampling and estimation

Overview

This chapter turns away from purely descriptive statistics to the problem of estimating the values of parameters by calculation and by simulation. It starts with the idea of observations as a random sample and the concept of an 'estimator' as a function of the sample. The desirable properties of estimators are discussed in detail and estimators for the mean, variance, and covariance with these properties are derived. The central importance of the normal distribution in statistics derives from the laws of large numbers and the powerful central limit theorem. These are derived and the working of the latter is illustrated by examples. Then, building on the introductory material on errors in Chapter 1, there is a more detailed discussion of the statistical interpretation of experimental errors, and it is shown how such errors can be propagated to find the errors on other functions. Finally, the Monte Carlo method of numerical simulation is introduced, and simple examples are given for numerical integration and simulation of simple systems.

The previous chapters have been concerned almost exclusively with descriptive statistics. The main properties of statistical distributions have been described, and some of the general principles associated with them established. In this chapter we will discuss how to use these ideas to make inferences about a population, given that in practice we usually only have access to a sample of the whole population. This raises several problems, including how to ensure that any sample is random, what the distribution of the function of the sample data chosen to make statistical inferences is, and how to define the desirable properties of such functions so that reliable estimates may be made about the corresponding population parameters. One thing that will emerge from this discussion is the explanation of why the normal distribution is so important in statistical applications in physical sciences. The link between theoretical statistics and experiments is made in Section 5.4. Finally, in Section 5.5 we give a short introduction to the Monte Carlo method.

5.1 Random samples and estimators

In this section we will consider how random samples are selected, what their probability distributions are, and what are the desirable properties of the functions of random variables that are used to make inferences about the underlying population.

5.1.1 Sampling distributions

A random sample of size n selected from a population was defined in Section 1.2 as resulting from a situation where every sample of this size had an equal chance of being selected while noting the intrinsic circularity of this definition. A more formal definition is as follows. If x is a random variable distributed according to a density $f(x)$, then the population is the totality of possible values of x. A sample $\mathbf{x}$ of size n, that is, $\mathbf{x} = x_1, x_2, ..., x_n$, defines a new sample space and is a random sample if the following two conditions are satisfied:

(a) all samples are taken from the same distribution as the population.
(b) the samples are independently selected.

Condition (b) implies that the joint probability density is given by

$$f(\mathbf{x}) = f(x_1)f(x_2)...f(x_n). \tag{5.1}$$

If $f(x)$ is known, random samples can be obtained in principle from (5.1), but if $f(x)$ is unknown, it is often difficult to ensure that the conditions for randomness are strictly met, particularly (a) above. It is possible to test whether a given sample is random, but since this is formally testing an hypothesis about the nature of the sample, we will defer discussion of this until hypothesis testing in general is discussed in Chapters 10 and 11. For the present, we will assume that samples have been selected randomly.

We are very often interested in a function y of the sample $x_1, x_2, ..., x_n$. Any such function is called a *statistic*, a term introduced in Chapter 1, and is itself a random variable. Because of this, the values of y will vary with different samples of the same size and will be distributed according to a new density function. The formal solution for finding the latter is via construction of the distribution function of y using (5.1), that is,

$$F(y) = \int \cdots \int \prod_{i=1}^{n} f(x_i) \, dx_i, \tag{5.2}$$

where the integral is taken over the region such that $y \leq y(x_1, x_2, ..., x_n)$. In practice it is often convenient to let $y(x_1, x_2, ..., x_n)$ be a new variable and then choose $n-1$ other variables (functions of x_i) such that the n-dimensional integrand in (5.2) takes a simple form. Example 5.1 below will illustrate this.

Example 5.1

Find the sampling distribution of the means $\bar{x}_n$ of samples of size n drawn from the Cauchy distribution

$$f(x) = \frac{1}{\pi} \frac{1}{1+x^2}, \quad -\infty \leq x \leq \infty.$$

Solution: If we choose new variables $u_i = x_i (i = 1, 2, ..., n-1)$ and $u_n = \bar{x}_n$, then the Jacobian of the transformation is

$$J = \frac{\partial(x_1, x_2, ..., x_n)}{\partial(u_1, u_2, ..., u_{n-1}, \bar{x}_n)},$$

and the distribution function of the means becomes

$$F(\bar{x}_n) = \int \cdots \int f(u_1) f(u_2) \dots f(u_{n-1}) f(\bar{x}'_n) J \, d\bar{x}'_n \prod_{i=1}^{n-1} du_i,$$

where the density functions are expressed in terms of the new set of variables and the integrals are taken over all x_i such that

$$\bar{x}'_n \geq \frac{1}{n} \sum x_i.$$

Thus,

$$F(\bar{x}_n) = \frac{n}{\pi^n} \int\limits_{-\infty}^{\bar{x}_n} du_n \int\limits_{-\infty}^{\infty} du_1 \cdots \int\limits_{-\infty}^{\infty} \left[1 + \left(nu_n - \sum_{i=1}^{n-1} u_i \right)^2 \right]^{-1} \prod_{j=1}^{n-1} \left(1 + u_j^2 \right)^{-1},$$

and the density function of $\bar{x}_n$ is given by differentiation of the $(n-1)$-fold integration in $u_j (j = 1, 2, \dots, n-1)$. The integral can be evaluated, although the algebra is rather lengthy. The result is the probability density

$$f(\bar{x}_n) = \frac{1}{\pi} \frac{1}{1 + \bar{x}_n},$$

which is the same form as the population density for any value of n. We will see later in this chapter that this result is unusual, and the sampling distribution of the sample mean for most distributions commonly met in physical science is a normal distribution, provided the sample size n is large.

Even in the simple case given in Example 5.1, the integral is complicated and in practice it is rarely possible to evaluate analytically the required multidimensional integrals. Instead, numerical evaluation is used, but even then, conventional techniques are usually far too time consuming, and a statistical method, known as the *Monte Carlo method*, mentioned in Section 3.4, is used. The Monte Carlo technique uses a sequence of random numbers to calculate probabilities and related quantities and is of rather general application. It will be discussed briefly in Section 5.5.

Another useful method for finding sampling distributions is to construct the moment generating function, or the characteristic function, for the statistic. If either of these is recognized as that of a known *pdf*, then this is the *pdf* of the sampling distribution. This technique is very practical, and we shall have occasion to use it later. Alternatively, the inversion theorem may be used to identify the density function. For example, we have shown in Section 4.2.3 that the *cf* of a Poisson distribution with a general term $e^{-\lambda} \lambda^k / k!$ is

$$\phi(t) = e^{-\lambda} \exp\left(\lambda e^{it} \right),$$

where λ, the Poisson parameter, is also the mean of the distribution. From this we can form the *cf* for the sum of a sample of size n as the product of terms of the form $\phi(t)$ and hence show that the *cf* for the sum is $\exp\{ n\lambda (e^{it} - 1) \}$. Since this is the *cf* of a Poisson distribution with parameter $n\lambda$, we can conclude that the sum of the sampling distribution is also a Poisson distribution, but whose general term is $e^{-n\lambda}(n\lambda)^k/k!$. (See Appendix A.3.)

Since the expected value of a Poisson distribution is its parameter, we can also conclude that the expected value of the average of the sample is the parameter λ of the original Poisson distribution of the elements making up the sample. The *pdf* of the sample average is easily determined by noting that the probability that the sum of a sample takes a value k is the same as the probability that the average has the value k/n. The result is

$$P[\bar{x}] = \frac{1}{(n\bar{x})!}e^{-n\lambda}(n\lambda)^{n\bar{x}},$$

which is not a Poisson distribution. The derivation of this result is given briefly in Appendix A.4.

A common situation is where the exact form of $f(x)$ is unknown, but one has a model (or hypothesis) for $f(x)$ that depends on an unknown parameter θ. The central problem is then to construct a function of the observations $x_1, x_2, ..., x_n$ that contains no unknown parameters, to *estimate* θ, that is, to give a value to θ, and hence determine $f(x)$. This situation is an example of *parametric statistics*. In these circumstances the statistic is referred to as a *point estimator* of θ and is written as $\hat{\theta}$. The word 'point' will be omitted when it is obvious that we are referring to the estimation of the value of a parameter by a single number. In general, there could be several unknown parameters $\boldsymbol{\theta} = \theta_1, \theta_2, ..., \theta_m$ and associated estimators. Since the estimator is a function of the random variables $\mathbf{x} = x_1, x_2, ..., x_n$, it is itself a random variable and its value will therefore vary with different samples of the same size and be distributed according to a new density function $g(\hat{\theta}; \theta)$. The merit of an estimator is judged by the properties of this distribution, and not by the values of a particular estimate. So we will now turn to consider the properties of 'good' estimators.

5.1.2 Properties of point estimators

It is intuitively obvious that a desirable property of an estimator is that the estimate tends to the value of the population parameter as the sample size increases. Any other result would be inconvenient, and even possibly misleading. This property is called *consistency*. Formally, an estimator $\hat{\theta}_n$, computed from a sample of size n, is said to be a *consistent* estimator of a population parameter θ if, for any positive ε, arbitrarily small,

$$\lim_{n \to \infty} P\left[|\hat{\theta}_n - \theta| > \varepsilon\right] = 0. \tag{5.3}$$

In these circumstances $\hat{\theta}_n$ is said to *converge in probability* to θ. Thus $\hat{\theta}_n$ is a consistent estimator of θ if it converges in probability to θ.

The property of consistency tells us the asymptotic $(n \to \infty)$ behavior of a suitable estimator, although the approach to consistency does not have to be monotonic as n increases.

Having found such an estimator, we may generate an infinite number of other consistent estimators

$$\widehat{\theta}'_n = p(n)\,\widehat{\theta}_n, \tag{5.4}$$

provided

$$\lim_{n \to \infty} p(n) = 1. \tag{5.5}$$

However, we may further restrict the possible estimators by requiring that for *all* n the expected value of $\widehat{\theta}_n$ is θ, that is, $E\left[\widehat{\theta}_n\right] = \theta$, or in full, using (5.1),

$$E\left[\widehat{\theta}(\mathbf{x})\right] = \int \widehat{\theta}(\mathbf{x})\, g\left(\widehat{\theta};\theta\right)\, \mathrm{d}\widehat{\theta} = \int \cdots \int \widehat{\theta}(\mathbf{x}) f(x_1) f(x_2)...f(x_n)\, \mathrm{d}x_1 \mathrm{d}x_2 ... \mathrm{d}x_n = \theta. \tag{5.6}$$

Estimators with this property are called *unbiased*, with the *bias* b given by

$$b = E\left[\widehat{\theta}_n\right] - \theta. \tag{5.7}$$

Estimators for which $b \to 0$ as $n \to \infty$ are said to be *asymptotically unbiased*. Despite the name, the fact that an estimator is biased is often not a serious problem, because there frequently exists a simple factor that converts such an estimator to an unbiased one. Unbiased estimators are more convenient to use in practice, although, as the following example shows, they are not unique.

Example 5.2

If $\widehat{\theta}_i$ $(i = 1, 2, ..., n)$ is a set of n unbiased estimators for the parameter θ, show that any linear combination

$$\widehat{\theta} = \sum_{i=1}^{m} \lambda_i \widehat{\theta}_i, \quad m \leq n$$

where λ_i are constants, is also an unbiased estimator for θ, provided

$$\sum_{i=1}^{m} \lambda_i = 1.$$

Solution: The expectation value of $\widehat{\theta}$ is

$$E\left[\widehat{\theta}\right] = E\left[\sum_{i=1}^{m} \lambda_i \widehat{\theta}_i\right] = \sum_{i=1}^{m} \lambda_i E\left[\widehat{\theta}_i\right] = \sum_{i=1}^{m} \lambda_i\, \theta = \theta.$$

Hence from (5.7), $\widehat{\theta}$ is an unbiased estimator for the parameter θ.

The requirements of consistency and lack of bias alone do not produce unique estimators. One can easily show that the sample mean is a consistent and unbiased estimator of the mean of a normal population with known variance. (This will be proved in Section 5.2.) But the same is true of the sample median. Further restrictions must be imposed if uniqueness is required. One of these is the *efficiency* of an estimator. An unbiased estimator with a small variance will produce estimates more closely grouped around the population value θ than one with a larger variance. If two estimators $\widehat{\theta}_1$ and $\widehat{\theta}_2$, both calculated from samples of size n, have variances such that $\mathrm{var}\widehat{\theta}_1 < \mathrm{var}\widehat{\theta}_2$, then $\widehat{\theta}_1$ is said to be more *efficient* than $\widehat{\theta}_2$ for samples of size n. For the normal distribution,

$$\mathrm{var}\,(\text{mean}) = \sigma^2/n,$$

for any n (this result is derived in Section 5.2 below). But for large n,

$$\mathrm{var}\,(\text{median}) = \pi\sigma^2/2n > \sigma^2/n.$$

The mean is thus the more efficient estimator for large n. (In fact this is true for all n.) Consistent estimators whose sampling variance for large samples is less than that of any other such estimators are called *most efficient*. Such estimators serve to define a scale of efficiency. Thus, if $\widehat{\theta}_2$ has variance v_2 and $\widehat{\theta}_1$, the most efficient estimator, has variance v_1, then the efficiency of $\widehat{\theta}_2$ is defined as

$$E_2 = v_1/v_2. \tag{5.8}$$

An estimator with efficiency of unity is called an efficient estimator. It may still be that there exist several efficient estimators $\widehat{\theta}$ for a population parameter θ. Can one choose a 'best' estimator from among them? The criterion of efficiency alone is not enough, since it is possible that for a given finite n, one estimator $\widehat{\theta}_n$, which is biased, is consistently closer to θ than an unbiased estimator $\widehat{\theta}'_n$. In this case the quantity to consider is not the variance but the second moment of $\widehat{\theta}_n$ about θ, which is

$$E\left[\left(\widehat{\theta}_n - \theta\right)^2\right].$$

Using (5.7) gives (showing this is Problem 5.1):

$$E\left[\left(\widehat{\theta}_n - \theta\right)^2\right] = \mathrm{var}\left(\widehat{\theta}_n\right) + b^2. \tag{5.9}$$

This quantity is called the *mean squared error* and we define $\widehat{\theta}_n$ to be a *best, or optimal, estimator* of the parameter θ if

$$E\left[\left(\widehat{\theta}_n - \theta\right)^2\right] \le E\left[\left(\widehat{\theta}'_n - \theta\right)^2\right],$$

where $\widehat{\theta}'_n$ is any other estimator of θ. Thus an optimal unbiased estimator $\widehat{\theta}_n$ is one with minimum variance. We will discuss how to obtain minimum variance estimators in more detail in Chapter 7, Section 7.4.

Example 5.3

If $\widehat{\theta}_i$ ($i = 1, 2$) are two independent and unbiased estimators for a parameter θ with variances σ_i^2, what value of the constant λ in the linear combination

$$\widehat{\theta} = \lambda \widehat{\theta}_1 + (1 - \lambda) \widehat{\theta}_2$$

ensures that $\widehat{\theta}$ is the optimal estimator for θ?

Solution: Because $\widehat{\theta}_1$ and $\widehat{\theta}_2$ are both unbiased, so is $\widehat{\theta}$ (see Example 5.2). The optimal estimator of θ is therefore one that minimizes the mean squared error

$$\text{var}\left(\widehat{\theta}\right) = E\left[\left(\widehat{\theta} - \theta\right)^2\right].$$

Now because $\widehat{\theta}_1$ and $\widehat{\theta}_2$ are independent,

$$\text{var}\left(\widehat{\theta}\right) = \lambda^2 \text{var}\left(\widehat{\theta}_1\right) + (1 - \lambda)^2 \text{var}\left(\widehat{\theta}_2\right) = \lambda^2 \sigma_1^2 + (1 - \lambda)^2 \sigma_2^2,$$

and the minimum of $\text{var}\left(\widehat{\theta}\right)$ is found from $d \, \text{var}\left(\widehat{\theta}\right)/d\lambda = 0$, that is,

$$\lambda = \frac{\sigma_2{}^2}{\sigma_1{}^2 + \sigma_2{}^2}.$$

The discussion above gives an idea of the desirable properties of estimators, but there is a more general criterion that can be used. Consider the case of estimating a parameter θ and let

$$f\left(\widehat{\theta}_1, \widehat{\theta}_2, \ldots, \widehat{\theta}_r; \theta\right)$$

be the joint density function of r independent estimators $\widehat{\theta}_i (i = 1, 2, \ldots, r)$. Then, from the definition of the multivariate conditional density, we have (compare Equation (3.23))

$$f\left(\widehat{\theta}_1, \widehat{\theta}_2, \ldots, \widehat{\theta}_r; \theta\right) = f^M\left(\widehat{\theta}_1; \theta\right) f^C\left(\widehat{\theta}_2, \widehat{\theta}_3, \ldots, \widehat{\theta}_r; \theta | \widehat{\theta}_1\right), \tag{5.10}$$

where $f^M\left(\widehat{\theta}_1; \theta\right)$ is the marginal density of $\widehat{\theta}_1$, and $f^C\left(\widehat{\theta}_2, \widehat{\theta}_3, \ldots, \widehat{\theta}_r; \theta | \widehat{\theta}_1\right)$ is the conditional density of all the other $\widehat{\theta}_i$, given $\widehat{\theta}_1$. Now if f^C is independent of θ, then clearly, once $\widehat{\theta}_1$ is specified, the other estimators contribute nothing to the problem of estimating θ, that is, $\widehat{\theta}_1$ contains *all* the information about θ. In these circumstances $\widehat{\theta}_1$ is called a *sufficient* statistic for θ. It is more convenient in practice to write (5.10) as a condition on the likelihood function introduced in Chapter 2.

Let $f(x;\theta)$ denote the density function of a random variable x, where the form of f is known, but not the value of θ, which is to be estimated. Then let $x_1, x_2, ..., x_n$ be a random sample of size n drawn from $f(x;\theta)$. The joint density function $f(x_1, x_2, ..., x_n; \theta)$ of the independent random variables $x_1, x_2, ..., x_n$ is given by

$$f(x_1, x_2, ..., x_n; \theta) = \prod_{i=1}^{n} f(x_i; \theta),\qquad(5.11)$$

where $f(x_i, \theta)$ is the density function for the ith random variable. The function $f(x_1, x_2, ..., x_n; \theta)$ is the likelihood function of θ and is written $L(x_1, x_2, ..., x_n; \theta)$. If L is expressible in the form

$$L(x_1, x_2, ..., x_n; \theta) = L_1(\widehat{\theta}; \theta) L_2(x_1, x_2, ..., x_n),\qquad(5.12)$$

where L_1 does not contain the x's other than in the form $\widehat{\theta}$, and L_2 is independent of θ, then $\widehat{\theta}$ is a *sufficient* statistic for the estimation of θ.

Example 5.4

Find a sufficient estimator for estimating the variance of a normal distribution with zero mean.
Solution: The probability density is

$$f(x) = \frac{1}{\sqrt{2\pi}} \frac{1}{\sigma} \exp\left(-\frac{x^2}{2\sigma^2}\right),$$

and the likelihood function is therefore

$$L(x_1, x_2, ..., x_n; \sigma^2) = \left(\frac{1}{\sigma\sqrt{2\pi}}\right)^n \exp\left(-\frac{1}{2\sigma^2} \sum_{i=1}^{n} x_i^2\right).$$

If we let $L_2 = 1$ in (5.12), we have $L_1 = L$ and L_1 is a function of the sample x_i only in terms of $\sum x_i^2$. Thus $\sum x_i^2$ is a sufficient estimator for σ^2. We will show in Section 5.2 that this estimator is biased.

5.2 Estimators for the mean, variance, and covariance

Estimators for the mean, variance, and covariance are of central importance in statistical analysis, so they are considered in more detail here. Let S denote a sample of n observations $x_i(i = 1, 2, ..., n)$ selected at random. The sample S is called a *random sample with replacement* (or a *simple random sample*) if, in general, the observation x_{n-1} is returned to the population before x_n is selected. If x_{n-1} is not returned, then S is called a *random sample without replacement*. Sampling with replacement implies, of course, that it is indeed possible to return the 'observation' to the population, as is the case when drawing cards from a deck. In most practical situations this is usually not possible, and the sampling is without replacement. Sampling from an infinite population is equivalent to sampling with replacement.

For any continuous population, finite or infinite, the sample mean $\bar{x}$ is an estimator for the population mean μ. This result follows simply from the definition of the sample mean, Equation (1.3). Thus

$$E[\bar{x}] = E\left[\frac{1}{n}\sum_{i=1}^{n} x_i\right] = \frac{1}{n}\sum_{i=1}^{n} E[x_i].$$

But using (5.1),

$$E[x_i] = \int \ldots \int x_i f(x_1)\ldots f(x_n)\, dx_1\ldots dx_n = \mu$$

and so

$$E[\bar{x}] = \frac{1}{n}\sum_{i=1}^{n} \mu = \mu. \tag{5.13}$$

We can also find the expectation of the sample variance s^2. This is

$$E\left[\frac{1}{n-1}\sum_{i=1}^{n}(x_i - \bar{x})^2\right] = \frac{1}{n-1}E\left[\sum_{i=1}^{n}\left(x_i - \frac{1}{n}\sum_{j=1}^{n} x_j\right)^2\right]$$

$$= \frac{1}{n-1}E\left[\frac{n-1}{n}\sum_{i=1}^{n}(x_i)^2 - \frac{1}{n}\sum_{i\neq j}^{n} x_i x_j\right] \tag{5.14}$$

$$= \mu_2' - (\mu_1')^2 = \sigma^2.$$

Thus the presence of the factor $1/(n-1)$ in the definition of the sample variance, which we noted in Chapter 1 differed from the analogous definition for the population variance, is to ensure that s^2 is an unbiased estimator of σ^2. Similarly, the sample covariance defined in (1.12b) is an unbiased estimator for the population covariance of Equation (1.12a).

Given any estimator $\hat{\theta}$, one can calculate its variance. For example, the variance of the sample mean drawn from an infinite population, or a finite population with replacement, is, by definition,

$$\sigma_{\bar{x}}^2 \equiv \text{var}(\bar{x}) = E\left[(\bar{x} - E[\bar{x}])^2\right] = E\left[(\bar{x} - \mu)^2\right], \tag{5.15}$$

which may be written

$$\text{var}(\bar{x}) = \frac{1}{n^2}E\left[\left(\sum_{i=1}^{n}(x_i - \mu)\right)^2\right].$$

If we expand the square bracket on the right-hand side and again use (5.1), there are n terms containing the form $(x_i - \mu)^2$, each of which gives a contribution

$$\int \cdots \int (x_i - \mu)^2 f(x_i) \ldots f(x_n) \, dx_i \ldots dx_n = \sigma^2.$$

The remaining terms are integrals over the forms $(x_i - \mu)(x_j - \mu)$ with $i < j$, each of which, using the definition of μ, is zero. Thus

$$\text{var}\,(\bar{x}) = \frac{1}{n^2} \sum_{i=1}^{n} \sigma^2 = \frac{\sigma^2}{n} = \frac{1}{n(n-1)} \sum_{i=1}^{n} (x_i - \bar{x})^2, \tag{5.16a}$$

where the result follows if σ^2 is replaced by its estimator s^2. If the sample is drawn from a finite population of N items without replacement, then this result is modified to

$$\text{var}\,(\bar{x}) = \frac{\sigma^2}{n} \left(\frac{N - n}{N - 1} \right). \tag{5.16b}$$

The square root of $\sigma_{\bar{x}}^2$, that is, the standard deviation $\sigma_{\bar{x}}$, is called the *(standard) error of the mean* and was introduced briefly when we discussed the normal distribution in Section 4.2. It is worth emphasizing the difference between the standard deviation σ and the standard error of the mean $\sigma_{\bar{x}}$. The former describes the extent to which a single observation is liable to vary from the population mean μ; the latter measures the extent that an estimate of the mean obtained from a sample of size n is liable to differ from the true mean.

The result (5.16a) is of considerable importance, because it shows that as the sample size n increases the variance of the sample mean decreases and hence the statistical error on a set of measurements decreases (like $1/\sqrt{n}$ in the case of (5.16a)) and thus the probability that the sample mean is a good estimation of the population mean increases, a result that was referred to in Chapter 1. The results (5.16) assume that the measurements are random samples and are uncorrelated. If this is not the case, then the nonzero correlation must be taken into account. They also assume the samples are obtained by simple random sampling from a single population. Better estimates can be obtained if we have additional information about the sample. One example is stratified sampling, mentioned briefly in Section 1.1. This technique requires that the population can be divided into several mutually exclusive subpopulations, with *known* fractions of the whole population in each. Then simple random samples, using, for example, sample sizes proportional to these fractions, lead to smaller estimates for $\text{var}(\bar{x})$ with the same total sample size. However, as this situation is not usually met in physical science, we will continue to consider only simple random sampling from a single homogeneous population.

We can go further, by using the general results for expectation values, and find the estimator of the variance of s^2, and hence the estimator of the standard deviation σ_s. The latter is not the square root of the former but, anticipating Equation (5.45), is given by

$$\text{var}\left(s^2\right) = \left(\frac{ds^2}{ds}\right)^2 \text{var}\left(s\right).$$

For a normal distribution, the result is[1]

$$\sigma_s = \frac{\sigma}{\sqrt{2(n-1)}}. \tag{5.17}$$

Just as in (5.16a), to use this result one would usually have to insert an estimate for σ obtained from the data. Providing n is large, there is little loss in precision in doing this, but for small n an alternative approach would have to be adopted. This will be discussed in Section 6.2. Alternatively, (5.17) can be used to predict how many events would be needed to measure σ to a given precision, under different assumptions about its value. This could be useful in the planning stages of an experiment.

Example 5.5

A random sample $x_i (i = 1, 2, ..., n)$ is drawn from a population with mean μ and variance σ^2. Two unbiased estimators for μ are

$$\widehat{\mu}_1 = \frac{1}{2}(x_1 + x_2) \quad \text{and} \quad \widehat{\mu}_2 = \bar{x}_n.$$

What is the relative efficiency of $\widehat{\mu}_1$ to $\widehat{\mu}_2$?
Solution: From (5.16a), var $(\widehat{\mu}_2) = $ var $(\bar{x}_n) = \sigma^2/n$. If the same steps to derive this result are used for $\widehat{\mu}_1$, then there is one term containing the form $(x_1 - \mu)^2$, and one containing the form $(x_2 - \mu)^2$, each of which gives a contribution σ^2 and a single term containing $(x_1 - \mu)(x_2 - \mu)$, which contributes zero. Thus var $(\widehat{\mu}_1) = \sigma^2/2$ and so

$$\text{relative efficiency} = \frac{\text{var}\left(\widehat{\mu}_2\right)}{\text{var}\left(\widehat{\mu}_1\right)} = \frac{n}{2}.$$

5.3 Laws of large numbers and the central limit theorem

The results of Section 5.2 may be stated formally as follows. Let x_i be a population of independent random variables with mean μ and finite variance and let $\bar{x}_n$ be the mean of a sample of size n. Then, given any $\varepsilon > 0$ and δ in the range $0 < \delta < 1$, there exists an integer n such that for all $m \geq n$.

$$P[|\bar{x}_m - \mu| \leq \varepsilon] \geq 1 - \delta. \tag{5.18}$$

This is the *weak law of large numbers*. It tells us that $|\bar{x}_n - \mu|$ will ultimately be very small but does not exclude the possibility that for some finite n it could be large. Since, in practice, we

---

[1] See, for example, Section 5.2.3 of the book by Barlow (1989) given in the Bibliography.

can only have access to finite samples, this possibility could be of some importance. Fortunately, there exists the so-called *strong law of large numbers*, which, in effect, states that the probability of such an occurrence is extremely small. The law of large numbers ensure that the frequency definition of probability adopted in Chapter 2 concurs in practice with the axiomatic one.

The weak law of large numbers may be proved by Chebyshev's inequality that we met in Chapter 1, Equation (1.11). It has several different, but equivalent, forms. A simple proof is given in Appendix (A.5). Provided the population distribution has a finite variance, Chebyshev's inequality may be written as

$$P\left[|\bar{x}_n - \mu| \geq \frac{k\sigma}{n^{1/2}}\right] \leq \frac{1}{k^2}, \tag{5.19}$$

so if we choose $k = \delta^{-1/2}$ and $n > \sigma^2/\delta\varepsilon^2$, and substitute in (5.19), then (5.18) results. As shown in Appendix (A.5), a more general form of the inequality applies to any random variable, not just the sampling distribution of the mean.

The bound given by (5.19) is usually weak, but if we continue to restrict ourselves to the sampling distribution of the mean, then we can derive the most important theorem in statistics, the *central limit theorem*, which may be stated as follows: Let the independent random variables x_i of unknown density function be identically distributed with mean μ and variance σ^2, both of which are finite. Then the distribution of the sample mean $\bar{x}_n$ tends to the normal distribution with mean μ and variance σ^2/n when n becomes large.

Thus, if $u(t)$ is the standard form of the normal density function, then for arbitrary t_1 and t_2,

$$\lim_{n \to \infty} P\left[t_1 \leq \frac{\bar{x}_n - \mu}{\sigma/n^{1/2}} \leq t_2\right] = \int_{t_1}^{t_2} u(t)\,dt. \tag{5.20}$$

The proof of this theorem illustrates the use of several earlier results and definitions and so is worth giving.

By applying the results on means of sums of random variables given in Chapter 3, and Equation (5.16a) to moment generating functions, it is easily shown that if the components of the sample are independent, then the mean and variance of their sum

$$S = \sum_{i=1}^{n} x_i$$

are given by

$$\mu_S = n\mu \quad \text{and} \quad \sigma_S^2 = n\sigma^2.$$

Now consider the variable

$$u = \frac{S - \mu_S}{\sigma_S} = \frac{1}{\sqrt{n}\sigma} \sum_{i=1}^{n} (x_i - \mu), \tag{5.21}$$

with characteristic function $\phi_u(t)$. If $\phi_i(t)$ is the *cf* of $(x_i - \mu)$, then

$$\phi_u(t) = \prod_{i=1}^{n} \phi_i\left(\frac{t}{\sqrt{n}\sigma}\right).$$

But all the $(x_i - \mu)$ have the same distribution and the same *cf*, $\phi = \phi_i$, and so

$$\phi_u(t) = \left[\phi\left(\frac{t}{\sqrt{n}\sigma}\right)\right]^n. \tag{5.22}$$

Just as the *mgf* can be expanded in an infinite series of moments, we can expand the *cf*, giving

$$\phi(t) = 1 + \sum_{r=1}^{\infty} \mu'_r \frac{(it)^r}{r!}, \tag{5.23}$$

and since the first two moments of $(x_i - \mu)$ are zero and σ^2, respectively, we have from (5.22) and (5.23)

$$\phi_u(t) = \left[1 - \frac{t^2}{2n} + o\left(\frac{1}{n^{3/2}}\right)\right]^n.$$

Expanding the square bracket, and then letting $n \to \infty$ but keeping fixed t, gives

$$\phi_u(t) \to e^{-t^2/2}, \tag{5.24}$$

which is the *cf* of a standardized normal distribution. So, by the inversion theorem, S is distributed as the normal distribution $n(S; \mu_S, \sigma_S^2)$, and hence $\bar{x}_n$ is distributed as $n(\bar{x}_n; \mu, \sigma^2/n)$.

In practice, the normal approximation is good for $n \geq 30$ regardless of the shape of the population distribution. For values of n less than about 30, the approximation is good only if the population distribution does not differ much from a normal distribution. The sampling distribution of the means when sampling from a normal population is also normal, independent of the size of n.

The form of the central limit theorem above is not the most general that can be given. Provided certain (weak) conditions on the third moments are obeyed, then the condition that the x_i all have the same distribution can be relaxed, and it is possible to prove that the sampling distribution of *any* linear combination of independent random variables having arbitrary distributions with finite means and variances tends to normality for large samples.

Example 5.6

Five hundred resistors are found to have a mean value of 10.3 ohms and a standard deviation of 0.2 ohms. What is the probability that a sample of 100 resistors drawn at random from this population will have a combined value between 1027 and 1035 ohms?

Solution: For the sampling distribution of the means, $\mu_{\bar{x}} = \mu = 10.3$ and, from Equation (5.16b), the standard deviation of this value is

$$\sigma_{\bar{x}} = \frac{\sigma}{\sqrt{n}} \sqrt{\frac{(N-n)}{(N-1)}} = \frac{0.2}{\sqrt{100}} \sqrt{\frac{(500-100)}{(500-1)}} = 0.018.$$

We seek the value of the probability such that $P[10.27 < x < 10.35]$. Using the central limit theorem, we can use the normal approximation. So, using standardized variables, this is equivalent to

$$P[-1.67 < z < 2.78] = N(2.78) + N(1.67) - 1 \approx 0.95.$$

The central limit theorem applies to both discrete and continuous distributions and is a remarkable theorem because nothing is said about the original density function, except that it has finite mean and variance. Although in practice these conditions are not usually restrictions, they are essential. For example, we have seen in Example 5.1 that the distribution of $\bar{x}_n$ for the Cauchy distribution is the same as for a single observation. The failure of the theorem in this case can be traced to the infinite variance of the Cauchy distribution, and there are other examples, such as the details of the scattering of particles from nuclei, where the long 'tails' of distributions cause the theorem to fail. It is the central limit theorem that gives the normal distribution such a prominent position both theoretically and in practice. It allows (approximate) quantitative probability statements to be made in experimental situations where the exact form of the underlying distribution is unknown. This was briefly mentioned in Section 1.4.

Just as we have been considering the sampling distribution of means, we can also consider the sampling distribution of sums $T = \sum x_i$ of random variables of size n. If the random variable x is distributed with mean μ, and variance σ^2, then the sampling distribution of T has mean

$$\mu_T = n\mu \tag{5.25}$$

and variance

$$\sigma_T^2 = \begin{cases} n\sigma^2 \left(\dfrac{N-n}{N-1} \right) \\ n\sigma^2 \end{cases} \tag{5.26}$$

where the first result is for sampling from a finite population of size N without replacement and the second result is otherwise.

We will conclude with some results on the properties of linear combinations of means, since up to now we have been concerned mainly with sampling distributions of a single sample mean. Let

$$l = \sum_{i=1}^{n} a_i x_i, \tag{5.27}$$

where a_i are real constants and x_i are random variables with means μ_i, variances σ_i^2, and covariances $\sigma_{ij}(i, j = 1, 2, \ldots n; i \neq j)$. (Note that the index i now indicates different random variables, not a sample of a single random variable.) Then,

$$\mu_l = \sum_{i=1}^{n} a_i \mu_i \tag{5.28}$$

and

$$\sigma_l^2 = \sum_{i=1}^{n} a_i^2 \sigma_i^2 + 2 \sum_{i<j} a_i a_j \sigma_{ij}, \tag{5.29}$$

which reduces to

$$\sigma_l^2 = \sum_{i=1}^{n} a_i^2 \sigma_i^2 \tag{5.30}$$

if the x's are mutually independent. Note that the constants are squared in (5.30), so, for example, the variance of $(x_1 + x_2)$ is the same as that of $(x_1 - x_2)$. (Problem 5.3 is to prove these results.)

A useful corollary to the above result is as follows. Let $\bar{x}_i$ ($i = 1, 2, \ldots, n$) be the means of random samples of size n_i drawn from an infinite populations with mean μ_i and variance σ_i. If $\bar{x}_1$ and $\bar{x}_2$ are independently distributed, then

$$\mu_{\bar{x}_1 \pm \bar{x}_2} = \mu_1 \pm \mu_2 \tag{5.31}$$

and

$$\sigma_{\bar{x}_1 \pm \bar{x}_2}^2 = \sum_{i=1}^{2} \left(\frac{\sigma_i^2}{n_i} \right). \tag{5.32}$$

These results follow immediately from (5.28) and (5.30), and the results (5.14) and (5.16a), by the substitutions $x_1 = \bar{x}_1$ and $x_2 = \bar{x}_2$, with $a_1 = a_2 = 1$ for the first case and $a_1 = -a_2 = 1$ for the second.

5.4 Experimental errors

In the preceding sections we have been concerned with theoretical statistics only. In this section we will provide the link between theoretical statistics and experimental situations. This continues the discussion started in Chapter 1.

In an experimental observation one can never measure the value of a quantity with absolute precision; that is, one can never reduce the statistical error on the measurement to zero, although we can reduce it by increasing n, that is, taking more data. Recall that in Section 1.4 we distinguished the *precision* of a measurement from its *accuracy*, that is, the deviation of the observation from the 'true' value, assuming that such a concept is meaningful. Thus there may exist, in addition to fluctuations in the measurement process that limit the precision, unknown systematic errors that limit the accuracy. In general, the only errors that we can deal with in detail here are the former type, and the conventional measure of this type of error is taken to be the standard error, defined above and which we have previously introduced in Section 4.2. This definition of the error is, of course, arbitrary, and formerly (but now only very rarely) the *probable error p*, defined by

$$\int_{\mu-p}^{\mu+p} f(x)\mathrm{d}x = 1/2,$$

was used. Multiplying errors by an arbitrary factor 'to be on the safe side' obviously renders statistical analyses meaningless.

Consider, for example, an idealized nuclear counting experiment for a scattering process. The number of trials is very large, because the numbers of particles in the beam and target are large, but the probability of a scatter, p is very small. In this situation the Poisson distribution is applicable, and as we have seen in Equation (4.51), if $N_e = np$ is the total number of counts recorded, then $\sigma = \sqrt{N_e}$. The result of the experiment would be given as

$$N = N_e \pm \Delta N, \tag{5.33}$$

where the standard error, in this case, is given by

$$\Delta N = \sqrt{N_e}. \tag{5.34}$$

If the population distribution is unknown, then we can consider the sampling distribution. For example, from a set of observations x_i, we know that an estimate of the mean is the sample mean

$$\bar{x} = \frac{1}{n}\sum_{i=1}^{n} x_i \tag{5.35}$$

and the laws of large numbers ensure that $\bar{x}$ is a good estimate for large n. The variance of $\bar{x}$ is

$$\sigma_{\bar{x}}^2 = \sigma^2/n, \tag{5.36}$$

so to calculate $\sigma_{\bar{x}}^2$, we need to estimate σ^2. We have seen that the sample variance is

$$s^2 = \frac{1}{n-1} \sum_{i=1}^{n} (x_i - \bar{x})^2, \tag{5.37}$$

and thus

$$\sigma_{\bar{x}}^2 = \frac{1}{n(n-1)} \sum_{i=1}^{n} (x_i - \bar{x})^2. \tag{5.38}$$

An experimental result would then be quoted as

$$x = \bar{x}_e \pm \Delta x, \tag{5.39a}$$

where the standard error is given by

$$\Delta x = \sigma_{\bar{x}} = \left[\frac{1}{n(n-1)} \sum_{i=1}^{n} (x_i - \bar{x})^2 \right]^{1/2}. \tag{5.39b}$$

This is the definition of the standard error commonly used to estimate the error in fitting a function to a set of experimental data.

Now by the central limit theorem, we know that for large sample sizes the distribution of the sample means is approximately normal, and therefore (5.39a, b) may be interpreted (compare Section 1.2) as

$$\begin{aligned} P[\bar{x}_e - \Delta x \le x \le \bar{x}_e + \Delta x] &\approx 68.3\%, \\ P[\bar{x}_e - 2\Delta x \le x \le \bar{x}_e + 2\Delta x] &\approx 95.4\%, \\ P[\bar{x}_e - 3\Delta x \le x \le \bar{x}_e + 3\Delta x] &\approx 99.7\%. \end{aligned} \tag{5.40}$$

So even though the form of the underlying distribution of x is unknown, the central limit theorem enables an approximate quantitative statement to be made about the probability of the true value of x lying within a specified range.

Since we have moved away from mathematical statistics into the real world of experimental data, it is worth commenting on a situation that commonly arises. In calculating $\sigma_{\bar{x}}$ from (5.39b) one often finds that a few data a long way from the mean (referred to as 'outliers') are making very significant contributions to the summation. What, if anything, should one do about this? A general comment is that transforming the data can reduce the effect of outliers. For example, taking logarithms shrinks large values much more than smaller ones, but this is not always practical. In light of (5.40), it might seem reasonable to ignore data that are, say, three standard deviations away from the mean, and tables exist giving criteria to select data for rejection. There are even 'rules' suggested for rejecting data. One version (called Chauvenet's criterion) states that if we have n data points, the point x_i should be rejected if $P[x_i > \bar{x}] < 1/2n$. However, common sense dictates that the more data taken, the more outliers will be found. So, if a rare (but real) event with a probability $1/2n$ is expected in a single trial, the probability of its occurrence at least once in n trials is

$$1 - \left(1 - \frac{1}{2n}\right)^n = 1 - \left\{\left(1 - \frac{1}{2n}\right)^{2n}\right\}^{1/2} \approx 1 - e^{-1/2} = 0.39$$

when n becomes large, which is not negligible. If outliers are rejected, for whatever reason, and then $\bar{x}_e$ and $\sigma_{\bar{x}}$ recalculated, because $\sigma_{\bar{x}}$ will now be smaller, new points may well be found that satisfy the recalculated rejection criterion, and logically these should also be rejected. But if this process were to be repeated, it could converge to a value that seriously distorts the information in the original data set. So blindly applying a rule, however reasonable it may appear, may result in misleading conclusions, and is to be discouraged.

The existence of outliers is an alert to possible problems with the data, so any outliers should be examined very carefully to see if there is any valid experimental reason why they should be rejected. But this should be done honestly, avoiding any temptation to 'massage the data', and should be defensible. In the absence of such reasons there are only two alternatives: either include the outliers and accept that such statistical fluctuations do rarely occur, or reject them and possibly miss the chance of finding some new phenomenon. Any rejection criterion should certainly never be used on a data set more than once. Also, if 'outside n standard deviations' is used as the criteria for rejection, as is commonly done, a value for n will have to be chosen, and $n = 3$ is often considered reasonable.[2] But whatever is used, it should be clearly stated when reporting the data.

5.4.1 Propagation of errors

If we have a function y of the p variables $\theta_i (i = 1, 2, ..., p)$, that is,

$$y \equiv y(\boldsymbol{\theta}) = y(\theta_1, \theta_2, ..., \theta_p),$$

then we are often interested in knowing the approximate error on y, given that we know the errors on θ_i. If the true values of θ_i are $\bar{\theta}_i$ (in practice, estimates of these quantities would usually have to be used) and the quantities $(\theta_i - \bar{\theta}_i)$ are small, then a Taylor expansion of $y(\boldsymbol{\theta})$ about the point $\boldsymbol{\theta} = \bar{\boldsymbol{\theta}}$ gives, to first order in $(\theta_i - \bar{\theta}_i)$,

$$y(\boldsymbol{\theta}) = y(\bar{\boldsymbol{\theta}}) + \sum_{i=1}^{p} (\theta_i - \bar{\theta}_i) \frac{\partial y(\boldsymbol{\theta})}{\partial \theta_i}\bigg|_{\boldsymbol{\theta}=\bar{\boldsymbol{\theta}}}. \tag{5.41}$$

---

[2] The acceptable value of n varies with different fields of interest, and how important the conclusions that might follow. For example, the 'gold standard' in experimental particle physics is often $n = 5$, an extremely high level that reflects the rigorous standards of the field.

Now

$$\text{var } y(\boldsymbol{\theta}) = E\left[(y(\boldsymbol{\theta}) - E[y(\boldsymbol{\theta})])^2\right] \simeq E\left[\{y(\boldsymbol{\theta}) - y(\overline{\boldsymbol{\theta}})\}^2\right], \tag{5.42}$$

and using (5.41) in (5.42) gives

$$\text{var } y(\boldsymbol{\theta}) \simeq \sum_{i=1}^{p} \sum_{j=1}^{p} \frac{\partial y(\boldsymbol{\theta})}{\partial \theta_i}\bigg|_{\theta=\overline{\theta}} E\left[(\theta_i - \overline{\theta}_i)(\theta_j - \overline{\theta}_j)\right] \frac{\partial y(\boldsymbol{\theta})}{\partial \theta_j}\bigg|_{\theta=\overline{\theta}}. \tag{5.43}$$

But

$$V_{ij} = E\left[(\theta_i - \overline{\theta}_i)(\theta_j - \overline{\theta}_j)\right]$$

is the variance matrix of the parameters θ_i. Thus if we set

$$(\Delta y)^2 = \text{var } y,$$

we have

$$(\Delta y)^2 = \sum_{i=1}^{p} \sum_{j=1}^{p} \left\{ \frac{\partial y(\boldsymbol{\theta})}{\partial \theta_i}\bigg|_{\theta=\overline{\theta}} V_{ij} \frac{\partial y(\boldsymbol{\theta})}{\partial \theta_j}\bigg|_{\theta=\overline{\theta}} \right\}. \tag{5.44}$$

Equation (5.44) is often referred to as the *law of propagation of errors*[3]. If the errors are un-correlated (i.e., $\text{cov}(\theta_i, \theta_j) = 0$), then

$$V_{ij} = \begin{cases} (\Delta \theta_i)^2 & i=j \\ 0 & i \neq j \end{cases}$$

and (5.44) reduces to

$$(\Delta y)^2 = \sum_{i=1}^{p} \left[\frac{\partial y(\boldsymbol{\theta})}{\partial \theta_i}\bigg|_{\theta=\overline{\theta}} \Delta \theta_i \right]^2. \tag{5.45}$$

When using these expressions, one should always ensure that the quantities $\Delta \theta_i \equiv \theta_i - \overline{\theta}_i$ are small enough to justify truncation of the Taylor series (5.41). Care should be taken with functions that are highly nonlinear in the vicinity of the mean, of a size comparable to the standard deviation of the parameters θ_i. Such situations are better dealt with by using the method of confidence intervals discussed in Chapter 9.

---

[3]Although this name is usually used in statistics, it is clear from (5.44), and other equations in this section, that mathematically, this is equivalent to a change of variables.

Example 5.7

If s and t are two random variables with variances σ_s^2 and σ_t^2, respectively, and a covariance σ_{st}^2, what are the approximate errors on the functions: (a) $x = as + bt$, (b) $x = ast$, and (c) $x = as/t$, where a and b are constants?

Solution: (a) Taking derivatives, we have $\partial x/\partial s = a$ and $\partial x/\partial t = b$. Also, the variance matrix is

$$V_{st} = \begin{bmatrix} \sigma_s^2 & \sigma_{st}^2 \\ \sigma_{ts}^2 & \sigma_t^2 \end{bmatrix}, \text{ with } \sigma_{ts}^2 = \sigma_{st}^2.$$

Then using (5.44) gives

$$\sigma_x^2 = a^2\sigma_s^2 + b^2\sigma_t^2 + 2ab\sigma_{st}^2,$$

and the approximate error on x is $\Delta x = \sigma_x$.

(b) Taking derivatives, $\partial x/\partial s = at$ and $\partial x/\partial t = as$, and using the same variance matrix as in (a) gives

$$\sigma_x^2 = (at\sigma_s)^2 + (as\sigma_t)^2 + 2a^2st\sigma_{st}^2.$$

(c) Taking derivatives, $\partial x/\partial s = a/t$ and $\partial x/\partial t = -as/t^2$, and using the same variance matrix as in (a) gives

$$\sigma_x^2 = \left(\frac{as}{t}\right)^2 \left[\frac{\sigma_s^2}{s^2} + \frac{\sigma_t^2}{t^2} - 2\frac{\sigma_{st}^2}{st}\right].$$

The results (5.44) and (5.45) are for the case of a single function $y(\boldsymbol{\theta})$ that is a function of the p parameters $\theta_i(i = 1, 2, ..., p)$. They are easily generalized to the case where there are n functions $y_k(k = 1, 2, ..., n) = \mathbf{y}(\boldsymbol{\theta})$ that are functions of the same p parameters. Then (5.44) becomes the set of equations

$$\text{var}\,(y_k) = (\Delta y_k)^2 = \sum_{i=1}^{p}\sum_{j=1}^{p}\left\{\frac{\partial y_k(\boldsymbol{\theta})}{\partial \theta_i}\bigg|_{\boldsymbol{\theta}=\bar{\boldsymbol{\theta}}} V_{ij} \frac{\partial y_k(\boldsymbol{\theta})}{\partial \theta_j}\bigg|_{\boldsymbol{\theta}=\bar{\boldsymbol{\theta}}}\right\}, k = 1, 2, ..., n. \tag{5.46}$$

A new feature is that the various functions y_k will be correlated, because they are all formed from the same set of parameters $\theta_i(i = 1, 2, ..., p)$. This will be true, independent of whether the parameters are themselves correlated. Their covariances may be found from the definition (3.28) and are

$$\text{cov}(y_k, y_l) = \sum_{i=1}^{n} \sum_{j=1}^{n} \left(\frac{\partial y_k}{\partial \theta_i}\right)\left(\frac{\partial y_l}{\partial \theta_j}\right) \text{cov}(\theta_i, \theta_j). \tag{5.47}$$

The two results (5.46) and (5.47) may be combined in the single matrix form

$$\mathbf{V_y} = \mathbf{G}\mathbf{V_0}\mathbf{G}^T, \tag{5.48}$$

where $\mathbf{V_y}$ is the $(n \times n)$ variance matrix of $\mathbf{y}$; $\mathbf{V_\theta}$ is the $(p \times p)$ variance matrix of $\mathbf{\theta}$; and $\mathbf{G}$ is an $(n \times p)$ matrix of derivatives with elements

$$G_{ki} = \frac{\partial y_k}{\partial \theta_i}.$$

Example 5.8

Measurements are made of a particle's position in two dimensions using the independent Cartesian coordinates (x, y), with measurement errors σ_x and σ_y. What is the variance matrix for the corresponding cylindrical polar coordinates (r, ϕ), where

$$r = \sqrt{x^2 + y^2} \quad \text{and} \quad \tan\phi = y/x?$$

Solution: From the relationship between cylindrical polar and Cartesian coordinates, we have

$$\mathbf{G} = \begin{bmatrix} \partial r/\partial x & \partial r/\partial y \\ \partial \phi/\partial x & \partial \phi/\partial y \end{bmatrix} = \begin{bmatrix} x/r & y/r \\ -y/r^2 & x/r^2 \end{bmatrix}.$$

Also,

$$\mathbf{V}_{\text{Cartesian}} = \begin{bmatrix} \sigma_x^2 & 0 \\ 0 & \sigma_y^2 \end{bmatrix},$$

and so, from (5.48)

$$\mathbf{V}_{\text{polar}} = \begin{bmatrix} x/r & y/r \\ -y/r^2 & x/r^2 \end{bmatrix} \begin{bmatrix} \sigma_x^2 & 0 \\ 0 & \sigma_y^2 \end{bmatrix} \begin{bmatrix} x/r & -y/r^2 \\ y/r & x/r^2 \end{bmatrix}.$$

Multiplying out gives

$$\mathbf{V}_{\text{polar}} = \begin{bmatrix} \left[(x/r)^2 \sigma_x^2 + (y/r)^2 \sigma_y^2\right] & \left[(xy/r^3)\left(\sigma_y^2 - \sigma_x^2\right)\right] \\ \left[(xy/r^3)\left(\sigma_y^2 - \sigma_x^2\right)\right] & \frac{1}{r^2}\left[(x/r)^2 \sigma_y^2 + (y/r)^2 \sigma_x^2\right] \end{bmatrix}.$$

Calculations involving a set of parameters that are uncorrelated are less complicated than those where correlations exist, because in the former case the variance matrix is diagonal. Given a set of variables θ_i that are correlated, it is always possible to find a new set of uncorrelated variables ω_i, for which the associated variance matrix is diagonal, in terms of the original set. This is achieved by a linear transformation of the form

$$\omega_i = \sum_{j=1}^{n} A_{ij}\theta_j, \tag{5.49}$$

which leads to a variance matrix U_{ij} for the set ω_i, given by

$$U_{ij} = \text{cov}(\omega_i, \omega_j) = \text{cov}\left(\sum_{k=1}^{n} A_{ik}\theta_k, \sum_{l=1}^{n} A_{jl}\theta_l \right)$$

$$= \sum_{k,l=1}^{n} A_{ik}A_{jl}\,\text{cov}(\theta_k, \theta_l) = \sum_{k,l=1}^{n} A_{ik}V_{kl}A_{lj}^{T}, \tag{5.50}$$

or in matrix notation $\mathbf{U} = \mathbf{AVA}^{T}$. Thus we need to find the matrix $\mathbf{A}$ that transforms the real symmetric matrix $\mathbf{V}$ into diagonal form. This is a standard technique in matrix algebra, but we will not pursue it further, because although this may simplify calculations, the transformed variables usually do not have a simple physical interpretation, and if there are more than three variables, numerical techniques are usually required anyway.

5.5 Monte Carlo method and simulations

Problem solving methods that rely on computer-generated random numbers to simulate elements of an experiment are generally called *Monte Carlo* methods, in reference to the famous gambling casino in Monaco. As mentioned in Sections 3.4 and 5.1.1, we will introduce such methods in this section.

One useful application of the Monte Carlo method is evaluation of integrals with complicated boundaries. We will discuss this and give an example in Section 5.5.1 to introduce the idea of using random variables in what are usually considered deterministic problems.

Then, in Section 5.5.2 we will briefly discuss how random variables are used in the important practical area of simulation, the technique that is used, for example, to test whether an apparatus can in practice achieve its design performance.

5.5.1 Monte Carlo method for integration

In Section 5.1.1 we discussed the problem of finding the joint probability density function of a statistic $y(\mathbf{x})$ that is a function of the n random variables $\mathbf{x} = x_1, x_2, ..., x_n$, independently and randomly selected from the same distribution $f(x)$. The formal solution to this problem was via construction of the joint distribution function of y, that is,

$$F(y) = \int \cdots \int \prod_{i=1}^{n} f(x_i) \, dx_i, \tag{5.51}$$

and an example was given for the case where $f(x)$ was a Cauchy distribution, but even in this very simple case the integrations were complex. In general, analytic evaluation of such integrals is rarely possible, and numerical methods must be used. Even then, conventional techniques are too time consuming to be practical and the Monte Carlo method is used. We will illustrate the technique by showing how the method can be used to evaluate a simple integral.

Consider the one-dimensional integral

$$I = \int_{0}^{1} f(x) dx, \tag{5.52}$$

where $f(x)$ is an arbitrary function. The simplest numerical method to evaluate the integral is to divide the range into n equal parts and to use as an estimate of I the sum

$$I = \frac{1}{n} \sum_{i=1}^{n} f(x_i),$$

where

$$x_i = (i - 1/2)1/n,$$

that is, the values at the midpoints of the intervals. More efficient numerical techniques are available, of course, such as Simpson's rule, which also uses equal spacing of the points in the range but approximates the value of the function between the points by a quadratic form. However, in a Monte Carlo evaluation, instead of calculating $f(x_i)$ at predetermined fixed points x_i, f is evaluated at the points $x_i = u_i$, where u_i are randomly chosen values uniformly distributed in the interval [0, 1]. Such numbers are called *random numbers* and may be generated, for example, by counting the number of decays observed from a radioactive source over a fixed period. In practice, random numbers u, uniformly distributed in the interval $0 \le u \le 1$, are readily available from computer programs called *random number generators* and are derived from an algorithm.[4]

If $u_1, u_2, \ldots, u_k$ are independent uniform random variables in the interval [0, 1], it follows that the random variables $f(u_1), f(u_2), \ldots, f(u_k)$ are independently and identically distributed random variables. It therefore follows from the strong law of large numbers that

---

[4] Random numbers generated by a computer are called *pseudorandom numbers*, because given enough numbers, the sequence will repeat, and this would have to be considered in calculations using very large numbers of random numbers. Nevertheless, for most purposes, pseudorandom numbers have properties that approximate very well the behaviors of true random numbers and are widely used.

$$\sum_1^k \frac{f(u_i)}{k} \to E[f(u)] = I \ \text{as} \ k \to \infty.$$

Hence we can estimate I by generating a large number of random variables u_i and taking the average value of $f(u_i)$. An analogous procedure can be followed for a general integral with lower and upper limits a and b, by making the simple substitution

$$y = (x-a)/(b-a), \text{and hence} \ dy = dx/(b-a),$$

so that random numbers in the range [0, 1] can still be used but now referred to the random variable y. The integral is then given by

$$I = \int_0^1 h(y)dy, \text{where} \ h(y) = (b-a)f(a+[b-a]y).$$

Alternately, random numbers uniform in the interval between the limits of integration can be used. As shown in Example 5.9, this requires including the uniform value of the *pdf* of the random distribution (which may not be unity) as a factor in the integrand.

For a single-variable integral such as (5.52), it may be shown that the error of a single integration using n equally spaced points in the domain of the integration is proportional to $1/n^2$, whereas for the Monte Carlo method using n randomly chosen points the error is proportional to $1/\sqrt{n}$, so less accurate. However, the power of Monte Carlo integrations becomes apparent when integrating functions in several dimensions. The accuracy of the Monte Carlo method remains proportional to $1/\sqrt{n}$, whereas simple methods using nonrandom variables have an accuracy approximately proportional to $n^{-2/d}$, where d is the number of dimensions. A more important reason for using the Monte Carlo method is that if the dimensionality is large and the integral complicated, it may not be possible to set up a network of points needed to evaluate the integrand using standard methods, or at least computationally very expensive to do so. For the Monte Carlo method it is only necessary to check whether the point is inside or outside the region of integration to decide whether to include or reject it. Finally, repeating a Monte Carlo calculation will produce a slightly different result depending on the exact sequence of random numbers used, which replicates the behavior of real experiments that invariably produce a finite amount of data of limited accuracy.

Example 5.9

Approximate the integral

$$I = \iiint_{x^2+y^2+z^2\le1} (xyz)^2 dxdydz$$

numerically, using the Monte Carlo method.

Solution: The exact solution, $I = 4\pi/945$, can be found, somewhat tediously, by transforming to polar coordinates and performing multiple integrations by parts. Nonetheless, it serves as a useful example for applying the Monte Carlo method to difficult integrals, in this case, where the boundary of the region of integration is not a surface of constant coordinate.

We can change the region of integration to the cubical volume $-1 \leq x, y, z \leq 1$ by defining

$$g(x, y, z) = \begin{cases} (xyz)^2 & \text{for } x^2 + y^2 + z^2 \leq 1 \\ 0 & \text{otherwise} \end{cases}$$

so that

$$I = \int_{z=-1}^{1} \int_{y=-1}^{1} \int_{x=-1}^{1} g(x, y, z) dx dy dz.$$

Now suppose that x, y, and z are independent, identically distributed random variables with *pdfs* that are uniform over the interval $[-1, 1]$ and zero outside of this interval. It is easy to see that each *pdf* has the value $1/2$ in the interval $[-1, 1]$ and is zero elsewhere. The expected value of the function g of these random variables is then

$$E[g] = \frac{1}{8} \int_{z=-1}^{1} \int_{y=-1}^{1} \int_{x=-1}^{1} g(x, y, z) dx dy dz,$$

so $I = 8E[g]$. We approximate the expected value of g as the mean of a sample of n realizations $g(x_i, y_i, z_i)$ where each x_i, y_i, and z_i is a realization of x, y, and z produced by a random number generator, respectively. Thus

$$I = 8E[g] \approx \frac{8}{n} \sum_{i=1}^{n} g(x_i, y_i, z_i).$$

As described in Chapter 4.1.1, we can use a uniform random variable on the interval $[0, 1]$ to generate any probability distribution. In the present case, if u is such a random number, then $2u - 1$ is a random variable identical to x, y, or z. Most modern computer languages are equipped with a random number generator for u, as well as many other distributions. In any such language an estimate of the integral I for a given value of n can be produced by the following simple steps:

1. Set a floating-point variable, s, say, to 0. This will hold the sum of the g_i.
2. Set an integer, i, say, to 1 as a counter for the number of values of g_i.
3. Get three values of the uniform distribution u_1, u_2, and u_3 and convert to values from the distribution uniform over $[-1, 1]$ by $x = 2u_1 - 1$, $y = 2u_2 - 1$, and $z = 2u_3 - 1$.
4. Check: Is $x^2 + y^2 + z^2 \leq 1$? If no, go to step 5. If yes, set $s = s + g(x_i, y_i, z_i)$.
5. Check: Is $i \geq n$? If no, set $i = i + 1$ and return to step 3.
6. Calculate estimate of integral: $I = 8s/n$.
7. End.

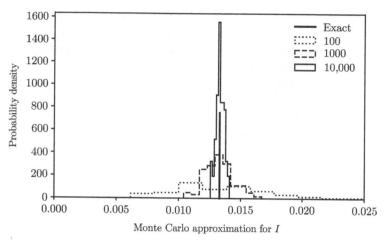

FIGURE 5.1   Results of Monte Carlo calculation of I for three values of n, the number of random positions within the region of integration at which the integrand is computed. The horizontal axis is the calculated value of the integral I. The vertical axis is the frequency of the value in 100 repetitions of the calculation for each value of n, normalized as a probability density.

Figure 5.1 shows the result of running this procedure for $n = 100, 1000$, and $10,000$. For each selection of n, the calculation was repeated 100 times and a histogram (normalized as a probability density) plotted. The exact result is shown as a vertical line.

In a real situation consideration would also have to be given to a measure of the overall accuracy. The simplest way to estimate this is to repeat the calculation several times with different sequences of random numbers and use the outcomes to calculate the mean and variance in the standard way used for any measurements. Alternatively, an analysis using n random numbers can be divided into 10 analyses, each using $n/10$ random numbers and the outcomes used to find the mean and variance as before. In the limit n estimates could be used, being obtained from a single point. In this case the final estimate will have an error $\sigma/\sqrt{n}$, where σ^2 is the variance of the distribution of the values of the function being integrated.

Monte Carlo calculations can use very large amounts of computer time, and so more efficient methods have been developed to reduce the variance of the estimate. These are analogous to methods that improve integrations based on using values at equally spaced points of the function, such as Simpson's rule mentioned earlier.[5]

Integration using the Monte Carlo method uses uniform random numbers distributed in the interval [0, 1], but other applications require random numbers distributed as other probability distributions. There are several methods of generating these. One is to use the transformation property of probability distributions referred to in Section 4.1. An example is given in Example 5.10

[5] There are several such *variance-reducing* methods, and interested readers can find the details in, for example, Chapter 11 of S.M. Ross *Introduction to Probability Models*, Elsevier (2019).

Example 5.10

Find an expression for random variables distributed with an exponential density in terms of random variables uniformly distributed with the density

$$p(u) = \begin{cases} 1 & 0 \leq u \leq 1 \\ 0 & \text{otherwise} \end{cases}.$$

Solution: The exponential density is, from Equation (4.26),

$$f(x) = \begin{cases} \lambda e^{-\lambda x} & \lambda > 0, x \geq 0 \\ 0 & \text{otherwise} \end{cases}.$$

In section 4.1 we showed that a random number x can be selected from any distribution F simply by setting $x = F^{-1}(u)$, where u is selected from the uniform distribution on [0, 1]. So $u = 1 - e^{-\lambda x}$, and $x = -\ln(1 - u)/\lambda$. As u is distributed uniformly, then so is $(1 - u)$. Thus, to select random numbers distributed with an exponential density, we generate a sequence u_i and from them generate a new sequence x_i using $x_i = -\ln(u_i)/\lambda$.

5.5.2 Simulation

A *simulation* is a computational model of a system. Simulations are usually used to determine whether a specific design for a system can in practice achieve the required performance, without incurring the cost of building and testing the actual system. Simulations can also be used to study the consequences of specific theories and provide evidence for or against the validity of a theory. Simulation has become an important and widely used technique in many areas and is crucial in fields such as particle physics and space exploration, where very large experiments are involved that are very expensive to construct and operate and can take years to complete. Simulations allow much of the design to be verified before expensive hardware is constructed.

In this section we will give a brief explanation of how in principle simulations are made, illustrated by two simple examples. Real situations are usually far more complex and often involve substantial amounts of high-speed computer time.

The first example concerns a particle scattering experiment, which usually involves highly specialized and expensive equipment that would merit simulation prior to construction. In this case the 'system' to be simulated could be individual components of the experiment, such as magnets to determine the momentum of particles by measuring the curvature of their trajectories, or electronic detectors to count and record the numbers and types of particles produced in collisions. In simulation of the former the computer model calculates the tracks (trajectories) of particles with specified momentum through the detector. After passing through each short segment of the track, the direction of each particle is modified by a small, randomly selected amount to simulate the effect of random multiple scatters. This is done by using a normal distribution with a variance that depends on the amount of material traversed.

The computed position of each particle may also be randomly modified to allow for the expected positional resolution. Finally, the complete set of computed coordinates of a track are used to calculate its estimated momentum, which can then be compared to the known input values, and thus a momentum resolution determined. If this value is too large to meet the aims of the experiment, a redesign of the detectors, or possibly other aspects of the experiment, such as the magnetic field, would have to be made. The redesign can proceed iteratively, with each version of the design simulated until a satisfactory result is obtained. The same ideas can be adapted to investigate systematic errors to determine whether they are within acceptable limits.[6]

Example 5.11

Simulate the trajectory of a 10 gigaelectron volt electron in a cloud chamber, at standard temperature and pressure, in a 10 Tesla uniform and constant magnetic field, accounting for random collisions of the electron with the gas molecules in the chamber. Determine the expected value and standard deviation of the radius of the trajectory.

Solution: A charged particle locally ionizes the gas in the cloud chamber resulting in a vapor trail that allows the particle trajectory to be observed. In the absence of the gas the force $\mathbf{F}$ on a particle of charge q and velocity $\mathbf{v}$ is due to the magnetic field $\mathbf{B}$ in accordance with Ampère's law. At any time t:

$$\mathbf{F}(t) = q\mathbf{v}(t) \times \mathbf{B}.$$

Since the force is perpendicular to the velocity, the speed, $v = |\mathbf{v}|$, of the electron is constant (we neglect the effect of radiation caused by the acceleration of the charge) and it is easily shown that the trajectory is a circle in the plane perpendicular to $\mathbf{B}$, of radius $r = mv/Bq$, where m is the electron mass and $B = |\mathbf{B}|$. The time required for the electron to complete one orbit in its trajectory is the distance traveled divided by the speed, that is, $2\pi m/Bq$. Given the radius of the trajectory and the charge of the electron, its momentum is determined.

For simplicity, we assume the gas molecules are sufficiently massive that a collision with the electron has negligible effect on it, so the electron's direction is changed, but not its speed. The velocity of the electron after a collision differs from that before the collision by an angle θ. This angle depends on the properties of the electron and the gas molecule as well as the distance between the electron and the molecule at nearest approach. Rather than attempting a deterministic model of the motions of all the gas molecules and the electron, we take advantage of the fact that the positions of the molecules are essentially random, and many collisions occur. Thus the angle θ for any collision is a random number with zero mean and a standard deviation σ_θ that depends on the characteristics of the molecules and the electron. For simplicity, we ignore any deflection out of the plane perpendicular to $\mathbf{B}$, so we assume the trajectory is a path in that plane, as it would be without collisions.

Over a short segment of the trajectory of the electron, the number of collisions n will be large enough that we can apply the central limit theorem. The sum of the angles of deflection of the electron velocity over this short segment is then a normal random variable with zero mean and standard deviation $\sqrt{n}\sigma_\theta$.

---

[6] Cloud chambers have not been 'state of the art' instruments for particle detection for several decades but are easy to understand, which is useful for this example.

The simulation of the trajectory proceeds in two steps for each time interval Δt. Firstly, the position and velocity of the electron at the end of the last time interval is used as the position and velocity at the start of the current time interval. The position and velocity at the end of the current time interval is first calculated, neglecting the effects of collisions. Then a normal distribution $N\left(0, \sqrt{n}\, \sigma_\theta^2\right)$ random number generator is used to select the collisional change in angle over the interval and the position and velocity at the end of the time step is adjusted accordingly.

To estimate the number of collisions n in a time step, we need an estimate of the distance between gas molecules. The number of molecules per unit volume η can be determined from the ideal gas law, usually expressed as $PV = nRT$, but in this situation more usefully expressed as:

$$\eta = \frac{P}{k_B T},$$

where P is the pressure, T is the absolute temperature, and k_B is the Boltzmann constant. The volume per molecule is $1/\eta$, which we take as the volume of the largest sphere centered on one molecule that, in an average sense, does not contain any others. The radius of this sphere is $\sqrt[3]{3/4\pi\eta}$, which we use as an approximation to the mean distance between molecules. The number of collisions in the time Δt is the distance the electron travels in this time divided by the distance between molecules:

$$n = \Delta t v \sqrt[3]{\frac{4\pi\eta}{3}}.$$

At a temperature of zero Celsius and atmospheric pressure, the distance between molecules calculated in this manner is 2.1 nanometers, indicating a very large number of collisions in very short segments of the trajectory.

Taking (x, y) Cartesian coordinates in the plane of the trajectory, we let x_k, y_k, v_{xk}, and v_{yk} be the components of the electron's position and velocity at the end of the kth time step. For convenience, we define Θ_k as the angle the electron velocity makes with the x axis at the end of the kth time step. Neglecting the effects of collisions, in the $(k+1)$th time step, the electron travels a circular arc of radius mv/Bq. It is straightforward to show that at the end of the time step, still neglecting collisions,

$$v_{x(k+1)} = v_{xk} \cos(\alpha\Delta t) - v_{yk} \sin(\alpha\Delta t)$$

$$v_{y(k+1)} = v_{xk} \sin(\alpha\Delta t) + v_{yk} \cos(\alpha\Delta t)$$

$$x_{(k+1)} = x_k + \frac{1}{\alpha}\left(v_{y(k+1)} - v_{yk}\right)$$

$$y_{(k+1)} = y_k - \frac{1}{\alpha}\left(v_{x(k+1)} - v_{xk}\right)$$

$$\Theta_{(k+1)} = \Theta_k + \alpha\Delta t$$

where

$$\alpha = \frac{Bq}{m}.$$

If the change in angle, $\delta\theta$, due to one collision is small, we can neglect terms of order $\delta\theta^2$ and use the approximation

$$\cos(\theta + \delta\theta) = \cos\theta - \delta\theta\sin\theta$$
$$\sin(\theta + \delta\theta) = \sin\theta + \delta\theta\cos\theta$$

To calculate the effect of all the collisions in the $(k+1)$th time step, we assume all the changes in angle occur at once at the end of the time step. We select a value $\Delta\Theta$ from the $N(0, \sqrt{n}\,\sigma_\theta^2)$ distribution and calculate the changes in components as

$$\Delta v_{x(k+1)} = -v\sin(\Theta_{k+1})\Delta\Theta$$
$$\Delta v_{y(k+1)} = v\cos(\Theta_{k+1})\Delta\Theta$$
$$\Delta x_{(k+1)} = \Delta v_{x(k+1)}\Delta t$$
$$\Delta y_{(k+1)} = \Delta v_{y(k+1)}\Delta t$$

Adding these changes to the components determined without collisions, and Adding $\Delta\Theta$ to the previously determined value of $\Theta_{(k+1)}$ completes the calculation for the $(k+1)$th time step.

Once the trajectory has been calculated over a sufficient number of time steps, the radius of the trajectory can be determined by finding the center and radius of the circle that best fits the calculated (x, y) coordinate pairs. Suppose (x_c, y_c) are the coordinates of the center of the 'true' trajectory and R is its radius. The distance from the kth calculated coordinate pair (x_k, y_k) to this center is

$$d_k = \sqrt{(x_k - x_c)^2 + (y_k - y_c)^2}.$$

A direct application of the least-squares method of estimating R is to find x_c, y_c and R that minimize the sum

$$\sum_k (d_k - R)^2$$

but the radical in the definition of d_k makes this a nonlinear problem that does not have an analytic solution. Instead, we minimize

$$\sum_k (d_k^2 - R^2)^2.$$

The solution is readily determined to be

$$R = \sqrt{D + x_c^2 + y_c^2}$$

where D, x_c, and y_c are the solutions of the matrix equation

$$\begin{pmatrix} 1 & 2\overline{x} & 2\overline{y} \\ \overline{x} & 2\overline{x^2} & 2\overline{xy} \\ \overline{y} & 2\overline{xy} & 2\overline{y^2} \end{pmatrix} \begin{pmatrix} D \\ x_c \\ y_c \end{pmatrix} = \begin{pmatrix} \overline{x^2 + y^2} \\ \overline{x^3 + xy^2} \\ \overline{yx^2 + y^3} \end{pmatrix}.$$

This simulation was implemented in a simple program in the Python programming language. In the simulation the cloud chamber temperature and pressure were set to 273.15 K and one atmosphere, respectively. The magnetic field was set to 10 T pointing in the negative z direction and the electron was given a kinetic energy of 10^{10} electron volts, from which the speed was calculated. Without collisions, the trajectory radius in this case is 2.38 cm.

The time step was set at $1/200$ of the time to complete the circular trajectory once, and the number of time steps was set to 100 so that without collisions, the electron would go halfway around the circular trajectory.

One thousand simulations of the trajectory were made for each of the three choices for σ_θ. The first 10 simulations of the trajectory, for $\sigma_\theta = 10^{-4}$ radians, are shown in Figure 5.2. Clearly, the effect of collisions in this case is too large for the trajectory to be of much use.

Figure 5.3 shows histograms of the calculated radius of each of the 1000 simulated trajectories for each of the three values of σ_θ. As expected, smaller values of σ_θ result in much narrower

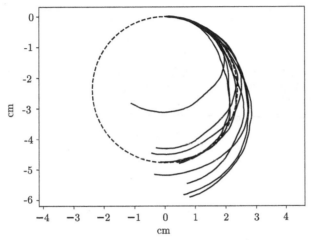

FIGURE 5.2 Simulations of 10 trajectories of an electron in a cloud chamber with a uniform 10 T magnetic field, positive pointing into the page, and $\sigma_\theta = 10^{-4}$ radians. The dotted circle is the trajectory that would be followed without collisions.

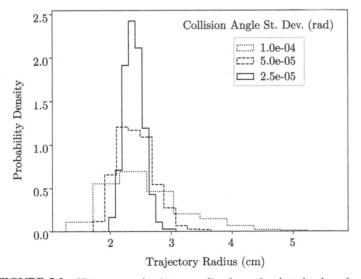

FIGURE 5.3 Histograms of trajectory radius for each selected value of σ_θ.

distributions of trajectory radius. Each histogram is normalized to an area of 1, making it an approximate probability density function for the radius.

As shown in Example 5.11, a simulation combines deterministic calculations with statistical calculations.[7] Simulation of entire experiments can help ensure the designs are appropriate, for example, to discriminate between the predictions of two competing theories.[8]

Using the Monte Carlo approach, a fixed number of Monte Carlo 'experiments' could be generated and distributed according to the theory being tested. The outcomes will allow the experimenters to decide whether the results can discriminate between the predictions of the two theories.[9] If not, more Monte Carlo 'experiments' would have to be generated, until sufficient data for successful discrimination is generated. The result indicates the number of events that would need to be measured in the real experiment.

Simulations can also be used to test whether an experiment to determine a parameter is adequate to establish whether the experimental result agrees with a theoretical prediction. A sequence of Monte Carlo 'experiments' are generated, each containing the same number of events as the proposed experiment and distributed according to the theory being tested. The design can then be adjusted until a sufficiently precise result is obtained in the simulation[10].

Another somewhat different example of simulation is the simple device illustrated in Figure 5.4, called a 'Galton Board'.[11] It consists of a board into which rows of pins are fixed, projecting perpendicular to its face. In each row the pins are equally spaced and positioned so that the pins in one row are halfway between the pins in the row above. If the pins were not present, the ball would simply drop into the central collection channel. Instead, the ball bounces in an unpredictable way to the left or the right and contacts the nearest pin on the next row down. A transparent cover, not shown, prevents the balls from bouncing away from the surface of the board.

Bouncing off one pin after another, the ball makes its way along a random path to one of the collection channels at the bottom of the board. Each bounce is a random contribution of half the distance between the pins to the left and the right. The sum of these contributions determines the collection channel in which the ball finally arrives. According to the central limit theorem, if enough balls are dropped, there are enough rows of pins, and each row contains enough pins, the distribution of balls in the channels will closely approximate a normal distribution.

---

[7] The least-squares method is discussed in detail in Chapter 8.

[8] For example, see I. Kasa 'A circle fitting procedure and its error analysis', IEEE Transactions on Instrumentation and Measurement (1976), vol. 25, pp. 8–14.

[9] How to use data to discriminate between theories, or hypotheses, is discussed in Chapters 10 and 11.

[10] Estimating parameters from data is discussed in Chapters 7 and 8.

[11] It is also known as a Quincunx, the Latin word for the pattern of five dots, with one in each corner of a square and one in the middle, which is the repeating pattern of the device.

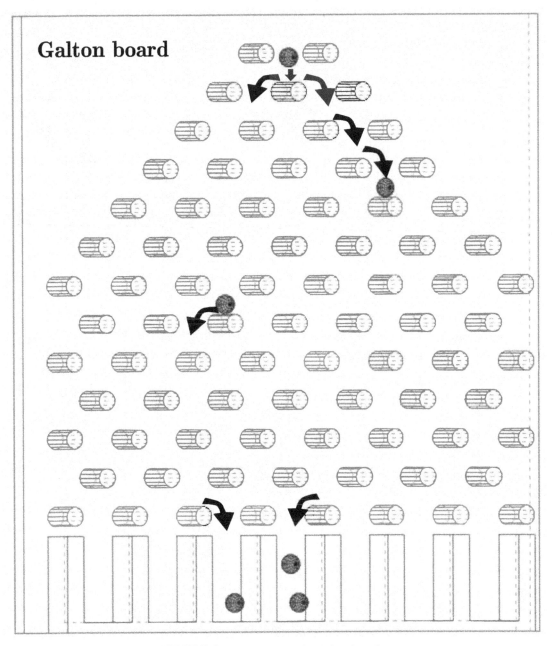

FIGURE 5.4   Illustration of a Galton board.

Example 5.12

Build a simulation of the Galton board and determine a design that will demonstrate the central limit theorem.

Solution: For simplicity, we assume we take the spacing between pins in each row as one unit of length. When a ball hits a pin, it bounces half a unit to either the left or the right and then contacts a pin in the next row down. The horizontal distance traveled is either $+1/2$, to the right, or $-1/2$, to the left. After bouncing off a pin in the last row, the ball is captured in the channel at distance from the center equal to the sum of all the $+1/2$ and $-1/2$ bounces.

The steps of the simulation are firstly to make a two-dimensional array with one row for each ball and one column for each row of pins. Secondly, the array is filled with $+1/2$ and $-1/2$ randomly by using a random number generator. A simple way to do this is, for each element of the array, to select a number x from a generator that produces a uniform distribution of random numbers in the range $[0, 1]$. If $0 \le x \le 0.5$, the element of the array is $-1/2$, otherwise it is $+1/2$.

Thirdly, create a one-dimensional array with an element for each ball. Set the value for each ball to the sum of the elements of the row in the two-dimensional array representing that ball, which is the position of the channel in which the ball is finally captured. To simulate a Galton Board with a limited number of pins in each row, check while summing the bounces to ensure the horizontal distance traveled has not gone past the last pin at either end of the row and if it has, set the distance at that point to the distance of the last channel.

A histogram of the values in the one-dimensional array, with the bin edges set to the positions of the pins in the last row, gives the number of balls in each channel. The result of 1 simulation for 1000 balls dropped through 100 rows of pins, with 15 pins on each side of the last row, is shown in Figure 5.5. The histogram is normalized to unit area so that it is a simulated approximation to the

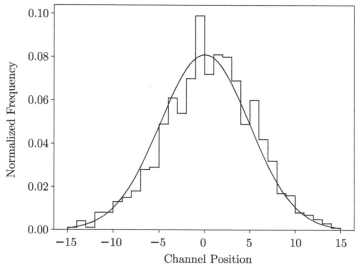

FIGURE 5.5   Histogram normalized to unit area for a Galton board with 100 rows of pins, spaced 1 unit of length apart and 1000 balls dropped.

pdf for ending up at any position along the width of the Galton Board. A normal *pdf* with the same mean and variance as the histogram is superimposed showing a sum of 100 random choices of bounce direction is a good fit to a normal *pdf*, as would be expected from the central limit theorem.

A similar type of test is used in deciding whether an enhancement observed in a mass distribution in the final spectrum of particles produced in a nuclear collision is due to the formation of a resonant state of the type described by the Breit-Wigner distribution discussed in Section 4.1.6, or a random statistical fluctuation.

The experiment mentioned in Section 1.6, where a measurement of the magnetic moment of the muon was made and compared to the theoretical prediction, is an excellent example of the Monte Carlo simulation method, and its importance in the planning and interpretation of modern experiments.

Problems 5.11 and 5.12 provide an opportunity to explore Monte Carlo simulation methods in very simple contexts.

Problems 5

5.1 Prove the relation (5.9).

5.2 A variable x is uniformly distributed in the interval $I \leq x \leq I + 1$. If $\bar{x}_n$ is the mean of a random sample of size n drawn from the population, find an unbiased estimator for I in terms of $\bar{x}_n$. What is the mean squared error of $\bar{x}_n$ if it is used as a biased estimator of I?

5.3 Prove the results given in (5.28) and (5.29).

5.4 A sample of size $n_1 = 6$ is drawn from a normal population with a mean $\mu_1 = 60$ and variance $\sigma_1^2 = 12$. A second sample, of size $n_2 = 4$, is selected, independent of the first sample, from a different normal population having a mean $\mu_2 = 50$ and variance $\sigma_2^2 = 8$. What is $P[(\bar{x}_1 - \bar{x}_2) < 7.5]$?

5.5 A particular living organ is repeatedly exposed to doses of radiation r_i that are normally distributed with mean 4 and variance 2 (in arbitrary units). It is found that the maximum cumulative dosage of radiation R that the organ can absorb without suffering permanent damage is normally distributed with mean 100 and variance 25 (in the same units). What is the maximum number of doses that the organ may absorb before the probability of damage exceeds 3%?

5.6 A power unit is manufactured by an identical process in several different factories A, B, C, etc. and the mean output of the unit across all factories is 50 watts with a standard deviation σ of 7 watts. A random sample of 100 units is taken from factory A and the sample mean is found to be 49 watts. Is the product from factory A consistent with the overall standard of manufacture?

5.7 A prospective purchaser of resistors decides to buy a sample R_n of size n from the manufacturer to check that their average value does not vary by more than 3% from the average value $\bar{R}$ of all resistors from the same manufacturer, with a probability of 0.05. If the values of the resistors are normally distributed with a standard deviation that is 15% of the value $\bar{R}$, how many would have to be bought?

5.8 $F(x, y, z)$ is a function of the three variables x, y, z with the form $F(x, y, z) = xy^2z^3$. If the variance matrix of x, y, z is

$$V = \frac{1}{1000}\begin{bmatrix} 1 & -1 & 1 \\ -1 & 1 & 0 \\ 1 & 0 & 1 \end{bmatrix},$$

what is the percentage error on F when $x = 2, y = 1$, and $z = 1$?

5.9 A beam of particles is incident on a target and F events are recorded where the particle scatters into the forward hemisphere and B events where it scatters into the backward hemisphere. What is the standard deviation of the 'forward-backward asymmetry' R, defined as $R \equiv (F - B)/(F + B)$?

5.10 Use the inversion method to find an expression for generating random numbers distributed as the Breit-Wigner (Cauchy) distribution

$$g(y) = \frac{2}{\pi\Gamma}\frac{\Gamma^2}{4(y - a)^2 + \Gamma^2},$$

where a and Γ are constants.

5.11 A large physics experiment requires a new type of detector and several vendors have submitted formal bids including not only price and delivery time but also explanations of why that vendor is the best choice for the job. A committee of five experts is formed to determine which vendor should get the job. Each committee member scores each bid on a scale of 0 (worst) to 10 (best) after considering several criteria. The scores are not limited to integer values. The scores of the individual committee members are combined to produce an agreed score for each vendor and the agreed scores are used to decide which vendor gets the job.

Create a simulation to decide which of two methods of combining the scores will produce the more accurate result. The first method, which we will call the average rule, is simply using the average of the scores of all the members as the score for the vendor. The second method, which we will call the truncation rule, is to leave out the highest and lowest member score and take the score for the vendor as the average of the remaining member scores. In the truncation rule, if there are multiple scores that are equally highest or lowest, only one of each is excluded from the average.

Assume the true score for each vendor is a random number uniformly distributed in the range [0, 10] and consider two scenarios:

1. Each expert is unbiased and produces a score that is the true score for each vendor plus a normally distributed random error with zero mean and standard deviation of 0.25.

2. There is a 10% probability that any expert is biased. A biased expert will, with equal probability, either increase the score of a favored vendor by 2.0 or decrease the score of an unfavored vendor by the same amount.

5.12 A bank wants to add a drive-through station for customers at a small branch office but has limited space for cars to wait in the queue. The drive-through will be operated by a single clerk who can serve one customer at a time. Assuming the time needed to process a single customer's transaction will vary uniformly from 2 minutes to 5 minutes and 10 customers are expected to arrive per hour to use the drive-through, write a simulation and create histograms to show the number of customers waiting in the queue and the time each customer will wait before being served.

CHAPTER

6

Sampling distributions associated with the normal distribution

Overview

In this chapter there is a description of the three most important sampling distributions encountered when sampling from populations that are normally distributed — the chi-squared (χ^2), Student's t, and F distributions — and the relations between them. Each is derived and their basic properties discussed. The main point of this chapter is to introduce these populations prior to a discussion of their use in applications in later chapters.

The special position held by the normal distribution, mainly by virtue of the central limit theorem, is reflected in the prominent positions of distributions resulting from sampling from the normal. In this chapter we will consider the basic properties of three frequently used sampling distributions: The chi-squared, the Student's t, and the F distributions. These are widely used in estimation problems, both for finding the best values of parameters and their optimal ranges and in testing hypotheses, topics that will be discussed in detail in Chapters 7–11.

6.1 Chi-squared distribution

If we wish to concentrate on a measure to describe the dispersion of a population, then we consider the sample variance. The chi-squared distribution is introduced for problems involving this quantity. It is defined as follows.

If $x_i (i = 1, 2, ..., n)$ is a sample of n random variables normally and independently distributed with means μ_i and variances σ_i^2, then the statistic

$$\chi^2 \equiv \sum_{i=1}^{n} \left(\frac{x_i - \mu_i}{\sigma_i} \right)^2 \tag{6.1}$$

is distributed with density function

$$f(\chi^2, n) = \frac{1}{2^{n/2}\Gamma(n/2)} \chi^{2[(n/2)-1]} \exp(-\chi^2/2), \quad \chi^2 > 0. \tag{6.2}$$

Probability and Statistics for Physical Sciences, Second Edition
https://doi.org/10.1016/B978-0-443-18969-2.00006-4

143

© 2024 Elsevier Inc. All rights reserved.

This is known as the χ^2-*distribution (chi-squared)* with *n degrees of freedom*. It is another example of the general gamma distribution discussed in Section 4.1.5 and defined in (4.32), this time with $x = \chi^2, \alpha = n/2$ and $\lambda = 1/2$. The symbol Γ in (6.2) is the gamma function, defined in Section 4.1.5, but repeated here for convenience:

$$\Gamma(x) \equiv \int_0^\infty e^{-u} u^{x-1}\, du, \quad 0 < x < \infty. \tag{6.3}$$

It is frequently encountered in sampling distributions associated with the normal distribution.

Example 6.1

Compute from the basic result (6.2) the quantity $P\left[\chi^2 < 4\right]$ for $n = 4$ degrees of freedom.
Solution: Setting $x = \chi^2$ in (6.2) gives

$$P[x < 4] = \frac{1}{4} \int_0^4 x \exp(-x/2) dx,$$

where we have used $\Gamma(2) = 1$ from the definition (6.3). The integral may then be evaluated by setting $y = x/2$, giving

$$[P < 4] = \int_0^2 y e^{-y} dy = -[e^{-y}(y+1)]_0^2 = -3e^{-2} + 1 = 0.594.$$

An approximate value consistent with this result can be obtained by using Table C.4 in Appendix C.

The χ^2-distribution may be derived using characteristic functions as follows. We first write χ^2 as

$$\chi^2 = \sum_{i=1}^n z_1^2,$$

where the quantities z_i are distributed as the standard normal distribution $N(z_i; 0, 1)$. The quantities $u_i = z_i^2$ therefore have density functions (see example 3.10)

$$n(u_i) = \frac{1}{(2\pi u_i)^{1/2}} \exp(-u_i/2),$$

and the *cf* of u_i is

$$\phi_i(t) = \int_0^\infty \frac{1}{(2\pi u_i)^{1/2}} e^{(-u_i/2)} e^{itu_i} du_i = (1 - 2it)^{-1/2}, \quad (u_i \geq 0). \tag{6.4}$$

If $\phi(t)$ is the *cf* of χ^2, and since the random variables u_i are independently distributed, we know from the work of Section 3.2.3 that

$$\phi(t) = \prod_{i=1}^{n} \phi_i(t) = (1 - 2it)^{-n/2}. \tag{6.5}$$

Finally, the density function of χ^2 is obtained from the inversion theorem

$$f(\chi^2, n) = \frac{1}{2\pi} \int_0^\infty (1 - 2it)^{-n/2} e^{-i\chi^2 t} \, dt.$$

Using the definition of the gamma function, this yields (6.2), although the evaluation of the integral is rather lengthy.

If the variables x_i are not independent, but have a joint n-dimensional normal distribution with an associated variance matrix $\mathbf{V}$, as discussed in Section 4.1.3, then the variable to consider is

$$z = (\mathbf{x} - \boldsymbol{\mu})^T \mathbf{V}^{-1} (\mathbf{x} - \boldsymbol{\mu}).$$

Example 6.2

Use the result $\Gamma(1/2) = \sqrt{\pi}$ to verify (6.4).

Solution: Change variables in the integrand to $x = u\left(\frac{1}{2} - it\right)$. This gives

$$\phi_i(t) = \int_0^\infty \frac{1}{\sqrt{\pi}(1 - 2it)^{1/2}} x^{-1/2} e^{-x} \, dx,$$

and using (6.3)

$$\int_0^\infty x^{-1/2} e^{-x} \, dx = \Gamma(1/2) = \sqrt{\pi}.$$

Therefore

$$\phi_i(t) = (1 - 2it)^{-1/2}.$$

The χ^2 distribution is one of the most important sampling distributions occurring in physical science. Its density and distribution functions are one-parameter families of curves. Examples of $f(\chi^2, n)$ and the distribution function $F(\chi^2, n)$ for $n = 1, 4,$ and 10 are shown in Figure 6.1. The distribution function $F(\chi^2, n)$ is also tabulated in Appendix C, Table C.4 for a range of values of n. An alternative useful table may be constructed by calculating the proportion α of the area under the χ^2 curves to the *right* of χ^2_α, that is, points such that

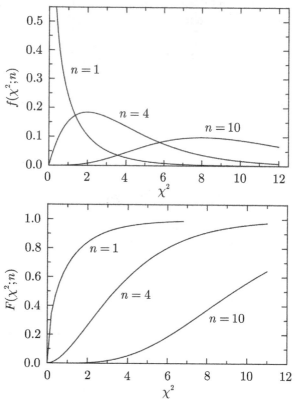

FIGURE 6.1   Graphs of the χ^2 density function $f(\chi^2, n)$ and its distribution function $F(\chi^2, n)$ for $n = 1, 4,$ and 10.

$$P\left[\chi^2 \geq \chi_\alpha^2\right] = \alpha = \int_{\chi_\alpha^2}^{\infty} f(\chi^2, n)\, d\chi^2. \tag{6.6}$$

Such points are called *percentage points,* or *critical values,* of the χ^2 distribution (recall the percentiles defined in Chapter 1) and may be deduced from Table C.4. They are shown graphically in Figure 6.2. A point of interest about these curves is that for a fixed value of P, the ratio $\chi_\alpha^2 / n \to 1$ as $n \to \infty$.

Example 6.3

(a) *What is $P\left[\chi^2 \leq 30\right]$ when χ^2 is a random variable with 26 degrees of freedom? (b) If χ^2 is a random variable with 15 degrees of freedom, what is its value that corresponds to $\alpha = 0.05$?*

Solution: (a) From Table C.4, we must find an entry close to, but not more than, 30 for $n = 26$. This is a little less than 0.75. The exact figure would have to be found by direct integration of the density function, as was done in Example 6.1.

(b) From the definition (6.6), we need to find a value χ_c^2 of χ^2 for 15 degrees of freedom such that $P\left[\chi^2 \geq \chi_c^2\right] = 0.05$, that is, a value χ_c^2 such that $P\left[\chi^2 \leq \chi_c^2\right] = 0.95$. From Table C.4, this is $\chi_c^2 = 25$.

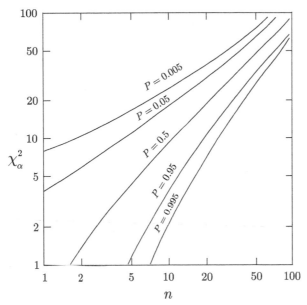

FIGURE 6.2   Percentage points of the chi-squared distribution, $P = P[\chi^2 \geq \chi_\alpha^2]$.

The *mgf* of the χ^2 distribution is obtainable directly from (6.4) and is

$$M(t) = (1 - 2t)^{-n/2}. \tag{6.7}$$

It follows that the mean and variance are given by

$$\mu = n \quad \text{and} \quad \sigma^2 = 2n. \tag{6.8}$$

The third and fourth moments about the mean may also be found from the *mgf*. They are

$$\mu_3 = 8n; \quad \mu_4 = 12n(n+4)$$

giving

$$\beta_1 = \frac{8}{n}; \quad \beta_2 = 3\left(1 + \frac{4}{n}\right),$$

which tend to the values for the normal distribution as $n \to \infty$, and the χ^2 distribution does indeed tend to normality for large samples. This can be demonstrated by constructing the *cf* for the standardized variable

$$y \equiv \left(\frac{\chi^2 - \mu}{\sigma}\right) = \left(\frac{\chi^2 - n}{\sqrt{2n}}\right),$$

which from (6.4) is

$$\phi_y(t) = \exp\left[-\frac{int}{(2n)^{1/2}}\right]\left[1 - \frac{2it}{(2n)^{1/2}}\right]^{-n/2}.$$

Then, taking logarithms gives

$$\ln\phi_y(t) \rightarrow -\frac{int}{(2n)^{1/2}} - \frac{n}{2}\ln\left[1 - \frac{2it}{(2n)^{1/2}}\right].$$

Finally, letting $n \rightarrow \infty$ and expanding the logarithm gives

$$\ln\phi_y(t) \rightarrow -\frac{int}{(2n)^{1/2}} - \frac{n}{2}\left[-\frac{2it}{(2n)^{1/2}} - \frac{1}{2}\left(\frac{2it}{(2n)^{1/2}}\right)^2 \cdots\right], \quad (n \rightarrow \infty),$$

implying

$$\phi_y(t) \rightarrow \exp(-t^2/2).$$

This is the *cf* of a standardized normal distribution and so, by the inversion theorem, the χ^2 distribution tends to normality as $n \rightarrow \infty$, although the rate of convergence is quite slow.

Because the χ^2 distribution is a one-parameter family of curves, it frequently happens that tabulated values do not exist for precisely the range one requires.[1] In such cases a very useful statistic is $(2\chi^2)^{1/2}$, which can be shown to tend rapidly to normality with mean $\mu = (2n - 1)^{1/2}$ and unit variance. The statistic

$$u = (2\chi^2)^{1/2} - (2n - 1)^{1/2} \tag{6.9a}$$

is therefore a standard normal variate for even quite moderate values of n, and so tables of the normal distribution may be used. Table 6.1 shows a comparison between the exact χ^2 distribution and the normal approximation based on the statistic $(2\chi^2)^{1/2}$ for a range of values of n and χ^2. Another statistic that converges to normality faster, but is more complicated to calculate, is $(\chi^2/n)^{1/3}$. This can be shown to tend very rapidly to normality with mean $1 - 2/(9n)$ and variance $2/(9n)$. Thus the statistic

---

[1] This is not as much of a problem as it once was. Statistical software packages and add-ons with functions to generate χ^2 and other probability density functions are readily available for personal computers.

$$u = \left\{ \left(\frac{\chi^2}{n} \right)^{1/3} + \frac{2}{9n} - 1 \right\} \left(\frac{9n}{2} \right)^{1/2} \tag{6.9b}$$

is a standard normal variate for even moderate values of n.

TABLE 6.1 Values of $P\left[\chi^2 \geq \chi_\alpha^2\right]$ for $n = 5, 10$, and 20 and $\chi_\alpha^2 = 2, 5, 10, 20$, and 30 using the exact χ^2 distribution function, and the normal approximation using the variable u of (6.9a).

n	5		10		20	
χ_α^2	**Exact**	**Approx.**	**Exact**	**Approx.**	**Exact**	**Approx.**
2	0.849	0.841	0.996	0.991		
5	0.416	0.436	0.891	0.885		
10	0.075	0.071	0.441	0.456	0.968	0.963
20	0.001	0.001	0.029	0.024	0.458	0.462
30			0.001	0.000	0.070	0.067

An important property of the χ^2 distribution is that the sum of m independent random variables $\chi_1^2, \chi_2^2, \ldots, \chi_m^2$, each having chi-squared distributions with $n_1, n_2, \ldots, n_m$ degrees of freedom, respectively, is itself distributed as χ^2 with $n = n_1 + n_2 + \ldots + n_m$ degrees of freedom. This is called the *additive property of* χ^2 and may be proved by using the characteristic function. (This Problem 6.3.)

Example 6.4

If x and y are two independent random chi-squared variables with $n = 3$ and $n = 6$ degrees of freedom, respectively, use the approximation given in (6.9b) to calculate the probability that their sum will exceed 10.

Solution:

By the additive property, the sum of two independent random chi-squared variables with n_1 and n_2 degrees of freedom, respectively, is itself an independent random chi-squared variable with $n = (n_1 + n_2)$ degrees of freedom. Thus the random variable $\chi^2 = (x+y) = 10$ is distributed as chi-squared with 9 degrees of freedom. Then using (6.9b), the random variable

$$u = \left\{ \left(\frac{\chi^2}{n} \right)^{1/3} + \frac{2}{9n} - 1 \right\} \left(\frac{9n}{2} \right)^{1/2}$$

is a standard normal variable. Finally, substituting the values above for χ^2 and n, and using Table C.1, gives

$$P\left[\chi^2 > 10\right] = P[u > 0.388] = 0.35.$$

There are two other important results that we shall need later. The first concerns a sample $x_1, x_2, \ldots, x_n$ of size n drawn from a normal population with mean zero and unit variance. Then the statistic

$$u = \sum_{i=1}^{n} (x_i - \bar{x})^2 \tag{6.10}$$

is distributed as χ^2 with $(n-1)$ degrees of freedom. In general, if the parent population has variance σ^2, then

$$\chi^2 = \frac{1}{\sigma^2} \sum_{i=1}^{n} (x_i - \bar{x})^2 \tag{6.11}$$

is distributed as χ^2 with $(n-1)$ degrees of freedom. Moreover, since the sample variance is

$$s^2 = \frac{\sigma^2 \chi^2}{n-1}, \tag{6.12}$$

it follows that $(n-1)s^2/\sigma^2$ is distributed as χ^2 with $(n-1)$ degrees of freedom, *independent* of the sample mean $\bar{x}$. Thus the sample mean and sample variance are independent random variables when sampling from normal populations. This somewhat surprising result is very important in practice, and we shall use it later to construct the sampling distribution known as the Student's t distribution.

If we assume that a sample is drawn at random from a single normal population with mean μ and variance σ^2, then from (6.1),

$$\chi^2 = \frac{1}{\sigma^2} \sum_{i=1}^{n} (x_i - \mu)^2. \tag{6.13}$$

However, since the mean of the population is rarely known, in these cases it is more useful to use the result that the quantity u in (6.10) is distributed as χ^2 with $(n-1)$ degrees of freedom. In that case χ^2 defined in (6.11) is distributed with $(n-1)$ degrees of freedom if $\bar{x}$ is used instead of μ.

The proof that u defined by (6.10) has $(n-1)$ degrees of freedom is fairly straightforward. We first note that the sample values $x_1, x_2, \ldots, x_n$ can be considered the Cartesian coordinates of a point in an n-dimensional space and u is the square of the distance between this point and the point $(\bar{x}, \bar{x}, \ldots, \bar{x})$, with n identical coordinates $\bar{x}$. It is easily shown that of all the possible points in the n-dimensional space, $(\bar{x}, \bar{x}, \ldots, \bar{x})$ is the point that minimizes the squared distance u. Thus, for any sample with the given value of $\bar{x}$, the sample point must lie in the $(n-1)$ dimensional hyperplane perpendicular to the line, passing through point $(\bar{x}, \bar{x}, \ldots, \bar{x})$ and the origin. In this sense the sample $x_1, x_2, \ldots, x_n$ has n degrees of freedom but u has $(n-1)$ degrees of freedom. The complete proof of (6.13) is problem 6.2. This illustrates an important general result: *The number of degrees of freedom must be reduced by one for each parameter estimated from the data.*

Example 6.5

Points are plotted randomly in a two-dimensional plane using Cartesian coordinates (x, y) and the distance from a fixed point (x_0, y_0) measured. If the differences

$$\Delta_x = x_0 - x \quad and \quad \Delta_y = y_0 - y$$

are independent random variables, normally distributed with zero means and standard deviations 2.1, what is the probability that the distance between the points (x, y) and (x_0, y_0) exceeds 3.5?

Solution: The distance d between the points (x, y) and (x_0, y_0) is given by

$$d^2 = \Delta_x^2 + \Delta_y^2,$$

and because the quantities $z_{x,y} = \Delta_{x,y}/\sigma = \Delta_{x,y}/2.1$ are standard normal variates,

$$P\left[d^2 > (3.5)^2 = 12.25\right] = P\left[z_x^2 + z_y^2 > (12.25/(2.1)^2 = 2.78\right]$$

$$= P\left[\chi^2 > 2.78\right] = 1 - P\left[\chi^2 < 2.78\right] = 0.25.$$

6.2 Student's *t* distribution[2]

The central limit theorem tells us that the distribution of the sample mean $\bar{x}$ is approximately normal with mean μ (the population mean) and variance σ^2/n (where σ^2 is the population variance and n is the sample size). Thus in standard measure the statistic

$$u = \left(\frac{\bar{x} - \mu}{\sigma_n}\right),$$

where $\sigma_n = \sigma/\sqrt{n}$, is approximately normally distributed with mean zero and unit variance for large n. However, in experimental situations neither the mean nor the population variance may be known, in which case they must be replaced by estimates from the sample. While σ^2 can be safely replaced by the sample variance s^2 for large $n \geq 30$, for small n, the statistic u will not be approximately normally distributed and serious loss of meaning in the interpretation will occur. Instead, we consider the distribution of the variable

$$t = \left(\frac{\bar{x} - \mu}{s/\sqrt{n}}\right),$$

where $s = \hat{\sigma}$ is an estimator for σ. If we write this as

---

[2] Despite its name, this is not a distribution specifically designed for use by students. The name refers to its originator, W.S. Gosset, who published under the pseudonym 'Student'. He worked as the Head Brewer at Guinness brewery and, as such, was obliged by his employers not to publish under his real name to protect Guinness' trade secrets. It is said he chose the name in recognition of the help that Karl Pearson had given him ('student') with the mathematics when he had worked for a year in Pearson's department.

$$t = \left(\frac{\bar{x} - \mu}{\sigma/\sqrt{n}}\right)\left(\frac{\hat{\sigma}}{\sigma}\right)^{-1},$$

then we can deduce from the central limit theorem that the numerator is distributed like a standard normal variable and, from (6.12) or (6.13), that the square of the denominator is distributed like a χ^2 variable with either $(n-1)$ or n degrees of freedom, divided by $(n-1)$ or n, depending whether or not μ is estimated from the data. The distribution of t is called the Student's t distribution. It enables one to use the sample variance, as well as the sample mean, to make statements about the population mean. The discussion will concentrate on three important results, but firstly we will derive the probability density function of t.

Let u have a normal distribution with mean zero and unit variance. Further, let w have a χ^2 distribution with n degrees of freedom, and let u and $\sqrt{w}$ be independently distributed. In this case the joint density of u and $\sqrt{w}$ is the product of their individual densities. Thus, from the form of the chi-squared distribution (6.2), and the standardized normal distribution (4.10), the joint density function of u and w is

$$f(u, w; n) = \frac{1}{(2\pi)^{1/2}}e^{-u^2/2}\frac{1}{\Gamma(n/2)2^{n/2}}w^{(n-2)/2}e^{-w/2}. \tag{6.14}$$

Then substituting

$$u = t\left(\frac{w}{n}\right)^{1/2},$$

(6.14) becomes

$$f(t, w; n) = \frac{e^{-t^2w/2n}e^{-w/2}w^{(n-2)/2}}{(2\pi)^{1/2}\Gamma(n/2)2^{n/2}}$$

and the marginal distribution of t, that is, $f(t; n)$, is given by

$$f(t; n) = \int_0^{\infty} f(t, w; n)\, dw.$$

This integral may be evaluated directly using the definition of the gamma function (6.3) with the result that the random variable

$$t = \frac{u}{(w/n)^{1/2}}$$

has a density function

$$f(t; n) = \frac{\Gamma[(n+1)/2]}{(\pi n)^{1/2}\Gamma(n/2)}\left[1 + \frac{t^2}{n}\right]^{-(n+1)/2}, \quad -\infty < t < \infty. \tag{6.15}$$

The statistic t is said to have a *Student's t distribution with n degrees of freedom*. It tends to a standard normal distribution for $n \to \infty$, as we will prove below, but for small n, the tails are wider than those of the latter, and for $n = 1$, the distribution is of the Cauchy form.

Like the χ^2 distribution, the Student's t distribution is a one-parameter family of curves. The distribution function is tabulated in Table C.5, and in using it one can use the fact that

$$P[t < -t_\alpha(n)] = P[t > t_\alpha(n)] = \alpha,$$

since the distribution is symmetrical about $t = 0$. Percentage points for the distribution are shown graphically in Figure 6.3.

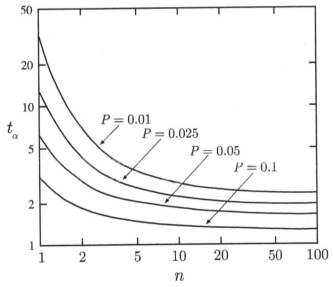

FIGURE 6.3   Percentage points of the Student's t distribution, $P = P[t > t_\alpha]$.

Example 6.6

(a) *What is $P[t \leq 0.7]$ when t is a random variable with 16 degrees of freedom? (b) If t is a random variable with 5 degrees of freedom, what is its value that corresponds to $\alpha = 0.05$?*

Solution: (a) From Table C.5, we need to find an entry close to, but not more than, 0.7 for $n = 16$. This is very close to 0.75.

(b) From the definition analogous to (6.6), we need to find a value t_c of t for 5 degrees of freedom such that $P[t \geq t_c] = 0.05$, that is, a value t_c such that $P[t \leq t_c] = 0.95$. From Table C.5, this is approximately $t_c = 2$.

The mean and variance as usual can be found from the *mgf*. From (6.13) and the definitions (3.11) and (3.13) we can show that moments of order r only exist for $r < n$ and are zero by symmetry for odd moments. For even moments, direct integration gives

$$\mu_{2r} = n^r \frac{\Gamma(r + 1/2)\Gamma(n/2 - r)}{\Gamma(1/2)\Gamma(n/2)}, \quad 2r < n. \tag{6.16}$$

The mean and variance follow from (6.16); they are

$$\mu = 0; \quad \sigma^2 = \frac{n}{n - 2}. \quad (n > 2) \tag{6.17}$$

We now return to the three basic results mentioned earlier. The first of these specifies the distribution of the difference of the sample mean and the population mean with respect to the sample variance. Let $x_i (i = 1, 2, ..., n)$ be a random sample of size n drawn from a normal population with mean μ and variance σ^2. Then the statistic

$$u = \left(\frac{\bar{x} - \mu}{\sigma/\sqrt{n}} \right)$$

is distributed as $N(u; 0, 1)$. Furthermore, from (6.12) we know that the statistic $w = (n - 1)s^2/\sigma^2$, where as usual s^2 is the sample variance, is distributed as χ^2 with $(n - 1)$ degrees of freedom. Therefore the statistic

$$t = \frac{u}{[w/(n - 1)]^{1/2}} = \frac{\sqrt{n}}{s}(\bar{x} - \mu) \tag{6.18}$$

is distributed as the Student's t distribution, with $(n - 1)$ degrees of freedom.

Example 6.7

A group of nine students entering the physics department of university A have a mean score of 78% in a national science examination, with a standard deviation of 5%. The national average for all students taking the same examination is 75%. What can be said about whether university A is getting significantly better than average students?

Solution: Using $\bar{x} = 78, s = 5$, and $\mu = 75$, we have $t = \sqrt{n}(\bar{x} - \mu)/s = 1.8$, and this value is for $n - 1 = 8$ degrees of freedom. From Table C.5, $F(t, n) = 0.95$ for $t = 1.86$ and $n = 8$. Thus the probability of getting a sample mean at least as large as this is approximately 5%.

The second result concerns the asymptotic behavior of the t distribution. As the number of degrees of freedom of the distribution approaches infinity, the distribution tends to the normal distribution in standard form. This follows by using the large-n approximation (called Stirling's approximation) in the gamma functions in (6.16). This gives

$$\Gamma(n+1) \to (2\pi)^{1/2} n^{n+1/2} e^{-n}, \quad n \to \infty.$$

Then the moments (6.16) become

$$\mu_{2r} \to \frac{(2r)!}{2^r r!}. \tag{6.19}$$

But (6.19) is the form for the moments of the normal distribution expressed in standard measure (see Equation (4.7)). Therefore the Student's t distribution tends to a normal distribution with mean zero and unit variance.

The final result concerns the t distribution when two normal populations are involved. Let random samples $x_{11}, x_{12}, \ldots, x_{1n_1}$ and $x_{21}, x_{22}, \ldots, x_{2n_2}$ of sizes n_1 and n_2, respectively, be independently drawn from two normal populations 1 and 2 with means μ_1 and μ_2, and the same variance σ^2; and define the statistic t by

$$t \equiv \frac{(\bar{x}_1 - \bar{x}_2) - (\mu_1 - \mu_2)}{\left[S_p^2(1/n_1 + 1/n_2)\right]^{1/2}}, \tag{6.20}$$

where

$$\bar{x}_i = \frac{1}{n_i} \sum_{j=1}^{n_i} x_{ij}, \quad i = 1, 2,$$

and S_p^2, the pooled sample variance, is given by

$$S_p^2 = \frac{\sum_{i=1}^{2} \sum_{j=1}^{n_i} \left(x_{ij} - \bar{x}_i\right)^2}{n_1 + n_2 - 2} = \frac{(n_1 - 1)s_1^2 + (n_2 - 1)s_2^2}{n_1 + n_2 - 2}. \tag{6.21}$$

Then, using (6.12) and the additive property of χ^2, the quantity

$$w = S_p^2(n_1 + n_2 - 2) / \sigma^2 \tag{6.22}$$

is distributed as χ^2 with $(n_1 + n_2 - 2)$ degrees of freedom. Furthermore, we know, from Equations (5.31) and (5.32), that $\bar{x} = \bar{x}_1 - \bar{x}_2$ is normally distributed with mean $\mu = \mu_1 - \mu_2$ and variance

$$\sigma_d^2 = \frac{\sigma^2}{n_1} + \frac{\sigma^2}{n_2}.$$

Thus the quantity

$$u = \frac{(\bar{x}_1 - \bar{x}_2) - (\mu_1 - \mu_2)}{\left[\sigma^2(1/n_1 + 1/n_2)\right]^{1/2}} = \frac{\bar{x} - \mu}{\sigma_d} \tag{6.23}$$

is normally distributed with mean zero and unit variance. But we showed in Section 6.1 that the sample mean and sample variance are independent variables when sampling randomly from a normal population, so $\bar{x}$ and u are independent random variables. Thus the quantity

$$t = \frac{u}{[w/(n_1 + n_2 - 2)]^{1/2}} \tag{6.24}$$

has a t distribution with $(n_1 + n_2 - 2)$ degrees of freedom. Substituting (6.22) and (6.23) into (6.24) gives (6.20) and completes the proof.

Example 6.8

The table shows the scores obtained in a certain examination by two groups of students, A and B. The seven students in group A attended revision classes to prepare for the exam. Has this significantly improved the mean score of the group, compared to that of the five students in group B?

n	1	2	3	4	5	6	7
A	71	75	79	71	70	73	72
B	70	68	72	73	67		

Solution: From the data we have $\bar{x}_A = 73, \bar{x}_B = 70, (n_A - 1)s_A^2 = 58$, and $(n_B - 1)s_B^2 = 26$, where $n_A = 7$ and $n_B = 5$. So from (6.21), $S_P^2 = 8.4$. We can now test whether $\mu_A = \mu_B$ by calculating t from (6.20). Thus, assuming $\mu_A = \mu_B$,

$$t = (\bar{x}_A - \bar{x}_B)\left[S_P^2\left(\frac{1}{n_A} + \frac{1}{n_B}\right)\right]^{-1/2} = 1.77.$$

From Table C.5, the probability of getting a value of t at least as great as this for 10 degrees of freedom is about 5%. In this example we are formally testing the hypothesis: The extra revision classes have improved the exams scores, against the alternative that they have not. Hypothesis testing will be discussed in more detail in Chapters 10 and 11.

Example 6.9

A factory produces ball bearings with diameters that are designed to be 1.00 in some units. This is checked regularly by measuring independent random samples of size 10. The table shows the results for two such samples. Use the t statistic with $\alpha = 0.05$ to check whether the manufacturing process is accurate.

Sample 1: 0.97, 0.98, 1.00, 1.02, 1.02, 1.02, 1.03, 1.03, 1.04, 1.08
Sample 2: 0.98, 0.99, 1.02, 1.02, 1.03, 1.03, 1.04, 1.04, 1.05, 1.08

Solution: From the data we calculate:

Run 1: Sample mean $\bar{x} = 1.019$ and sample standard deviation $s = 0.0311$
Run 2: Sample mean $\bar{x} = 1.028$ and sample standard deviation $s = 0.0286$

and so for Run 1, $t = \dfrac{1.019 - 1}{0.0311/\sqrt{10}} = 1.932$, and for Run 2, $t = \dfrac{1.028 - 1}{0.0286/\sqrt{10}} = 3.096$.

Finally, since $t_{n-1,\alpha/2} = 2.262$, we must conclude that the results of Run 1 are acceptable, but those of Run 2 are not.

6.3 F distribution

The F distribution[3] is designed for use in situations where we wish to compare two variances, or more than two means, situations for which the χ^2 and the Student's t distributions are not appropriate.

We begin by constructing the form of the F density function. Let two independent random variables $u = \chi_1^2$ and $v = \chi_2^2$ be distributed as χ^2 with n and m degrees of freedom, respectively. Then the joint density of u and v is, from (6.2),

$$g(u,v) = \frac{u^{(n-2)/2}v^{(m-2)/2}}{\Gamma(n/2)\Gamma(m/2)2^{(n+m)/2}} \exp\left[-\frac{1}{2}(u+v)\right].$$

The statistic F is defined by[4]

$$F = F(n,m) \equiv \frac{\chi_1^2/n}{\chi_2^2/m} = \frac{u/n}{v/m}. \tag{6.25}$$

We can transform the density function from the representation in variables u and v to F and v using the formulae (3.32a) and (3.32b). Thus using these and substituting

$$u = \left(\frac{n}{m}\right)vF$$

into $g(u,v)$ gives the joint density function of F and v as

$$f(F,v) = \frac{v^{(m-2)/2}}{\Gamma(n/2)\Gamma(m/2)2^{(m+n)/2}} \left(\frac{mv}{n}\right)\left(\frac{nvF}{m}\right)^{(n-2)/2} \exp\left[-\frac{v}{2}\left(1+\frac{n}{m}F\right)\right].$$

The density function of F is then obtained by integrating out the dependence on v. Thus

---

[3] This distribution is also known by other names, including the Fisher distribution, after one of the distinguished founders of modern statistics, Sir Ronald Fisher.

[4] The notation for F and its *pdf* is not standardized. Some authors write the degrees of freedom as arguments as in (6.25), but other authors prefer indices, as in F_{nm}. In either case the first degree of freedom is that of the numerator and the second is that of the denominator. Also, in constructing the F statistic, one must choose which χ^2 statistic to be the numerator and which to be the denominator in (6.25). Mathematically, the choice is arbitrary, but traditionally the choice is made so the numerator is larger than the denominator and the value of the statistic is always greater than one.

$$f(F; n, m) = \frac{F^{(n-2)/2}}{\Gamma(n/2)\Gamma(m/2)\,2^{(n+m)/2}}\left(\frac{n}{m}\right)^{n/2} I(F; n, m),$$

where, using the definition of the gamma function,

$$I(F; n, m) = \int_0^\infty v^{(n+m-2)/2}\exp\left[-\frac{v}{2}\left(1+\frac{n}{m}F\right)\right]dv = \frac{\Gamma[(n+m)/2]\,2^{(n+m)/2}}{(1+nF/m)^{(n+m)/2}}.$$

So, finally, the density function for the statistic F is

$$f(F; n, m) = \frac{\Gamma[(n+m)/2]}{\Gamma(n/2)\Gamma(m/2)}\left(\frac{n}{m}\right)^{n/2}\frac{F^{(n-2)/2}}{(1+nF/m)^{(n+m)/2}}, \quad F > 0 \qquad (6.26\text{a})$$

with n and m degrees of freedom, which may also be written in the form

$$f(F; n, m) = \frac{\Gamma[(n+m)/2]}{\Gamma(n/2)\Gamma(m/2)}\frac{n^{n/2}m^{m/2}F^{(n-2)/2}}{(m+nF)^{(n+m)/2}}, \quad F > 0. \qquad (6.26\text{b})$$

The *mgf* may be deduced in the usual way from its definition. The moments of order r exist only for $2r < n$ and are given by

$$\mu_r' = \left(\frac{m}{n}\right)^r \frac{\Gamma(r+n/2)\Gamma(m/2-r)}{\Gamma(n/2)\Gamma(m/2)}. \qquad (6.27)$$

The mean and variance follow directly from (6.27) and are

$$\mu = \frac{m}{m-2}, \quad m > 2,$$

and

$$\sigma^2 = \frac{2m^2(m+n-2)}{n(m-2)^2(m-4)}, \quad m > 4.$$

Equation (6.27) may also be used to calculate β_1 and β_2, and the result shows that the F distribution is always skewed. The *pdf* of the F distribution is more complicated than those of the χ^2 and t distributions in being a two-parameter family of curves.

The distribution function of F is tabulated in Table C.6. Percentage points are defined in the same way as for the χ^2 distribution. Thus

$$P[F \geq F_\alpha] = \alpha = \int_{F_\alpha}^\infty f(F; n, m)\,\mathrm{d}F.$$

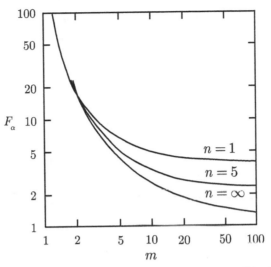

FIGURE 6.4   Percentage points of the *F* distribution, $P = P[F < F_\alpha] = 0.95$.

Right-tailed percentage points may be obtained from Table C.6, and should left-tailed percentage points be needed, they may be obtained from the relation

$$F_{1-\alpha}(n, m) = [F_\alpha(m, n)]^{-1}. \tag{6.28}$$

As an example, the percentage points for $P = 0.05$. are shown graphically in Figure 6.4.

Example 6.10

(a) *Find the critical value $F_{0.05}$ for $n = 6$ and $m = 14$. **(b)** Find the critical value $F_{0.975}$ for $m = 9$ and $n = 30$.*

   Solution: (a) This is the value for which $P[F > F_{0.05}] = 0.05$, or equivalently $P[F \leq F_{0.05}] = 0.95$. From Table C.6, with $n = 6$ and $m = 14$, this is $F_{0.05} = 2.85$.

(b) We first find the critical value $F_{0.025}$ for $m = 30$ and $n = 9$. This is the value for which $P[F > F_{0.025}] = 0.025$, or equivalently $P[F \leq F_{0.025}] = 0.975$. From Table C.6, with $m = 30$ and $n = 9$, this is $F_{0.025} = 3.56$. Now we can use (6.28) to give $F_{0.975}(30, 9) = [F_{0.025}(9, 30)]^{-1} = 0.281$.

Example 6.11

Use the tables in Appendix C to find the following quantities at the given critical points:
The values of χ^2 for (a) $n = 20$ and $\alpha = 0.025$, and (b) $n = 8$ and $\alpha = 0.95$; and the values of F for (c)
$n = 15, m = 10$, and $\alpha = 0.05$ and (d) $n = 6, m = 8$, and $\alpha = 0.95$.
Solution: (a) 34.2, (b) 2.73, (c) 2.85, (d) 0.241 (use (6.28)).

One use of the F distribution is to compare two variances, for example, to see whether two variances are compatible, so that the conditions for applying the Student's t test are satisfied. Let s_1^2 and s_2^2 be the variances of two independent random samples of sizes n and m, respectively, whose populations are assumed to be normal with variances σ_1^2 and σ_2^2. Then from the definition of the sample variance, we may write

$$s_1^2 = \frac{1}{(n-1)\sigma_1^2} \sum_{i=1}^{n} (x_i - \bar{x})^2 = \frac{\chi^2}{n-1},$$

where χ^2 is a chi-squared random variable with $(n-1)$ degrees of freedom. Thus, if we assume that $\sigma_1^2 = \sigma_2^2$, the ratio of s_1^2 to s_2^2 is distributed as an F random variable with $(n-1)$ and $(m-1)$ degrees of freedom.[5] As noted above, it is conventional to always designate the population with the larger variance as population 1 and put its χ^2 value in the numerator of the ratio. This ensures that the F-test statistic will always be greater or equal to 1, which makes it simpler to use published tables.

Example 6.12

Do the data of Example 6.6 justify the use of the t distribution to compare the mean scores of the two groups of students?

Solution: From the results of Example 6.6, we have $s_A^2 = 9.67$ and $s_B^2 = 6.50$ so that $F(6,4) = 1.49$. From Table C.6, the value of $F(6.4)$ is 4.01 for a critical value of 10%. As the value found from the data is well below this, the use of the Student's t distribution is compatible with the data.

6.4 Relations between χ^2, t, and F distributions

The F distribution is related to the χ^2 distribution as follows. Because χ^2 approaches a normal distribution as n, the number of degrees of freedom, goes to infinity and the mean and standard deviation are n and $\sqrt{2n}$, respectively, it is straightforward to show that as $n \to \infty$,

$$P[|\chi^2/n - 1|] \to 0.$$

---

[5] Because of this relation, F is also sometimes called the *variance ratio statistic*.

Thus

$$F(n, \infty) = \chi^2/n. \tag{6.29}$$

(See, for example, Figure 6.2.) The distribution of χ^2/n with n degrees of freedom is a special case of the F distribution with n and ∞ degrees of freedom. So for any α,

$$F_\alpha(n, \infty) = \frac{\chi^2_\alpha(n)}{n}, \tag{6.30}$$

which is directly verifiable by the use of a set of tables. If we consider the limit as $n \to \infty$, we have

$$F(\infty, m) = \frac{m}{\chi^2(m)}, \tag{6.31}$$

and so

$$F_\alpha(\infty, m) = \frac{m}{\chi^2_{1-\alpha}(m)}. \tag{6.32}$$

Thus the left-tailed percentage points of χ^2/m are special cases of the right-tailed percentage points of $F(\infty, m)$.

The F distribution is also related to the Student's t distribution. This can be seen by noting that when $n = 1$, then $\chi^2/n = u^2$, where u is a standard normal variate. We may thus write

$$F(1, m) = \frac{u^2}{(\chi^2/m)}. \tag{6.33}$$

in which χ^2 has m degrees of freedom.

But the variate

$$t = \frac{u}{(\chi^2/m)^{1/2}}, \tag{6.34}$$

is distributed as the Student's t distribution with m degrees of freedom, so (6.33) may be written as

$$F(1, m) = t^2(m) \tag{6.35}$$

Using (6.35), the relation

$$P[F(1, m) < F_\alpha(1, m)] = 1 - \alpha$$

is equivalent to

$$P\left[-(F_\alpha(1,m))^{1/2} < t(m) < (F_\alpha(1,m))^{1/2} \right] = 1 - \alpha,$$

and using the symmetry of the t distribution about $t = 0$ gives

$$P\left[t(m) < -(F_\alpha(1,m))^{1/2} \right] = P\left[t(m) > (F_\alpha(1,m))^{1/2} \right] = \alpha/2. \qquad (6.36)$$

But

$$P\left[t(m) > t_{\alpha/2}(m) \right] = \alpha/2$$

and so

$$t_{\alpha/2}(m) = F_\alpha[(1,m)]^{1/2},$$

or

$$F_\alpha(1,m) = t^2_{\alpha/2}(m) \qquad (6.37)$$

Similarly, we can show that for $m = 1$,

$$F(n,1) = \left[t^2(n) \right]^{-1} \qquad (6.38)$$

and

$$F_\alpha(n,1) = \left[t^2_{(1-\alpha)/2}(n) \right]^{-1}. \qquad (6.39)$$

Finally, if $n = 1$ and $m \to \infty$,

$$F_\alpha(1,\infty) = u^2_{\alpha/2}, \qquad (6.40)$$

and if $m = 1$ and $n \to \infty$,

$$F_\alpha(\infty,1) = \left[u^2_{(1-\alpha/2)/2} \right]^{-1}, \qquad (6.41)$$

where u_α is a point of the standard normal variate such that

$$P[u > u_\alpha] = \alpha.$$

The various relationships above are summarized in Table 6.2.

The relationships between these three distributions in certain limiting situations are shown in Figure 6.5, together with their relationships to the three most important population distributions discussed in Chapter 4.

TABLE 6.2 Percentage points F_α of the $F(n,m)$ distribution and their relation to the χ^2 and Student's t distributions.

		n	
m	1	n	∞
1	$t^2_{\alpha/2}(1) = \dfrac{1}{t^2_{(1-\alpha)/2}(1)}$	$\dfrac{1}{t^2_{(1-\alpha)/2}(n)}$	$\dfrac{1}{u^2_{(1-\alpha)/2}}$
m	$t^2_{\alpha/2}(m)$	$F_\alpha(n,m)$	$\dfrac{m}{\chi^2_{1-\alpha}(m)}$
∞	$u^2_{\alpha/2}$	$\dfrac{\chi^2_\alpha(n)}{n}$	1

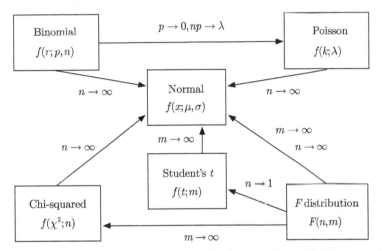

FIGURE 6.5 The relationships between the population distributions (binomial, Poisson, and normal) and the sampling distributions (chi-squared, Student's t, and F) as their parameters tend to certain limits.

Problems 6

6.1 Find the value of the 90th percentage point of the χ^2 distribution for $n = 100$ degrees of freedom using the normal distribution data given in Table C.1.

6.2 Prove the result stated in the text: That for a sample $x_1, x_2, ..., x_n$ of size n drawn from a normal population with mean zero and unit variance, the statistic

$$u = \sum_{i=1}^{n} (x_i - \bar{x})^2$$

is distributed as χ^2 with $(n-1)$ degrees of freedom. Complete the proof by using the following steps:

(Step 1) Show that for a given sample, $\bar{x}$ minimizes u. This demonstrates that for a given $\bar{x}$ all points representing a sample with mean $\bar{x}$ must lie in the $(n-1)$-dimensional hyperplane that contains the point M, having n components $(\bar{x}, \bar{x}, ..., \bar{x})$, and is perpendicular to the line L through the origin and M.

(Step 2) A coordinate transformation that rotates the Cartesian coordinate system $(\widehat{x}_1, \widehat{x}_2, ..., \widehat{x}_n)$ into another Cartesian system $(\widehat{y}_1, \widehat{y}_2, ..., \widehat{y}_n)$, preserving the distance between any pair of points, is an orthogonal transformation

$$\widehat{y}_i = \sum_{j=1}^{n} a_{ij} \widehat{x}_j$$

in which the matrix of elements a_{ij} has the property that its inverse is its transpose:

$$\sum_{i=1}^{n} a_{ji} a_{ki} = \delta_{ik}$$

where $\delta_{ik} = 1$ if $i = k$ and is zero otherwise. Show the following:

(Step 2.a) A rotation to a Cartesian coordinate system $(\widehat{y}_1, \widehat{y}_2, ..., \widehat{y}_n)$ with coordinate axis $\widehat{y}_n$ along the line L transforms u to

$$u = \sum_{i=1}^{n-1} y_i$$

in which the y_i are the transformed coordinates of the sample point.

(Step 2.b) That each y_i is normally distributed with mean zero and variance one, thus proving that u has the χ^2_{n-1} distribution.

(Step 3) Show that the matrix of a rotation has the property

$$\sum_{i=1}^{n} a_{ji} a_{ki} = \delta_{ik}.$$

6.3 Prove the additive property of χ^2.

6.4 Projectiles are fired at points in the general direction of a target that has fixed coordinates in three dimensions. The distances by which the projectiles miss the target are given by $x_i (i = 1, 2, 3)$ and are random normal variables with means 0 and standard deviations 3 (in arbitrary units). Find the approximate probability that the distance d from the target exceeds 6 units.

6.5 Twenty measurements taken from a normal population that was assumed to have mean 14.5 yielded values of sample mean and sample variance of 15 and 2.31, respectively. Use the Student's t distribution to test the assumption.

6.6 A target is to be located in an n-dimensional hyperspace by measuring its coordinates from a fixed point. If the coordinate errors are normal variates with mean zero and variance 4/3, what is the largest value of n for which the method can be used if the probability that the distance from the fixed point to the target D exceeds 4 is 10%?

6.7 Prove the relation (6.28).

6.8 A random sample of size 20 is selected from a normal distribution and the t statistic calculated. Find the value of k that satisfies $P[k < t < -1.328] = 0.075$.

6.9 Find the number h such that $P[-h < t < h] = 0.90$ where t is a random variable with a Student's t distribution with 15 degrees of freedom.

6.10 Resistors of a given value are manufactured by two machines A and B, and for consistency the outputs from both machines should have equal variances. A random sample of $n_A = 16$ resistors from machine A has $s_A^2 = 15$ and an independent random sample of size $n_B = 21$ from machine B has $s_B^2 = 5$. Are the machines making the resistors consistently?

6.11 Barrels of oil are sold marked as containing 300 liters. This is checked monthly by sampling the contents of 25 barrels and is considered acceptable if they yield a t-value between $t_{-0.05}$ and $t_{0.05}$. Assuming the actual volume of oil is a random variable distributed as a normal distribution with mean $\mu = 300$ liters, what can be deduced about the filling mechanism if a batch yields a mean volume $\bar{x} = 310$ liters and a sample standard deviation $s = 20$ liters?

6.12 Two mathematics teachers, A and B, each select a random sample of students from their class, and both are given an identical test. Teacher A (B) selects 20 (30) students and finds the variance in the marks of 8 (3). Is it reasonable to assume these results are compatible? Assume that the samples are from populations with equal variances, and use $\alpha = 0.05$.

Parameter estimation I: Maximum likelihood and minimum variance

Overview

This chapter is the first of two that address one of the most important topics in statistics as applied to the physical sciences: Estimation of the values of parameters from random samples of data, so-called 'point estimation'. In this chapter, the general method of maximum likelihood is discussed in some detail, both for cases where an exact solution is possible and the more practical cases where approximate methods must be used, including how to calculate the variance of the estimator. This is done both for estimating the value of a single parameter and for estimating several parameters simultaneously. The maximum likelihood technique is an example of a method of estimation that can result in the estimator having minimum variance, and so the last part of the chapter discusses the concept of minimum variance in general and derives a theoretical bound on the size of the variance.

In previous chapters we have encountered the problem of estimating the values of the parameters of a population from a sample. For example, we have used the sample mean and the sample variance as estimators of the corresponding population parameters. These choices satisfy the desirable general properties of point estimators discussed in Chapter 5 and are supported by the laws of large numbers. The likelihood function was introduced briefly in Chapters 2 and 5, but in this chapter we will consider in detail its use in estimation problems.

Of all the possible methods of parameter estimation, that of maximum likelihood (ML) is, in a sense to be discussed below, the most general, and is widely used in practice in physical sciences. The name is a little misleading and might suggest that an ML estimate is 'most likely', but as mentioned in Chapter 2, an ML estimate is the estimate that makes the observed data the most likely, i.e., most probable, which is not the same thing. This is illustrated in Example 7.1 below.

Example 7.1

An urn contains six balls, of which an unknown number N are black. Samples of size three are drawn without replacement. Calculate the probabilities of drawing two black balls in the sample, and hence the ML estimator of the number of black balls in the urn.

Solution: If x is the probability of getting two black balls in the sample, and is assumed to be from a sample independently and identically distributed from identical distributions, then the probability of getting two black balls if $N = 2$ is

$$P[x = 2|N = 2] = \frac{{}_2C_2\,{}_4C_1}{{}_6C_3} = \frac{4}{20}.$$

Similarly, for $N = 3$,

$$P[x = 2|N = 3] = \frac{{}_3C_2\,{}_3C_1}{{}_6C_3} = \frac{9}{20},$$

and for $N = 4$ and $N = 5$,

$$P[x = 2|N = 4] = \frac{12}{20}, \quad P[x = 2|N = 5] = \frac{10}{20}.$$

The ML estimate is therefore $N = 4$, because this value maximizes the probability of getting two black balls in the sample. This result agrees with that obtained by maximizing the likelihood function.

7.1 Estimation of a single parameter

The likelihood function was defined in (5.2). For a sample of size n, drawn from a distribution with density function $f(x; \theta)$, where θ, is a parameter,

$$L(\theta) = \prod_{i=1}^{n} f(x_i; \theta), \tag{7.1}$$

and we have suppressed the dependence of L on x. In (7.1) we have assumed that $f(x; \theta)$ is a function of the single random variable x, but all the results of this section, and Section 7.2, may be generalized in a straightforward way to the case of estimating a single parameter from a multivariate distribution. It is worth remarking, once again, that $L(\theta)$ is not itself a probability density function. The ML *estimator* of a population parameter θ is defined as that statistic $\widehat{\theta}$ which maximizes $L(\theta)$ for variations of θ that is, the solution (if it exists) of the equation[1]

$$\frac{\partial L(\theta)}{\partial \theta} = 0, \tag{7.2a}$$

with the condition

$$\frac{\partial^2 L(\theta)}{\partial \theta^2} < 0. \tag{7.2b}$$

---

[1] A brief review of maxima and minima is given in Appendix A, Section A.2.

The latter condition should always be checked to confirm that (7.2a) does produce a maximum. The ML *estimate* is found by evaluating the ML estimator using the values of the data sample. Since $L(\theta) > 0$, the first equation is equivalent to

$$\frac{1}{L}\frac{\partial L(\theta)}{\partial \theta} = \frac{\partial \ln L(\theta)}{\partial \theta} = 0, \tag{7.3}$$

which is the form more often used in practice, and again the equivalent condition to (7.2b) should be checked.

It is clear from (7.2a) that the solution obtained by estimating the parameter θ is the same as estimating a function of θ, for example $F(\theta)$, since

$$\frac{\partial \ln L}{\partial \theta} = \frac{\partial \ln L(F)}{\partial F}\cdot\frac{\partial F}{\partial \theta}, \tag{7.4}$$

and the two sides vanish together. This is a useful invariance property of ML estimators, but it does not extend to their variances. This is readily seen by considering the probability

$$P[\theta_1 \leq \theta \leq \theta_2] = \left[\int_{\theta_1}^{\theta_2} L(\theta)d\theta\right]\left[\int_{-\infty}^{\infty} L(\theta)d\theta\right]^{-1}, \tag{7.5}$$

i.e., the probability that an interval (θ_1, θ_2) will contain the true value θ. For example, if this probability is chosen to be 0.68, then for a normal distribution it would correspond to a standard error of one standard deviation in the usual sense. If we now use a function $F(\theta)$ as an estimator with $F_1 = F(\theta_1)$ and $F_2 = F(\theta_2)$, then

$$P[F_1 \leq F \leq F_2] = \left[\int_{F_1}^{F_2} L(F)dF\right]\left[\int_{-\infty}^{\infty} L(F)dF\right]^{-1},$$

which in general will not be equal to the value obtained from (7.5). Another reason to be careful about using (7.4) is that it does not always follow that the estimator obtained by maximizing a function of θ is also unbiased. One must balance the convenience of the invariance property of ML estimators against the fact that the resulting estimator may not be unbiased. This is illustrated in Example 7.2 for the exponential distribution.

Example 7.2

Find the ML estimator for the parameter $\lambda = 1/\tau$ in a population with an exponential density function

$$f(x; \lambda) = \lambda \exp(-\lambda x).$$

The quantity λ could be, for example, the inverse of the lifetime, i.e., the decay rate, of an unstable atomic state. Then the quantities x_i would be a sample of measurements t_i of the lifetime.

Solution: The likelihood function is

$$L(x, \lambda) = \prod_{i=1}^{n} \lambda^n \exp\left(-\lambda \sum_i x_i\right),$$

and so

$$\ln L(x, \lambda) = n \ln \lambda - \lambda \sum x_i.$$

Differentiating gives

$$\frac{d \ln L(x, \lambda)}{d\lambda} = \frac{n}{\lambda} - \sum_i x_i = \frac{n}{\lambda} - n\bar{x},$$

and setting this to zero gives the ML estimator $\widehat{\lambda} = 1/\bar{x}$. Thus, the mean of the measurements t_i is an unbiased ML estimator for the lifetime τ, i.e.,

$$\widehat{\tau} = \frac{1}{n} \sum_{i=1}^{n} t_i \tag{7.6}$$

for all n. However, the estimator for any function of τ may be found by evaluating the function using $\widehat{\tau}$. So, if we were to take the function to be the rate of decay $R = 1/\tau$, then from (7.6),

$$\widehat{R} = \frac{1}{\widehat{\tau}} = n \left(\sum_{i=1}^{n} t_i\right)^{-1},$$

but is it straightforward to show (for example, by the method used above) that

$$E\left[\widehat{R}\right] = \frac{n}{n-1} R,$$

and so $\widehat{R}$ is not an unbiased estimator for R, except asymptotically, when n is large. Fortunately, this latter condition is usually satisfied in practical applications of the ML method.

A practical consideration when using the ML method is that the data do not have to be binned. However, this strength can become a weakness in the case of very large samples, because a complicated function may have to be evaluated at many points. In this case it is usual to apply the method to binned data. If N observations, distributed with a density $f(x; \theta)$, are divided into m bins, with $n_j(i = 1, 2, ..., m)$ entries in bin j, then

$$e_j(\theta) = N \int_{x_j^{\min}}^{x_j^{\max}} f(x; \theta) dx$$

is the expectation value of the number of entries in the jth bin having lower and upper limits $x_j^{\min}$ and $x_j^{\max}$, respectively. If we take the probability to be in bin j as (e_j / N), the joint probability is given by the multinomial distribution discussed in Section 4.2.2,

$$f_{\text{joint}} = \frac{N!}{n_1!\,n_1!\ldots n_m!}\left(\frac{e_1}{N}\right)^{n_1}\left(\frac{e_2}{N}\right)^{n_2}\cdots\left(\frac{e_m}{N}\right)^{n_m}$$

and

$$\ln L(\theta) = \sum_{j=1}^{m} n_j \ln e_j(\theta) + C, \tag{7.7}$$

where C does not depend on the parameter θ. The ML estimator $\hat{\theta}$ is now obtained by maximizing (7.7) with respect to θ, usually by numerical means. It is evident from this expression that the method has no difficulty in accommodating bins that have no data.

The importance of ML estimators stems from their properties. It can be shown that they are generally consistent and have minimum variance. If a sufficient estimator for a parameter exists, then it is a function of the ML estimator. The latter follows directly from the factorization condition (5.12), because maximizing L is equivalent to choosing $\hat{\theta}$ to maximize $L_1\left(\hat{\theta};\theta\right)$ in that equation. Another important property is that for large samples, ML estimators have a distribution that tends to normality. There are situations where these results do not hold and the ML estimator is a poor estimator, but for the common distributions met in practice they are valid.

To prove the normality property, we set $\ln L(\theta) = h(\theta)$, so that the ML estimator is defined by the solution of $\mathrm{d}h(\theta)/\mathrm{d}\theta \equiv h'(\theta) = 0$. Then, providing $h'(\theta)$ can be differentiated further, we can expand it about the point $\hat{\theta}$ to give

$$h'(\theta) = h'\left(\hat{\theta}\right) + \left(\theta - \hat{\theta}\right) h''\left(\hat{\theta}\right) + \cdots \tag{7.8}$$

where, setting $f_i = f(x_i;\theta)$ and using (7.1),

$$h'\left(\hat{\theta}\right) = \sum_{i=1}^{n}\left(\frac{f_i'}{f_i}\right)\Bigg|_{\theta=\hat{\theta}} \quad \text{and} \quad h''\left(\hat{\theta}\right) = \sum_{i=1}^{n}\left[\left(\frac{f_i'}{f_i}\right)'\right]\Bigg|_{\theta=\hat{\theta}},$$

and the primes denote differentiation with respect to θ. For large samples we know that $E\left[h'\left(\hat{\theta}\right)\right] = 0$, i.e.,

$$E\left[h'\left(\hat{\theta}\right)\right] = \int_{-\infty}^{\infty} \frac{f'(x)}{f(x)} f(x)\,\mathrm{d}x = 0.$$

Differentiating this result again, and writing out in full gives

$$\int_{-\infty}^{\infty}\left[\frac{f'^2(x)}{f(x)} + f(x)\left(\frac{f'(x)}{f(x)}\right)'\right]\mathrm{d}x = \int_{-\infty}^{\infty}\left[\left(\frac{f'(x)}{f(x)}\right)^2 + \left(\frac{f'(x)}{f(x)}\right)'\right]f(x)\,\mathrm{d}x = E\left[\left(\frac{f'(x)}{f(x)}\right)^2\right] + E\left[\left(\frac{f'(x)}{f(x)}\right)'\right] = 0,$$

that is,

$$E\left[\left\{h'\left(\hat{\theta}\right)\right\}^2\right] = -E\left[h''\left(\hat{\theta}\right)\right] = 1/c^2, \tag{7.9}$$

where c^2 depends on the density f and the estimator $\hat{\theta}$. Substituting (7.9) into (7.8) and integrating gives

$$h(\theta) - h\left(\hat{\theta}\right) = -\frac{1}{2}\left(\frac{\theta - \hat{\theta}}{c}\right)^2.$$

Finally, taking exponentials, and using $h'\left(\hat{\theta}\right) = 0$,

$$L(\theta) = k \exp\left[-\frac{1}{2}\left(\frac{\theta - \hat{\theta}}{c}\right)^2\right],$$

which completes the proof that ML estimators are asymptotically distributed as a normal distribution.

Some examples of the ML method for estimating one parameter are illustrated by the following examples, starting with the ML estimator for the mean of a normal population. The related problem of finding the ML estimator for the variable σ^2 in a normal population is left to Problem 7.1.

Example 7.3

Find the ML estimator $\hat{\mu}$ for the parameter μ in the normal population

$$f(x; \mu, \sigma^2) = \frac{1}{(2\pi\sigma^2)^{1/2}} \exp\left[-\frac{1}{2}\left(\frac{x - \mu}{\sigma}\right)^2\right],$$

for samples of size n, where σ is known and $-\infty \leq x \leq \infty$. Is the estimator unbiased?
Solution: Using (7.1), we have

$$\ln L(\mu, \sigma^2) = -n \ln\left[(2\pi\sigma^2)^{1/2}\right] - \frac{1}{2\sigma^2}\sum_{i=1}^{n}(x_i - \mu)^2.$$

The ML estimator of μ is found by maximizing $\ln L$ with respect to μ, that is, the solution of

$$\frac{\partial \ln L(\mu)}{\partial \mu} = \frac{1}{\sigma^2}\sum_{i=1}^{n}(x_i - \mu) = 0.$$

Thus,

$$\hat{\mu} = \frac{1}{n}\sum_{i=1}^{n}x_i = \bar{x}.$$

Therefore, the sample mean is the ML estimator of the parameter μ. In a straightforward way, one could show that $\hat{\mu}$ is an unbiased estimator for μ by calculating its expectation value using the joint probability distribution for the x_i. This is done in full for a related exercise in Problem 7.11. It also follows from the general result that the sample mean is an unbiased estimator of the mean for any probability density function.

Example 7.4

Find the ML estimator for the parameter θ in a population with a density function

$$f(x;\theta) = (1+\theta)x^{\theta}, \qquad 0 \le x \le 1.$$

Solution: The likelihood function is

$$L(\theta) = \prod_{i=1}^{n}(1+\theta)x_i^{\theta},$$

with

$$\ln L(\theta) = n\ln(1+\theta) + \theta\Sigma, \quad \text{where} \quad \Sigma \equiv \sum_{i=1}^{n}\ln x_i.$$

Taking the derivative gives

$$\partial\ln L(\theta)/\partial\theta = n/(1+\theta) + \Sigma = 0,$$

and so

$$\widehat{\theta} = -\frac{(n+\Sigma)}{\Sigma} = -\left(\frac{n + \sum_{i=1}^{n}\ln x_i}{\sum_{i=1}^{n}\ln x_i}\right).$$

Example 7.5

A discrete random variable x has a geometric distribution if its probability mass function is

$$f(x;p) = (1-p)^{x-1}p,$$

for parameter p, where x is a positive integer. Otherwise $f(x;p) = 0$. *Using* $p = 1/N$, *calculate the maximum likelihood estimator* $\widehat{N}$ *for N, using a sample* $x_i (i = 1,2,3,...,n)$.

Solution: The likelihood estimator is the product of the probability functions for x_i i.e.,

$$L(N) = \prod_{i}\frac{1}{N}\left(1-\frac{1}{N}\right)^{x_i-1} = \left(\frac{1}{N}\right)^{n}\left(1-\frac{1}{N}\right)^{\left(\sum_{i=1}^{n}x_i-n\right)} = \left(\frac{1}{N}\right)^{n}\left(1-\frac{1}{N}\right)^{n(\bar{x}_n-1)}.$$

Hence, taking logarithms,

$$\ln[L(N)] = -n\ln N + \left(\sum_{i}^{n}x_i - n\right)\ln\left(1-\frac{1}{N}\right),$$

and then differentiating gives

$$\frac{\partial \ln L(N)}{\partial N} = -\frac{n}{N} + (n\bar{x}_n - n)\left(\frac{1}{N(N-1)}\right).$$

Finally, setting $\partial \ln N / \partial N = 0$, gives $N = \bar{x}_n$ and it is straightforward to show, by differentiating again, that this is the ML estimator $\widehat{N}$ for N.

In the above discussion, we have made the usual assumption that n is a fixed known number. However, there are often circumstances where the number of events observed in an experiment is itself a random variable, typically with a Poisson distribution with mean λ. In these cases, the overall likelihood function is the product of the probability of finding a given value of n (given by Equation (4.47)) and the usual likelihood function for the n values of x. The combined likelihood function is therefore

$$\mathcal{L}(n, \theta) = \frac{\lambda^n}{n!} e^{-\lambda} \prod_{i=1}^{n} f(x_i; \theta),$$

and is called the *extended likelihood function*. It differs from the usual likelihood function only in that it is taken to be a function of both n and the sample values x_i. Much of the standard formalism of the ML method carries over to $\mathcal{L}(n, \theta)$ and we will not pursue it further here, except to say that the extended likelihood method usually results in smaller variances for estimators $\widehat{\theta}$ because the method uses the statistical information contained in n as well as that in the sample.

7.2 Variance of an estimator

Although the likelihood function $L(\theta)$ is not a density function, when appropriately scaled, it may be *formally* regarded as such for the parameter θ viewed as a random variable. Thus, we can define the variance of the estimator as

$$\mathrm{var}\,\widehat{\theta} = \int_{-\infty}^{\infty} \left(\theta - \widehat{\theta}\right)^2 L(\theta)\, d\theta \Big/ \int_{-\infty}^{\infty} L(\theta)\, d\theta, \qquad (7.10)$$

and, by analogy with the work of Section 5.4, an estimate from experimental data would be quoted as

$$\theta = \widehat{\theta}_e \pm \Delta\widehat{\theta}_e, \qquad (7.11a)$$

where $\widehat{\theta}_e$ is the ML estimator obtained from the data, and

$$\Delta\widehat{\theta}_e = \left(\mathrm{var}\,\widehat{\theta}_e\right)^{1/2}. \qquad (7.11b)$$

The interpretation of (7.11a) is that if the experiment were to be repeated many times, with the same number of measurements in each experiment, one would expect the standard deviation of the distribution of the estimates of θ to be $\Delta\widehat{\theta}_e$.

From the normality property of ML estimators, it follows that for large samples, the form of $L(\theta)$ is

$$L(\theta) = \frac{1}{(2\pi v)^{1/2}} \exp\left[-\frac{1}{2}\frac{\left(\theta - \widehat{\theta}\right)^2}{v}\right],\qquad(7.12)$$

where $v = \text{var}\widehat{\theta}$. Then

$$\ln L(\theta) = -\ln\left[(2\pi v)^{1/2}\right] - \frac{1}{2}\frac{\left(\theta - \widehat{\theta}\right)^2}{v},\qquad(7.13)$$

and

$$\frac{\partial^2 \ln L(\theta)}{\partial\theta^2} = -\frac{1}{v}.$$

Thus

$$\text{var}\widehat{\theta} = \left[-\frac{\partial^2 \ln L(\theta)}{\partial\theta^2}\right]^{-1}\Bigg|_{\theta=\widehat{\theta}}.\qquad(7.14)$$

This is the usual form used for the variance of an ML estimator when making numerical calculations.

Example 7.6

Find the ML estimator $\widehat{\mu}$ and its variance for the parameter μ in the same normal population used in Example 7.3, but now for a set of experimental observations of the same quantity x_i with associated experimental errors Δx_i.

Solution: The density function is

$$f(x, \Delta x; \mu) = \frac{1}{\sqrt{2\pi}\Delta x}\exp\left[-\frac{1}{2}\left(\frac{x-\mu}{\Delta x}\right)^2\right],$$

from which

$$\ln L(\mu) = -\ln\left[(2\pi)^{1/2}\sum_{i=1}^n \Delta x_i\right] - \frac{1}{2}\sum_{i=1}^n \left(\frac{x_i-\mu}{\Delta x_i}\right)^2,$$

and

$$\frac{\partial \ln L(\mu)}{\partial \mu} = \sum_{i=1}^{n} \left[\frac{x_i - \mu}{(\Delta x_i)^2} \right].$$

Setting this last expression to zero gives

$$\hat{\mu} = \sum_{i=1}^{n} (x_i/\Delta x_i^2) \Big/ \sum_{i=1}^{n} (1/\Delta x_i^2).$$

This result is called the *weighted mean* of a set of observations, where each data value is weighted by the inverse of its squared error. The variance of $\hat{\mu}$ may be found using the second derivative

$$\frac{\partial^2 \ln L(\mu)}{\partial \mu^2} = -\sum_{i=1}^{n} \left[\frac{1}{(\Delta x_i)^2} \right]$$

in (7.14), with the result

$$\mathrm{var}\,\hat{\mu} = (\Delta \hat{\mu})^2 = \left[\sum_{i=1}^{n} \left(\frac{1}{\Delta x_i} \right)^2 \right]^{-1}.$$

The formula in Example 7.6 for the weighted mean $\hat{\mu}$, although formally correct, should be used with care. This is because the experimental errors Δx_i are only estimates of the population standard deviation and we must be sure that they are mutually consistent; that is, we must be sure that the measurements all come from the same normal distribution. We will return to this question in Chapter 11, where we discuss ways of testing whether a set of data do indeed come from the same population distribution.

7.2.1 Approximate methods

If an experiment has 'good statistics' then the likelihood function will indeed be a close approximation to a normal distribution and the method above for estimating the variance will be valid. However, many effects may be present which could produce a function that is clearly not normal and in this case the use of (7.14) usually produces an underestimate for $\Delta \hat{\theta}$. In these circumstances a more realistic estimate is to average $\partial^2 \ln L(\theta)/\partial \theta^2$ over the likelihood function, so that

$$\frac{1}{\left(\Delta \hat{\theta}\right)^2} = -\overline{\frac{\partial^2 \ln L(\theta)}{\partial \theta^2}} = \left[\int_{-\infty}^{\infty} \left(-\frac{\partial^2 \ln L(\theta)}{\partial \theta^2} \right) L(\theta) d\theta \right] \left[\int_{-\infty}^{\infty} L(\theta) d\theta \right]^{-1}, \qquad (7.15)$$

where, as usual, the overbar denotes an average.

A related method that partially deals with the problem of non-invariance is to use the function

$$S(\theta) = \left[-\overline{\frac{\partial^2 \ln L(\theta)}{\partial \theta^2}} \right]^{-1/2} \frac{\partial \ln L(\theta)}{\partial \theta},$$

called the *Bartlett S function*,[2] which can be shown to have a mean $\mu = 0$ and variance $\sigma^2 = 1$. In the case of a normal distribution, $S(\theta)$ is a straight line passing through zero when $\theta = \widehat{\theta}$ and the values at $\pm n$ standard deviations are found from the points where $S = \pm n$. For non-normal functions, the solutions of the equations $S(\theta_\pm) = \mp 1$ determine the 'one-standard deviation' quantities $\theta_\pm$ so that the result would be quoted as

$$\theta = \widehat{\theta} \, {}^{+\theta_+}_{-\theta_-}.$$

Alternatively, a direct graphical method can be used to estimate the variance. Thus, a plot of $L(\theta)$ is made and the two values found where it falls to $e^{1/2}$ of its maximum value, i.e., the two values that would correspond to one standard deviation in the case of a normal distribution. Reverting to using $\ln L(\theta)$, we can expand this in a Taylor series about the ML estimate $\widehat{\theta}$ to give

$$\ln L(\theta) = \ln L\left(\widehat{\theta}\right) + \left[\frac{\partial \ln L}{\partial \theta}\right]\bigg|_{\theta=\widehat{\theta}} \left(\theta - \widehat{\theta}\right) + \frac{1}{2!}\left[\frac{\partial^2 \ln L}{\partial \theta^2}\right]\bigg|_{\theta=\widehat{\theta}} \left(\theta - \widehat{\theta}\right)^2 + \cdots \qquad (7.16)$$

From the definition of $\widehat{\theta}$, we know that $\ln L(\widehat{\theta}) = \ln L_{\max}$, and the second term is zero because $\partial \ln L(\theta)/\partial \theta = 0$ for $\theta = \widehat{\theta}$. So, if we ignore terms of higher order than those shown in (7.16), we have

$$\ln L(\theta) = \ln L_{\max} - \frac{(\theta - \widehat{\theta})^2}{2 \mathrm{var}\widehat{\theta}}, \qquad (7.17a)$$

where we have used (7.14). Equation (7.17a) implies that

$$\ln L\left(\widehat{\theta} \pm \widehat{\sigma}\right) - \ln L_{\max} - 1/2. \qquad (7.17b)$$

Thus a change in the value of θ of one standard deviation from its ML estimate $\widehat{\theta}$ corresponds to a decrease in the value of $\ln L(\theta)$ of $1/2$ from its maximum value. Likewise, a decrease of 2 defines the points where θ changes by two standard deviations from its ML estimate, and so on. In the case where the likelihood function has an approximate normal distribution, $\ln L(\theta)$ will be approximately parabolic.

This is illustrated in Figure 7.1 for a case where $\widehat{\theta} = 10.0$ and $\ln L_{\max} = -50$. In this case the two points where $\widehat{\theta}$ changes by one-standard deviation can be found from the figure and lead to error estimates $\widehat{\sigma}_- = 0.52$ and $\widehat{\sigma}_+ = 0.58$. These are close enough that it is reasonable

[2] See, for example, Roe, 1992 listed in the bibliography.

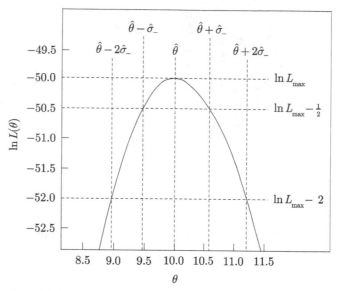

FIGURE 7.1   Graphical method for finding errors on an ML estimate.

to average them and quote the result as $\hat{\sigma} = 10.0 \pm 0.55$. Alternatively, the asymmetric errors could be quoted, i.e., $\hat{\sigma} = 10.0\,^{+0.58}_{-0.52}$, as would have to be done if the curve was not approximately parabolic.

   Another useful formula for $\Delta\hat{\theta}$ may be derived for situations where one wants to answer the question: How many data are required to establish a particular result to a specified accuracy? The problem is to find a value for $\partial^2 \ln L(\theta)/\partial\theta^2$ averaged over many repeated experiments consisting of n events each. Since

$$\ln L(x; \theta) = \sum_{i=1}^{n} \ln f(x_i; \theta),$$

we have

$$\overline{\frac{\partial^2 \ln L(\theta)}{\partial\theta^2}} = n \int \frac{\partial^2 \ln f(x; \theta)}{\partial\theta^2} f(x; \theta)\, dx = nE\left[\frac{\partial^2 \ln f(x; \theta)}{\partial\theta^2}\right]. \tag{7.18a}$$

   This form may be used in (7.15) directly, or it may be expressed in terms of first derivatives by writing

$$\frac{\partial^2 \ln f}{\partial\theta^2} = \frac{1}{f}\frac{\partial^2 f}{\partial\theta^2} - \frac{1}{f^2}\left(\frac{\partial f}{\partial\theta}\right)^2 = \frac{1}{f}\frac{\partial^2 f}{\partial\theta^2} - \left(\frac{\partial \ln f}{\partial\theta}\right)^2,$$

and then taking expectation values to give

$$E\left[\frac{\partial^2 \ln f}{\partial\theta^2}\right] = -E\left[\frac{\partial\ln f}{\partial\theta}\right]^2 + E\left[\frac{1}{f}\frac{\partial^2 f}{\partial\theta^2}\right] = -\int\frac{1}{f}\left(\frac{\partial f}{\partial\theta}\right)^2 \mathrm{d}x, \qquad (7.18\mathrm{b})$$

since the second term is

$$E\left[\frac{1}{f}\frac{\partial^2 f}{\partial\theta^2}\right] = \frac{\partial^2}{\partial\theta^2}\int f(\theta)\mathrm{d}x = 0$$

From (7.8a) and (7.8b),

$$\int\frac{\partial^2 \ln f(x;\theta)}{\partial\theta^2}f(x;\theta)\,\mathrm{d}x = -\int\left(\frac{\partial f(x;\theta)}{\partial\theta}\right)^2\frac{1}{f(x;\theta)}\,\mathrm{d}x, \qquad (7.19)$$

and so

$$\Delta\widehat{\theta} = \frac{1}{\sqrt{n}}\left[\int\left(\frac{\partial f(x;\theta)}{\partial\theta}\right)^2\frac{1}{f(x;\theta)}\,\mathrm{d}x\right]^{-1/2}. \qquad (7.20)$$

This result also confirms an earlier statement, that to increase the precision of the experiment n-fold requires n^2 as many events.

Example 7.7

Consider the density function

$$f(x;\theta) = \frac{1}{2}(1+\theta x), \qquad -1 \le x \le 1$$

How many events would be required to determine θ to a precision of 1% for a value $\widehat{\theta} = 0.5$?
Solution: We have

$$\frac{\partial f(x;\theta)}{\partial\theta} = \frac{x}{2},$$

and

$$\int_{-1}^{1}\left(\frac{\partial f(x;\theta)}{\partial\theta}\right)^2\frac{1}{f(x;\theta)}\,\mathrm{d}x = \frac{1}{2\theta^3}\left[\frac{1}{2}(1+\theta x)^2 - 2(1+\theta x) + \ln(1+\theta x)\right]_{-1}^{1}$$

$$= \frac{1}{2\theta^3}\left[\ln\left(\frac{1+\theta}{1-\theta}\right) - 2\theta\right].$$

Thus from (7.20),

$$\left(\frac{\Delta\widehat{\theta}}{\widehat{\theta}}\right) = \left(\frac{2\widehat{\theta}}{n}\right)^{1/2}\left[\ln\left(\frac{1+\widehat{\theta}}{1-\widehat{\theta}}\right) - 2\widehat{\theta}\right]^{-1/2}.$$

Setting $\left(\Delta\widehat{\theta}\big/\widehat{\theta}\right) = 0.01$ for $\widehat{\theta} = 0.5$ gives $n \approx 1.01 \times 10^5$.

7.3 Simultaneous estimation of several parameters

If we wish to estimate simultaneously several parameters, then the preceding results generalize in a straightforward way. The ML equation becomes the set of simultaneous equations

$$\frac{\partial \ln L(\theta_1, \theta_2, \cdots, \theta_i, ..., \theta_n)}{\partial \theta_i} = 0, \qquad i = 1, 2, ..., n \tag{7.21}$$

Analogous properties of ML estimators for a single parameter also hold. As an example, consider the generalization of the normality property. This states that the ML estimators $\widehat{\theta}_i (i = 1, 2, ..., \theta_p)$ for the parameters of a density function $f(x; \theta_1, \theta_2, ..., \theta_p)$ from samples of size n are, for large samples, approximately distributed as the multivariate normal distribution with means $\theta_1, \theta_2, ..., \theta_p$ and a variance matrix $\mathbf{V}$, where

$$M_{ij} = \left(V_{ij}\right)^{-1} = -nE\left[\frac{\partial^2 \ln f(x; \theta_1, \theta_2, ..., \theta_p)}{\partial \theta_i \, \partial \theta_j}\right]. \tag{7.22}$$

The use of (7.21) and (7.22) is illustrated in the following example.

Example 7.8

Find the simultaneous ML estimators for the parameters μ and σ of the normal population

$$f(x; \mu, \sigma) = \frac{1}{(2\pi\sigma^2)^{1/2}} \exp\left[-\frac{1}{2}\left(\frac{x-\mu}{\sigma}\right)^2\right],$$

and find the form of the joint distribution of the estimators for large samples.
Solution: From (7.21),

$$\frac{\partial \ln L(\mu, \sigma)}{\partial \mu} = \frac{1}{\sigma^2}\sum_{j=1}^{n}(x_j - \mu) = 0,$$

and

$$\frac{\partial \ln L(\mu, \sigma)}{\partial \sigma} = \frac{1}{2\sigma^4}\sum_{j=1}^{n}(x_j - \mu)^2 - \frac{n}{2\sigma^2} = 0,$$

giving

$$\hat{\mu} = \bar{x}; \quad \hat{\sigma}^2 = \frac{1}{n} \sum_{j=1}^{n} (x_j - \bar{x})^2.$$

Note that $\hat{\sigma}^2$ is a biased estimator of σ^2. This is often the case with ML estimators, but fortunately there usually exists a constant c, in this case $n/(n-1)$, such that multiplying the ML estimator by c produces an unbiased estimator.

Because the two estimators $\hat{\mu}$ and $\hat{\sigma}$ are approximately normally distributed with mean μ and a matrix $\mathbf{M}$ given by (7.22), we have, with $\theta_1 = \mu$ and $\theta_2 = \sigma$,

$$M_{11} = -nE\left[-\frac{1}{\sigma^2}\right] = \frac{n}{\sigma^2}, \quad M_{22} = -nE\left[-\frac{3(x-\mu)^2}{\sigma^4} + \frac{1}{\sigma^2}\right] = \frac{2n}{\sigma^2}$$

and

$$M_{12} = M_{21} = -nE\left[-\frac{2(x-\mu)}{\sigma^3}\right] = 0.$$

Thus, the variance matrix is

$$V_{ij} = \left(M^{-1}\right)_{ij} = \begin{pmatrix} \sigma^2/n & 0 \\ 0 & \sigma^2/2n \end{pmatrix},$$

and the variance and covariances are given by

$$\sigma_{ij}/n = V_{ij}.$$

Finally, from (4.16) the form of the distribution of the estimators is

$$S(\hat{\mu}, \hat{\sigma}) = \frac{\sqrt{2}n}{2\pi\sigma^2} \exp\left\{ -\frac{n}{2}\left[2\left(\frac{\hat{\sigma}-\sigma}{\sigma}\right)^2 + \left(\frac{\hat{\mu}-\mu}{\sigma}\right)^2\right]\right\}.$$

There is one point that should be remarked about the simultaneous estimation of several parameters, which is illustrated by reference to Example 7.4. If we know μ, the estimation of σ^2 alone gives (see Problem 7.1)

$$\hat{\sigma}^2 = \frac{1}{n} \sum_{i=1}^{n} (x_i - \mu)^2, \tag{7.23a}$$

whereas in Example 7.8, we found from the *simultaneous* estimation of μ and σ^2 that

$$\hat{\sigma}^2 = \frac{1}{n} \sum_{j=1}^{n} (x_j - \bar{x})^2. \tag{7.23b}$$

However, from the results found in Example 7.3 we see that we can estimate μ, independent of any possible knowledge of σ^2, to be $\bar{x}$. Thus, if we now find the estimator of σ^2 that maximizes the likelihood for all samples giving the estimated value of $\mu = \bar{x}$, it might be thought that the result (7.23b) would result, whereas in fact in this latter case

$$\widehat{\sigma}^2 = \frac{1}{n-1} \sum_{i=1}^{n} (x_i - \bar{x})^2. \tag{7.23c}$$

The difference between (7.23b) and (7.23c) is that in the former case we have considered the variations of $\ln L(\mu, \sigma)$ over *all* samples of size n, whereas in the latter the constraint that $\sum x$ is a constant has been imposed, and thus the number of degrees of freedom has been lowered by one. For large n the difference is of little importance, but it is a useful reminder that every parameter estimated from the sample (i.e., every constraint applied) lowers the number of degrees of freedom by one.

The ML method has the disadvantage that to estimate a parameter, the form of the distribution must be known. Furthermore, it often happens that $L(\theta)$ is a highly nonlinear function of the parameters, and so maximizing the likelihood function may be a difficult problem.[3] Finally, if the data under study are normally distributed, then maximizing $L(\theta)$ is equivalent to minimizing

$$\chi^2 = \sum_{i=1}^{n} \left(\frac{x_i - \mu_i}{\sigma_i} \right)^2,$$

which may be more useful in practice, as we shall illustrate when we consider the method of minimum chi-squared in Chapter 8.

We will conclude with a few brief remarks on the interpretation of ML estimators. Bayes' theorem tells us that maximizing the likelihood does not necessarily maximize the posterior probability of an event. This is only true if the prior probabilities are equal or somehow 'smooth'. Thus, ML estimators (and of course other estimators) should always be interpreted in the light of prior knowledge. In Chapter 8 we shall see how such knowledge can formally be included in the estimation procedure. However, because in general it is difficult to reduce prior knowledge to the required form, the actual method of estimation is not always of practical use. An alternative method is to form a likelihood function that is the product of the likelihood functions for all previous related experiments and use this function to make a new estimate of the parameter.

7.4 Minimum variance

The requirement that an estimator has minimum variance can also, in principle, be used as a criterion for parameter estimation and this is illustrated below.

---

[3] A brief discussion of optimizing non-linear functions is given in Appendix B.

7.4.1 Parameter estimation

Consider the problem of estimating the population parameter μ, where samples are drawn from n populations, each with the same mean μ but with different variances. The estimate will be obtained by combining the sample means $\bar{x}_i$ and the corresponding sample variances. Since $\bar{x}_i$ is an unbiased estimate of μ, we have seen in Example 5.2 that the quantity

$$\bar{x} = \sum_{i=1}^{n} a_i \bar{x}_i, \tag{7.24a}$$

with

$$\sum_{i=1}^{n} a_i = 1, \tag{7.24b}$$

is also an unbiased estimate, regardless of the values of the coefficients a_i, so the problem is one of selecting a suitable set of a_i. This will be done by choosing the set a_i such that $\bar{x}$ has minimum variance. Thus, we seek to minimize

$$\text{var}\,(\bar{x}) = \text{var}\left(\sum_{i=1}^{n} a_i \bar{x}_i \right) = \sum_{i=1}^{n} a_i^2 \text{var}(\bar{x}_i) = \sum_{i=1}^{n} a_i^2 \sigma_i^2,$$

subject to the constraint (7.24b). To do this we use the method of Lagrange multipliers.[4] If we introduce a multiplier λ, then the variational function is

$$L = \sum_{i=1}^{n} a_i^2 \sigma_i^2 + \lambda \left(\sum_{i=1}^{n} a_i - 1 \right),$$

and

$$\frac{\partial L}{\partial a_i} = 0 = 2 a_i \sigma_i^2 + \lambda.$$

Thus $a_i = -\lambda / 2\sigma_i^2$, and since the sum of the a_i is unity,

$$\lambda = -2 \left[\sum_{j=1}^{n} \left(1/\sigma_j^2 \right) \right]^{-1}.$$

Hence

---

[4] Readers unfamiliar with this technique are referred to Appendix A.

$$a_i = (1/\sigma_i^2) \left[\sum_{j=1}^{n} \left(1/\sigma_j^2\right) \right]^{-1},$$

giving

$$\bar{x} = \left[\sum_{i=1}^{n} \left(\bar{x}_i / \sigma_i^2\right) \right] \left[\sum_{i=1}^{n} \left(1/\sigma_i^2\right) \right]^{-1} \quad \text{and} \quad \text{var}\,(\bar{x}) = \left[\sum_{j=1}^{n} \left(1/\sigma_j^2\right) \right]^{-1}.$$

In this example, the minimum variance estimator is the weighted mean, identical to the estimator obtained using the ML method (cf Example 7.6), *where the population distribution was assumed to be normal.* For other population densities, the results of the two methods will differ.

7.4.2 Minimum variance bound

In many cases it is not possible to find the variance of an estimator analytically (i.e., exactly) and even to do so numerically, for example, by using the Monte Carlo method discussed in Chapter 5, involves a great deal of computation. In this situation, a very useful result that puts a lower limit on the variance may be used. This has various names, such as the *Cramér-Rao*, or *Fréchet, inequality*, or simply the *minimum variance bound*, and is true in general and not just for ML estimators, although it will be illustrated for ML estimators in what follows.

Consider the estimator $\hat{\theta}$ of a parameter θ that is a function of the sample $\mathbf{x} = x_1, x_2, \ldots, x_n$, with a joint pdf given by (7.1). The expectation value of $\hat{\theta}$ is

$$E\left[\hat{\theta}\right] = \int \hat{\theta} L(\mathbf{x}; \theta)\, d\mathbf{x}, \tag{7.25}$$

which differentiating with respect to θ, gives

$$\frac{dE\left[\hat{\theta}\right]}{d\theta} = \int \hat{\theta} \frac{dL(\mathbf{x}; \theta)}{d\theta}\, d\mathbf{x} = \int \hat{\theta} \frac{d\ln L(\mathbf{x}; \theta)}{d\theta} L(\mathbf{x}; \theta)\, d\mathbf{x} = E\left[\hat{\theta} \frac{d\ln L(\theta)}{d\theta}\right]. \tag{7.26}$$

But in general,

$$E\left[\hat{\theta}\right] = \theta + b(\theta), \tag{7.27}$$

where b is the bias. Differentiating (7.27) and suppressing the dependence of L on $\mathbf{x}$, gives

$$1 + \frac{db(\theta)}{d\theta} = E\left[\hat{\theta} \frac{d\ln L(\theta)}{d\theta}\right]. \tag{7.28}$$

The right-hand side of (7.28) may be evaluated by firstly differentiating the normalization condition

$$\int L(\mathbf{x}; \theta) \, d\mathbf{x} = 1$$

to give

$$\int \frac{dL(\theta)}{d\theta} \, d\mathbf{x} = \int \frac{d \ln L(\theta)}{d\theta} L(\theta) \, d\mathbf{x} = E\left[\frac{d \ln L(\mathbf{x}; \theta)}{d\theta}\right] = 0. \tag{7.29}$$

Multiplying (7.29) by $E\left[\widehat{\theta}\right]$ and subtracting the result from (7.28) gives

$$E\left[\widehat{\theta}\frac{d \ln L(\theta)}{d\theta}\right] - E\left[\widehat{\theta}\right]E\left[\frac{d \ln L(\theta)}{d\theta}\right] = E\left[\left(\widehat{\theta} - E\left[\widehat{\theta}\right]\right)\frac{d \ln L(\theta)}{d\theta}\right] = 1 + \frac{db(\theta)}{d\theta}. \tag{7.30}$$

The second step is to use a form of the so-called *Schwarz inequality*, which for two random variables x and y, such that x^2 and y^2 have finite expectation values, takes the form[5]

$$\{E[xy]\}^2 \leq E[x^2]E[y^2]. \tag{7.31}$$

Applying (7.31) to (7.30) gives

$$\left(1 + \frac{db(\theta)}{d\theta}\right)^2 \leq E\left[\left(\widehat{\theta} - E\left[\widehat{\theta}\right]\right)^2\right] E\left[\left(\frac{d \ln L(\theta)}{d\theta}\right)^2\right]. \tag{7.32}$$

The first factor on the right-hand side is $\mathrm{var}\left(\widehat{\theta}\right) = \sigma^2\left(\widehat{\theta}\right)$. To evaluate the second factor, we have

$$E\left[\left(\frac{d \ln L(\theta)}{d\theta}\right)^2\right] = nE\left[\left(\frac{df(x; \theta)/d\theta}{f(x; \theta)}\right)^2\right] \equiv I(\theta) \geq 0, \tag{7.33}$$

---

[5] The result (7.31) may be proved as follows. For any real value λ,

$$E[(\lambda x + y)^2] = \lambda^2 E[x^2] + 2\lambda E[xy] + E[y^2] \geq 0$$

and the solutions for λ in the case of the equality are

$$\lambda_{\pm} = -\frac{E[xy]}{E[x^2]} \pm \left\{\left(\frac{E[xy]}{E[x^2]}\right)^2 - \left(\frac{E[y^2]}{E[x^2]}\right)\right\}^{1/2},$$

so the inequality holds only if $\{E[xy]\}^2 \leq E[x^2]E[y^2]$.

where the quantity $I(\theta)$ is called the *information of the sample with respect to θ*, or simply the *information*. It may also be written in the form (cf (7.19)),

$$I(\theta) = E\left[-\frac{d^2 \ln L(\theta)}{d\theta^2}\right].$$

So, finally, for a single parameter θ having an estimator $\widehat{\theta}$ with a bias b, the bound is

$$\mathrm{var}\left(\widehat{\theta}\right) = \sigma^2\left(\widehat{\theta}\right) \geq \left(1 + \frac{db}{d\theta}\right)^2 \frac{1}{I(\theta)}. \tag{7.34}$$

It is worth noting that in the derivation no assumption has been made about the estimator. If the equality holds in (7.34), the estimator is efficient.

There remains the question of what conditions are necessary for the MVB to be attained, that is, for the equality in (7.31) to hold. This is valid if $(\lambda x + y) = 0$, because only then is $E[(\lambda x + y)^2] = 0$ for all values of λ, x, and y. Applying this to (7.31) with $x = \widehat{\theta} - E[\widehat{\theta}]$, $y = d \ln L(\theta)/d\theta$, and $\lambda = A(\theta)$, where $A(\theta)$ does not depend on the sample $x_1, x_2, ..., x_n$, gives

$$\frac{d \ln L(\theta)}{d\theta} = A(\theta)\left(\widehat{\theta} - E\left[\widehat{\theta}\right]\right). \tag{7.35}$$

Finally, integrating (7.35) gives the condition

$$\ln L(\theta) = B(\theta)\,\widehat{\theta} + C(\theta) + D, \tag{7.36}$$

where B and C are functions of θ, and D is independent of θ. Thus, an estimator $\widehat{\theta}$ will have a variance that satisfies the minimum bound if the associated likelihood function has the structure (7.36). The actual value of the MVB may be found by using (7.35) in (7.34). For an unbiased estimator, this gives

$$\sigma^2\left(\widehat{\theta}\right) = \frac{1}{A^2(\theta)E\left[\left(\widehat{\theta} - E\left[\widehat{\theta}\right]\right)^2\right]} = \frac{1}{A^2(\theta)\sigma^2\left(\widehat{\theta}\right)},$$

and so

$$\sigma^2\left(\widehat{\theta}\right) = |A(\theta)|^{-1}. \tag{7.37}$$

Example 7.9

Find an ML estimator for the parameter p of the binomial distribution and show that it is an unbiased minimum variance estimator.

Solution: The binomial probability is given by Equation (4.34) as

$$f(r;p,n) = \binom{n}{r} p^r (1-p)^{n-r},$$

and gives the probability of r successes, each with probability p, out of n trials. If we conduct N sets of n trials, with r_i successes in the ith set, the likelihood function is

$$L(p) = \prod_{i=1}^{N} \binom{n}{r_i} p^{r_i} (1-p)^{n-r_i}.$$

Then, taking logarithms,

$$\ln L(p) = \sum_{i=1}^{N} \left\{ r_i \ln p + (n-r_i)\ln(1-p) + \ln\binom{n}{r_i} \right\}$$

and so

$$\frac{d \ln L(p)}{dp} = \frac{h}{p} - \frac{Nn - h}{1-p},$$

where

$$h = \sum_{i=1}^{N} r_i.$$

Setting $d \ln L(p)/dp = 0$ shows that $\hat{p} = h/Nn$ is an unbiased estimator for the parameter p. Note that the average number of successes in the N sets of n trials is $\bar{r} = h/N$ so $n\hat{p} = \bar{r}$ as would be expected intuitively. Because $\bar{r}$ is a mean, and therefore an unbiased estimator, $\hat{p}$ is also unbiased. Moreover, using $n\hat{p} = \bar{r}$ we can write $\ln L(p)$ as

$$\ln L(p) = Nn\hat{p}[\ln p - \ln(1-p)] + Nn \ln(1-p) + \sum_{i=1}^{N} \ln\binom{n}{r_i},$$

which is of the form (7.36), so $\bar{r}/n$ is a minimum variance estimator for p. The value of the variance is found by writing

$$\frac{d \ln L(p)}{dp} = \frac{h}{p} - \frac{Nn - h}{1-p} = \frac{Nn(\hat{p} - p)}{p(1-p)},$$

which is of the form (7.35), with $A(p) = Nn/[p(1-p)]$, and so

$$\sigma^2(\hat{p}) = A^{-1}(p) = p(1-p)/Nn.$$

There are other ways of expressing the requirements of the Cramér-Rao bound. An example is: A minimum variance unbiased estimator of a parameter θ must satisfy

$$\text{var}\left(\widehat{\theta}\right) = \left(nE\left[\left(\frac{\partial \ln f(x;\theta)}{\partial\theta}\right)^2\right]\right)^{-1} = -\left(nE\left[\frac{\partial^2 \ln f(x;\theta)}{\partial\theta^2}\right]\right)^{-1}. \tag{7.38}$$

Certain conditions must be satisfied by the function $f(x,\theta)$ for this result to be true, but they will not be discussed here, as they are satisfied by the distributions met in physical science.

Example 7.10

Use the result (7.38) to find the minimum variance, unbiased estimator of the parameter λ in the exponential density $f(x) = \lambda e^{-\lambda x}$.

Solution: For n trials,

$$L(\lambda) = \prod_i^n \lambda e^{-\lambda x_i}$$

and

$$\ln L(\lambda) = \sum_{i=1}^n (\ln \lambda - \lambda x_i).$$

Differentiating with respect to λ and setting the result equal to zero, we find the ML estimator $\widehat{\lambda} = 1/\bar{x}$. We know the mean of the exponential distribution is $1/\lambda$ and $\bar{x}$ is an unbiased estimate of the mean for any distribution. It is easy to show from this that $\widehat{\lambda}$ is an unbiased estimate of λ. With this result, we have

$$\ln L(\lambda) = n \ln \lambda - n\frac{\lambda}{\bar{\lambda}},$$

which is of the form (7.35), so the estimator attains the MVB.

From the density formula $\ln f(x;\lambda) = \ln \lambda - \lambda x$ and its derivative

$$\partial \ln f(x;\lambda)/\partial\lambda = 1/\lambda - x,$$

we have

$$nE\left[\left(\frac{\partial \ln f(x;\lambda)}{\partial\lambda}\right)^2\right] = nE\left[\frac{1}{\lambda^2} - \frac{2x}{\lambda} + x^2\right] = \frac{n}{\lambda^2},$$

where we have used the easily verified results for the exponential distribution that $E[x] = 1/\lambda, E[x^2] = 2/\lambda^2$, and the general result for any distribution that

$$E[x^2] = \text{var}(x) + (E[x])^2.$$

Thus $\text{var}(\widehat{\lambda}) = \lambda^2/n$, and hence $1/\bar{x}$ is a minimum variance unbiased estimator for λ in an exponential distribution.

Example 7.11

Show that $\overline{X}$ is a minimum variance unbiased estimator for the parameter μ in a normal distribution $N(\mu, \sigma^2)$.

Solution: From earlier work we know that $\overline{X}$ is an unbiased estimator for the mean μ of a normal distribution,

$$f(x) = \frac{1}{\sqrt{2\pi}\sigma} \exp\left[-\frac{(x-\mu)^2}{2\sigma^2} \right],$$

and from the *pdf* we have

$$\ln f(x) = -\ln(\sqrt{2\pi}\,\sigma) - \frac{(x-\mu)^2}{2\sigma^2},$$

and

$$\ln L(\mu) = \frac{n}{\sigma^2}\overline{X}\mu - \frac{n\mu^2}{2\sigma^2} - \frac{1}{2\sigma^2}\sum_{i=1}^{n}x_i^2 - n\ln(\sqrt{2\pi}\,\sigma),$$

which is of the form (7.36) so $\overline{X}$ is a minimum variance estimator. To determine the variance we differentiate $\ln f(x)$:

$$\frac{\partial \ln f(x)}{\partial \mu} = \frac{(x-\mu)}{\sigma^2}.$$

Then

$$nE\left[\left(\frac{\partial \ln(f(x))}{\partial \mu} \right)^2 \right] = nE\left[\frac{(x-\mu)^2}{\sigma^4} \right] = \frac{n}{\sigma^2}E\left[\frac{(x-\mu)^2}{\sigma^2} \right],$$

which for a normal distribution is

$$nE\left[\left(\frac{\partial \ln(f(x))}{\partial \mu} \right)^2 \right] = \frac{n}{\sigma^2}.$$

Thus, from the Cramér-Rao bound (7.38),

$$\text{var}(\overline{X}) = \left\{ \frac{1}{nE[(\partial \ln(f(x)/\partial \mu)]^2} \right\} = \frac{\sigma^2}{n},$$

and $\overline{X}$ is a minimum variance unbiased estimator for μ.

Problems 7

7.1 Find the ML estimator for the parameter σ^2 in the normal population

$$f(x;\mu,\sigma^2) = \frac{1}{(2\pi\sigma^2)^{1/2}}\exp\left[-\frac{1}{2}\left(\frac{x-\mu}{\sigma}\right)^2\right],$$

for samples of size n. Is the estimator unbiased?

7.2 Find the ML estimator for the parameter k for a sample of size n from a population having a density function

$$f(x;k) = \begin{cases} a(k+2)^3 x^k & 0 \le x \le 1 \\ 0 & \text{otherwise} \end{cases},$$

where a is a constant.

7.3 A data set is subject to two independent scans. In the first scan, n_1 events of a given type x are identified, and in the second, n_2 events of the same type are found. If there are n_{12} events in common in the two scans, what is the efficiency E_1 of the first scan and what is its standard deviation? Estimate the total number of events of type x.

7.4 A set of n independent measurements $E_i(i = 1,2,...,n)$ are made of the energy of a quantum system in the vicinity of an excited state of energy E_0 and width Γ described by the Breit-Wigner density (Cauchy distribution) of Section 4.1.6. If $|E_i - E_0| \ll \Gamma$, show that the mean energy $\bar{E}$ is the ML estimator of E_0.

7.5 A sample of size 10 is taken from a population with a known density function

$$f(x;\theta) = \begin{cases} \theta/x^{(\theta+1)} & x > 1 \\ 0 & \text{elsewhere} \end{cases},$$

where $\theta > 0$, with the results:

i	1	2	3	4	5	6	7	8	9	10
x_i	10	11	12	9	10	13	12	10	11	9

Calculate the ML estimate for θ.

7.6 The discrete *Bernoulli* distribution describes a situation where there are n independent trials, each of which has a probability p of success. A random Bernoulli variable x has a probability function

$$f(x;p) = p^x(1-p)^{1-x}, \quad \text{with } x = 1, \text{ if the trial is a success}$$
$$\text{and } x = 0, \text{ otherwise}$$

Calculate the ML estimator for the parameter p.

7.7 Find equations for the ML estimators of the constants α and β in the Weibull distribution of Section 4.1.5.

7.8 Find the ML estimator for the parameter λ of the gamma distribution given in Equation (4.32), for a fixed value of the parameter α.

7.9 Find the ML estimator for the parameter λ of the Poisson distribution (see Equation (4.47)) and show that it is a minimum variance estimator.

7.10 Find the unbiased minimum variance bound (MVB) for the parameter θ in the distribution

$$f(x;\theta) = \frac{1}{\pi} \frac{1}{\left[1 + (x-\theta)^2\right]}.$$

Note the integral:

$$\int_0^\infty \frac{(1-x^2)}{(1+x^2)^3}\,dx = \frac{\pi}{8}.$$

7.11 Show that the ML estimator $\hat{\tau}$ of the lifetime τ for the exponential distribution found in Example 7.2 is an unbiased estimator.

Parameter estimation II: Least-squares and other methods

Overview

This second chapter on point estimation is largely about the method of least-squares, the most popular technique for estimating the values of parameters. The method for finding parameters and their variances is discussed in detail for the case where the fitting functions are linear in the parameters, both in general using matrix notation and also for the simple, but practical, case of fitting data with a straight line. There are short sections on the quality of the fit and the use of orthogonal polynomials in the fitting process. It is also shown how the method can be used to combine the results of several experiments. The linear least-squares technique is then extended to cover cases in which either the parameters or the data are subject to linear constraints. There is also a short section on the far more difficult situation where the fitting functions are nonlinear in the parameters. Finally, there are brief accounts of three other methods of point estimation that are less commonly used: Minimum chi-squared, the method of moments, and Bayes' estimation.

The method of least-squares is an application of minimum variance estimators, which were introduced in Section 7.4, and is widely used in situations where a functional form is known (or assumed) to exist between the observed quantities and the parameters to be estimated. This may be dictated by the requirements of a theoretical model of the data or may be chosen arbitrarily to provide a convenient interpolation formula for use in other situations. We will firstly consider the technique for the situation where it is most used; where the data depend *linearly* on the parameters to be estimated. In this form the least-squares method is frequently used in curve-fitting problems.

8.1 Unconstrained linear least-squares

The method will be formulated as a general procedure for finding estimators $\widehat{\theta}_i(i = 1, 2, ..., p)$ of parameters $\theta_i(i = 1, 2, ..., p)$ that minimize the function

$$S = \sum_{i=1}^{n} \left(y_i - \widehat{\eta}_i\right)^2 = \sum_{i=1}^{n} r_i^2, \tag{8.1}$$

Probability and Statistics for Physical Sciences, Second Edition
https://doi.org/10.1016/B978-0-443-18969-2.00008-8

where

$$\widehat{\eta}_i = f\left(x_{1i}, x_{2i}, ..., x_{ki}; \ \widehat{\theta}_1, \widehat{\theta}_2, ..., \widehat{\theta}_p\right), \tag{8.2}$$

and $x_{1i}, x_{2i}, ..., x_{ki}, y_i$ denotes the ith set of observations on $(k+1)$ variables, of which only y_i has any appreciable random variation, typically due to measurement error. The relation (8.2) is called the *equation of the regression curve of best fit*, or simply the *best fit curve*. The word 'regression' comes from an early investigation[1] that showed that tall fathers tended to have tall sons, although not on average as tall as themselves — referred to as 'regression to the norm'. Some authors prefer that regression is used to describe situations such as this, where only qualitative statements can be made about the relationship between two variables.

We shall consider firstly the general case where the observations are correlated and have different 'weights' that are related to their experimental errors.[2] Later we will look at simpler cases, which follow easily from the general situation. Suppose we make observations of a quantity y that is a function $f(x; \ \theta_1, \theta_2, ..., \theta_p)$ of one variable x and p parameters $\theta_i(i = 1, 2, ..., p)$. Note that x is *not* a random variable and f is *not* a density function. The observations y_i are made at points x_i and differ from the function f due to experimental error and any difference between f and the true values of the y_i. This difference is the residual r_i. If the n observations y_i depend *linearly* on the p parameters, then the observational equations may be written

$$y_i = \sum_{k=1}^{p} \theta_k \phi_k(x_i) + r_i, \quad i = 1, 2, ..., n, \tag{8.3}$$

where $\phi_k(x_i)$ are any linearly independent functions of x_i. The word 'linear' in 'linear least squares' or 'linear regression', refers to the coefficients θ_k, that is, they contain no powers, square roots, trigonometric functions, etc. Many situations that at first sight look nonlinear can be transformed so that the linear least-squares method may be used. For example, taking logarithms of the equation $y = ae^{\lambda x}$ gives $\ln y = \ln a + \lambda x$, which is a relationship which is linear in λ. (See Example 8.1.) On the other hand, the fitting functions $\phi_k(x)$ can be nonlinear provided they only depend on the variables x_i. In matrix notation[3] (8.3) may be written

$$\mathbf{Y} = \mathbf{\Phi}\,\mathbf{\Theta} + \mathbf{R}, \tag{8.4}$$

where $\mathbf{Y}$ and $\mathbf{R}$ are $(n \times 1)$ column vectors, $\mathbf{\Theta}$ is a $(p \times 1)$ column vector and $\mathbf{\Phi}$ is an $(n \times p)$ matrix (called the *design matrix* or *model matrix*)

---

[1] F. Galton and J.D.H Dickson *Family Likeness in Stature*, Proceedings of Royal Society of London (1886), vol. 40, pp. 42–73.

[2] The least-squares method can also be formulated for random fluctuations in both x and y but is more complicated. As it is not the usual situation met in practice, it will not be discussed here.

[3] A review of matrix algebra is given in Appendix A, Section A.1.

$$\Phi = \begin{pmatrix} \phi_1(x_1) & \phi_2(x_1) & \cdots & \phi_p(x_1) \\ \phi_1(x_2) & \phi_2(x_2) & \cdots & \phi_p(x_2) \\ \vdots & \vdots & & \vdots \\ \phi_1(x_n) & \phi_2(x_n) & \cdots & \phi_p(x_n) \end{pmatrix}.$$

The components r_i of the residual vector $\mathbf{R}$ are the deviations of the measured values $y_i(x_i)$ from $f(x_i;\ \theta_1, \theta_2, ..., \theta_p)$, which is the sum of three parts. One part of the deviation is the difference between the true value of y and the value predicted by f and the second is any systematic error that might be present, as discussed in Chapter 1. These two parts of the deviation are not random but may depend on the value of x. The third part of the deviation is the random error, which may be assumed to have zero mean since any nonzero mean value may be treated as part of the systematic error rather than the random error.

8.1.1 General solution for the parameters

The problem is to obtain estimates $\widehat{\theta}_k$ for the parameters. For $n = p$ a unique solution exists and is obtainable directly from (8.4) by a simple matrix inversion, but for the more practical case where $n > p$ the system of equations is over determined. In this situation no general unique solution exists, and so what we seek is a 'best approximate solution' in a sense that will be discussed later. Thus, we seek to approximate the experimental points y_i by a series of degree p, that is,

$$f_i = f(x_i;\ \theta_1, \theta_2, ..., \theta_p) = \sum_{k=1}^{p} \theta_k \phi_k(x_i). \tag{8.5}$$

It is usually assumed that f_i represents the true value of $y(x_i)$ and no systematic error is present. In this case, the only deviation of the measurements from f_i is a random error with zero mean and obviously

$$E[\mathbf{Y}] = \Phi\,\Theta. \tag{8.6}$$

We will usually make this assumption but will not do so in this and most of the following section so that the impact of systematic error on the results of the method of least squares can be understood.

The error has an associated variance matrix

$$V_{ij} = \begin{pmatrix} \sigma_1^2 & \sigma_{12} & \cdots & \sigma_{1n} \\ \sigma_{21} & \sigma_2^2 & \cdots & \sigma_{2n} \\ \vdots & \vdots & & \vdots \\ \sigma_{n1} & \sigma_{n2} & \cdots & \sigma_n^2 \end{pmatrix}, \tag{8.7}$$

where

$$\sigma_i^2 = \text{var}(y_i),$$

and

$$\sigma_{ij} = \sigma_{ji} = \text{cov}\left(y_i, y_j\right).$$

Two observations are useful here. Firstly, note that we have only assumed that the population distribution of the errors has a *finite second moment*. It is *not* necessary to assume that the distribution is normal. However, *if* the errors are normally distributed, as is often the case, then the method of least squares gives the same results as the method of maximum likelihood.

Secondly, we observe that the variance matrix depends only on the random error. If a systematic error is present, it does not change the variance matrix. This is easily demonstrated by noting that if there is a systematic error, then $y_i = f_i + \varepsilon_i + \omega_i$ where ε_i is the random error with zero mean, and ω_i is the nonrandom systematic error. By definition,

$$\sigma_{ij} = \text{cov}\left[y_i y_j\right] = E\left[\left(f_i + \varepsilon_i + \omega_i - E[f_i + \varepsilon_i + \omega_i]\right)\left(f_j + \varepsilon_j + \omega_j - E[f_j + \varepsilon_j + \omega_j]\right)\right].$$

Since the random error has zero mean,

$$E\left[f_i + \varepsilon_i + \omega_i\right] = f_i + \omega_i$$

and

$$\sigma_{ij} = E\left[\varepsilon_i \varepsilon_j\right].$$

Similarly, $\sigma_i^2 = E\left[\varepsilon_i^2\right]$. Thus, only the random part of the error affects the value of variance matrix.

Returning to the task of finding a least-squares estimate of the parameters, we note that from (8.3),

$$r_i \equiv y_i - f_i = y_i - \sum_{k=1}^{p} \theta_k \phi_k(x_i), \tag{8.8}$$

and we will minimize the weighted sum

$$S = \sum_{i=1}^{n} \sum_{j=1}^{n} r_i r_j \left(V^{-1}\right)_{ij} = \mathbf{R}^T \mathbf{V}^{-1} \mathbf{R}. \tag{8.9}$$

To minimize S with respect to $\mathbf{\Theta}$, we set $\partial S / \partial \mathbf{\Theta} = 0$, giving the solution

$$\widehat{\mathbf{\Theta}} = \left(\mathbf{\Phi}^T \mathbf{V}^{-1} \mathbf{\Phi}\right)^{-1} \mathbf{\Phi}^T \mathbf{V}^{-1} \mathbf{Y}, \tag{8.10}$$

or, in nonmatrix notation,

$$\widehat{\theta}_k = \sum_{l=1}^{p} \left(E^{-1}\right)_{kl} \sum_{i=1}^{n} \sum_{j=1}^{n} \phi_l(x_i)\left(V^{-1}\right)_{ij} y_j, \tag{8.11}$$

where

$$E_{kl} = \sum_{i=1}^{n} \sum_{j=1}^{n} \phi_k(x_i)\left(V^{-1}\right)_{ij} \phi_l(x_j). \tag{8.12}$$

These are the so-called *normal equations* for the parameters. Note that to find the estimators for the parameters only requires knowledge of the *relative* errors on the observations, because any scale factor in **V** would cancel in (8.10). However, this is not true for the variances of the parameters, as we shall see in Section 8.1.2 below.

As mentioned above, one advantage of weighting the residual sum in (8.9) with V^{-1} is if the distribution of the errors is normal, the values determined by (8.11) for $\widehat{\theta}_k$ are the same as would be determined by the maximum likelihood method. A second advantage is that the quantities $\widehat{\theta}_k$ are unbiased estimators when there is no systematic error. This can be shown easily from (8.10) by using (8.4) to replace Y:

$$\widehat{\Theta} = \left(\mathbf{\Phi}^T \mathbf{V}^{-1} \mathbf{\Phi}\right)^{-1} \mathbf{\Phi}^T \mathbf{V}^{-1} (\mathbf{\Phi}\,\Theta + \mathbf{R}) = \Theta + \left(\mathbf{\Phi}^T \mathbf{V}^{-1} \mathbf{\Phi}\right)^{-1} \mathbf{\Phi}^T \mathbf{V}^{-1} \mathbf{R}.$$

Since the expected value of the residual in this case is zero, we have $E\left[\widehat{\Theta}\right] = \Theta$.

Example 8.1

The table below shows the values of data $y_i(i = 1, 2, ..., 7)$ with uncorrelated errors σ_i taken at the points x_i. Use the general formulation of the least-squares method to find estimators for the parameters a and b in a fit to the data of the form $y = a\exp(bx)$ and calculate the predictions for $\widehat{y}_i$. Plot the data and the best-fit line.

i	1	2	3	4	5	6	7
x_i	1	2	3	4	5	6	7
y_i	4	5	8	16	30	38	70
σ_i	2	2	3	3	4	4	5

Solution: By taking logarithms of the fitting function, the problem can be converted to the linear form $y' = a' + b'x'$, where $y' = \ln y, a' = \ln a, b' = b$ and $x' = x$. The errors on y' follow from (5.45) for the propagation of errors, that is

$$\sigma' = \frac{d \ln y}{dy} \sigma = \frac{\sigma}{y}.$$

A new table can then be constructed as follows:

i	1	2	3	4	5	6	7
x_i'	1	2	3	4	5	6	7
y_i'	1.386	1.609	2.079	2.773	3.401	3.638	4.248
σ_i'	0.500	0.400	0.375	0.188	0.133	0.105	0.071

Using the notation above, the various matrices needed for the primed quantities are:

$$\mathbf{\Phi}'^{T} = \begin{pmatrix} 1 & 1 & 1 & 1 & 1 & 1 & 1 \\ 1 & 2 & 3 & 4 & 5 & 6 & 7 \end{pmatrix}$$

$$\mathbf{Y}'^{T} = (1.386 \quad 1.609 \quad 2.079 \quad 2.773 \quad 3.401 \quad 3.638 \quad 4.248),$$

and

$$\mathbf{V}' = 10^{-3} \begin{pmatrix} 250.00 & 0 & 0 & 0 & 0 & 0 & 0 \\ 0 & 160.00 & 0 & 0 & 0 & 0 & 0 \\ 0 & 0 & 140.63 & 0 & 0 & 0 & 0 \\ 0 & 0 & 0 & 35.15 & 0 & 0 & 0 \\ 0 & 0 & 0 & 0 & 17.78 & 0 & 0 \\ 0 & 0 & 0 & 0 & 0 & 11.08 & 0 \\ 0 & 0 & 0 & 0 & 0 & 0 & 5.10 \end{pmatrix}.$$

These can be used to calculate the matrices $\left(\mathbf{\Phi}'^{T}\mathbf{V}'^{-1}\mathbf{\Phi}'\right)^{-1}$ and $\left(\mathbf{\Phi}'^{T}\mathbf{V}'^{-1}\mathbf{Y}'\right)$ and hence $\widehat{\mathbf{\Theta}}$ from (8.11), where $\widehat{\theta}_1 = \widehat{a}' = \ln\widehat{a}$, and $\widehat{\theta}_2 = \widehat{b}' = \widehat{b}$. The result is $\widehat{a} = 2.2163$ and $\widehat{b} = 0.491$. From these we can calculate the fitted values from $\widehat{y}_i = \widehat{a}\exp\left(\widehat{b}x_i\right)$ and they are given below.

i	1	2	3	4	5	6	7
$\widehat{y}_i$	3.62	5.92	9.67	15.81	25.83	42.23	69.01

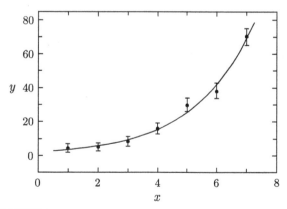

FIGURE 8.1   Best fit to the data using $y = 2.22\exp(0.49x)$.

A plot of the data and the fitted function is shown in Figure 8.1.

In (8.9) the sums are over all the data points, but the least-squares method can also be applied to binned data. In this case we will assume that the fitting function *is* a probability density, and for simplicity is a function of a single parameter θ to be estimated. Using the notation in the analogous discussion in Section 7.1 about fitting binned data using the maximum likelihood method, we assume there are N observations of a random variable x independently distributed with a density function $f(x; \theta)$, and divided between m bins. If the observed number of entries in the jth bin is o_j, then the predicted (expected) number of entries for that bin is

$$e_j(\theta) = E[o_j] = N \int_{x_j^{xmin}}^{x_j^{xmax}} f(x; \theta)\,dx = Np_j(\theta), \tag{8.13}$$

where x_j^{xmin} and x_j^{xmax} define the bin limits and $p_j(\theta)$ is the probability of having an entry in the jth bin. Then, by analogy with (8.9), the least-squares estimators are found by numerically minimizing the quantity

$$S = \sum_{j=1}^{m} \frac{\left[o_j - e_j(\theta)\right]^2}{\sigma_j^2}, \tag{8.14}$$

where σ_j^2 are the variances on the observed number of entries in the jth bin. If the mean number of entries in each bin is small compared to N, the entries in each bin are defined by the Poisson distribution, for which the variance is equal to the mean, $\sigma_j^2 = e_j$, so in this case

$$S = \sum_{j=1}^{m} \frac{\left[o_j - e_j(\theta)\right]^2}{e_j(\theta)} = \sum_{j=1}^{m} \frac{\left[o_j - Np_j(\theta)\right]^2}{Np_j(\theta)}. \tag{8.15}$$

Sometimes, for reasons of computational simplicity, the variance of the number of entries in a bin is replaced by the number of observed entries o_j, rather than the predicted number e_j, so that S becomes

$$S = \sum_{j=1}^{m} \frac{\left[o_j - e_j(\theta)\right]^2}{o_j} = \sum_{j=1}^{m} \frac{\left[o_j - Np_j(\theta)\right]^2}{o_j}, \tag{8.16}$$

but this is only valid if the number of entries in each bin is large; if for example any of the o_j were zero, clearly S would be undefined.

The estimates $\widehat{\theta}_k$ from (8.11) have the advantage of being unbiased estimates when there is no systematic error, as discussed previously. However, a greater importance of weighted least-squares estimates stems from their minimum variance properties, which are summarized by the statement that the estimator of any linear combination of the parameters θ_k with minimum variance is the linear combination of the least-squares estimates $\widehat{\theta}_k$. To prove this, consider the general linear combination of parameters

$$L = \mathbf{C}^T \mathbf{\Theta}, \tag{8.17}$$

where $\mathbf{C}$ is a $(p \times 1)$ vector of known constant coefficients. An estimator is a function of the experimental measurements. In this case, and in many others, it is sufficient to take the estimator to be a linear combination of the y_i which may be represented in matrix form as $\mathbf{G}^T\mathbf{Y}$ where $\mathbf{G}$ is an $(n \times 1)$ vector. Multiplying the terms of (8.4) on the left by $\mathbf{G}^T$ we have

$$\mathbf{G}^T\mathbf{Y} = \mathbf{G}^T\mathbf{\Phi}\mathbf{\Theta} + \mathbf{G}^T\mathbf{e}.$$

Clearly, $\mathbf{G}^T\mathbf{Y}$ is an estimator of L if

$$\mathbf{C}^T = \mathbf{G}^T\mathbf{\Phi}. \tag{8.18}$$

Because $\mathbf{C}$ is a $(p \times 1)$ vector and $\mathbf{G}$ is an $(n \times 1)$ vector, (8.18) does not have a unique solution. The problem of minimizing the variance of the estimator $\mathbf{G}^T\mathbf{Y}$ of L is now equivalent to selecting $\mathbf{G}$ to minimize the variance of $\mathbf{G}^T\mathbf{Y}$, subject to the constraint (8.18). To do this we use the method of Lagrange multipliers, as used in Section 7.4.1 when discussing the maximum likelihood method. Since $\mathbf{G}$ is a constant vector,

$$\text{var}(\mathbf{G}^T\mathbf{Y}) = \mathbf{G}^T(\text{var}\,\mathbf{Y})\,\mathbf{G} = \mathbf{G}^T\mathbf{V}\mathbf{G},$$

which follows from the definition of the variance matrix, and we can construct a variational function

$$F = \mathbf{G}^T\mathbf{V}\mathbf{G} - \mathbf{\Lambda}^T(\mathbf{\Phi}^T\mathbf{G} - \mathbf{C}), \tag{8.19}$$

where $\mathbf{\Lambda}$ is a $(p \times 1)$ vector of Lagrange multipliers. Differentiating F with respect to $\mathbf{G}$ and setting the result equal to zero, gives

$$\mathbf{G}^T = \mathbf{\Lambda}^T\mathbf{\Phi}^T\mathbf{V}^{-1}. \tag{8.20}$$

Applying (8.20) to (8.18) and solving for $\mathbf{\Lambda}^T$ we have

$$\mathbf{\Lambda}^T = \mathbf{C}^T(\mathbf{\Phi}^T\mathbf{V}^{-1}\mathbf{\Phi})^{-1}. \tag{8.21}$$

Eliminating $\boldsymbol{\Lambda}^T$ between (8.20) and (8.21) gives

$$\mathbf{G}^T = \mathbf{C}^T \left(\boldsymbol{\Phi}^T \mathbf{V}^{-1} \boldsymbol{\Phi}\right)^{-1} \boldsymbol{\Phi}^T \mathbf{V}^{-1}. \tag{8.22}$$

If we now multiply (8.22) on the right by $\mathbf{Y}$ and use (8.10), we have

$$\mathbf{G}^T \mathbf{Y} = \left(\mathbf{G}^T \boldsymbol{\Phi}\right) \widehat{\boldsymbol{\Theta}} = \mathbf{C}^T \widehat{\boldsymbol{\Theta}}. \tag{8.23}$$

Thus, we have shown the important result that the value of $\widehat{\boldsymbol{\Theta}}$ which minimizes the variance of the estimate of any linear combination of the parameters is the least-squares estimate, a result originally due to Gauss.

8.1.2 Errors on the parameter estimates

Having obtained the least-squares estimates $\widehat{\theta}_k$, we can now consider their variances and covariances. As mentioned above, this cannot be done with only knowledge of the relative errors on the observations, but instead requires the absolute values of these quantities. It is therefore convenient at this stage to allow for the possibility that the variance matrix may only be determined up to a scale factor w by writing

$$\mathbf{V} = w\mathbf{W}^{-1}, \tag{8.24}$$

where $\mathbf{W}$ is the so-called *weight matrix* of the observations. In this case (8.9) becomes

$$S = \frac{1}{w}(\mathbf{Y} - \boldsymbol{\Phi}\boldsymbol{\Theta})^T \mathbf{W}(\mathbf{Y} - \boldsymbol{\Phi}\boldsymbol{\Theta}) \tag{8.25}$$

and the solution of the normal equations is

$$\widehat{\boldsymbol{\Theta}} = \left(\boldsymbol{\Phi}^T \mathbf{W} \boldsymbol{\Phi}\right)^{-1} \boldsymbol{\Phi}^T \mathbf{W} \mathbf{Y}. \tag{8.26}$$

We have previously used the result that for any linear combination of y_i, say $\mathbf{P}^T \mathbf{Y}$, with $\mathbf{P}$ a constant vector

$$\mathrm{var}\left(\mathbf{P}^T \mathbf{Y}\right) = \mathbf{P}^T \mathrm{var}(\mathbf{Y}) \mathbf{P}. \tag{8.27}$$

Applying (8.27) to $\widehat{\boldsymbol{\Theta}}$ as given by (8.26), we have

$$\mathrm{var}\left(\widehat{\boldsymbol{\Theta}}\right) = \left(\boldsymbol{\Phi}^T \mathbf{W} \boldsymbol{\Phi}\right)^{-1} \boldsymbol{\Phi}^T \mathbf{W} \, \mathrm{var}(\mathbf{Y}) \mathbf{W} \boldsymbol{\Phi} \left(\boldsymbol{\Phi}^T \mathbf{W} \boldsymbol{\Phi}\right)^{-1},$$

and using

$$\mathrm{var}(\mathbf{Y}) = \mathbf{V} = w \, \mathbf{W}^{-1}$$

and the definition (8.12) of the components of the matrix $\mathbf{E}$, gives

$$\mathbf{E}^{-1} = \text{var}\big(\widehat{\mathbf{\Theta}}\big) = w\big(\mathbf{\Phi}^T\mathbf{W}\mathbf{\Phi}\big)^{-1}. \tag{8.28}$$

This is the variance matrix of the parameters and is given by a quantity that appears in the solution (8.26) for the parameters themselves. The matrix $\mathbf{E}^{-1}$ is also called the *error matrix*, and the errors (as usual this means their uncertainties) on the parameters are

$$\Delta\widehat{\theta}_i = \widehat{\sigma}_i = \big((\mathbf{E}^{-1})_{ii}\big)^{1/2}.$$

It is sometimes useful to know which linear combinations of parameter estimates have zero covariances. Since $\mathbf{E}^{-1}$ is a real, symmetric matrix, it can be diagonalized by an orthogonal[4] matrix $\mathbf{U}$. This same matrix then transforms the parameter estimates into the required linear combination.

Finally, if w is unknown, we need to find an estimate for it. This may be done by using $\widehat{\mathbf{\Theta}}$ as an estimate of $\mathbf{\Theta}$ in (8.25) and finding the expected value of this estimate of the weighted sum of residuals S,

$$wE[S] = E\big[\mathbf{R}^T\mathbf{W}\mathbf{R}\big]. \tag{8.29}$$

When $\mathbf{\Theta} = \widehat{\mathbf{\Theta}}$, the right-hand side of (8.29) becomes

$$E\big[\mathbf{R}^T\mathbf{W}(\mathbf{Y} - \mathbf{\Phi}\,\widehat{\mathbf{\Theta}})\big] = E\big[\mathbf{R}^T\mathbf{W}\mathbf{Y}\big],$$

since in this case

$$\mathbf{R}^T\mathbf{W}\mathbf{\Phi}\mathbf{\Theta} = 0$$

is equivalent to the statement of the normal equations. Furthermore,

$$\mathbf{R}^T\mathbf{W}\mathbf{Y} = \big(\mathbf{Y}^T - \widehat{\mathbf{\Theta}}^T\mathbf{\Phi}^T\big)\mathbf{W}\mathbf{Y} = \big(\mathbf{Y}^T\mathbf{W}\mathbf{Y}\big) - \big(\widehat{\mathbf{\Theta}}^T\mathbf{N}\widehat{\mathbf{\Theta}}\big), \tag{8.30}$$

where

$$\mathbf{N} = \mathbf{\Phi}^T\mathbf{W}\mathbf{\Phi}.$$

By using the normal equations once again, the right-hand side of (8.30) may be reduced to

$$\big(\mathbf{Y} - \mathbf{Y}^0\big)^T\mathbf{W}\big(\mathbf{Y} - \mathbf{Y}^0\big) - \big(\widehat{\mathbf{\Theta}} - \mathbf{\Theta}\big)^T\mathbf{N}\big(\widehat{\mathbf{\Theta}} - \mathbf{\Theta}\big),$$

where $\mathbf{Y}^0 = \mathbf{\Phi}\mathbf{\Theta}$. Thus, we have arrived at the result that

---

[4] A matrix is orthogonal if its inverse is equal to its transpose. In cartesian coordinates, multiplying a vector by an orthogonal matrix is a rotation of the vector about the origin or an opposite rotation of the coordinates.

$$E[S] = E\left[\mathbf{R}^T\mathbf{V}^{-1}\mathbf{R}\right] = E\left[\left(\mathbf{Y} - \mathbf{Y}^0\right)^T\mathbf{V}^{-1}\left(\mathbf{Y} - \mathbf{Y}^0\right) - \left(\widehat{\boldsymbol{\Theta}} - \boldsymbol{\Theta}\right)^T\mathbf{M}^{-1}\left(\widehat{\boldsymbol{\Theta}} - \boldsymbol{\Theta}\right)\right], \qquad (8.31)$$

where

$$\mathbf{M} = w\,\mathbf{N}^{-1}.$$

Comparing with (8.28), we see that $\mathbf{M}$ is the variance matrix of the parameter estimate $\widehat{\boldsymbol{\Theta}}$. Up to this point, the derivation is independent of whether the error has a systematic component. To arrive at a standard and quite useful result, we must now assume the error has zero mean, so that

$$E\left[\left(\mathbf{Y} - \mathbf{Y}^0\right)^T\left(\mathbf{Y} - \mathbf{Y}^0\right)\right] = \mathbf{V},$$

In this case

$$E\left[\left(\mathbf{Y} - \mathbf{Y}^0\right)^T\mathbf{V}^{-1}\left(\mathbf{Y} - \mathbf{Y}^0\right)\right] = \mathrm{Tr}\left\{E\left[\left(\mathbf{Y} - \mathbf{Y}^0\right)\left(\mathbf{Y} - \mathbf{Y}^0\right)^T\right]\mathbf{V}^{-1}\right\} = \mathrm{Tr}\left(\mathbf{V}\mathbf{V}^{-1}\right) = n,$$

where Tr denotes the trace of a matrix. Similarly, since $\mathbf{M}$ is the variance matrix of $\widehat{\boldsymbol{\Theta}}$,

$$E\left[\left(\widehat{\boldsymbol{\Theta}} - \boldsymbol{\Theta}\right)^T\mathbf{M}^{-1}\left(\widehat{\boldsymbol{\Theta}} - \boldsymbol{\Theta}\right)\right] = p.$$

Thus, from (8.31) we have

$$E\left[\mathbf{R}^T\mathbf{V}^{-1}\mathbf{R}\right] = n - p,$$

and so, from (8.29) an unbiased estimate for w is

$$\widehat{w} = \frac{\mathbf{R}^T\mathbf{W}\mathbf{R}}{n - p},$$

and consequently, an unbiased estimate for the variance matrix of $\widehat{\boldsymbol{\Theta}}$ is

$$\mathbf{E}^{-1} = \frac{\mathbf{R}^T\mathbf{W}\mathbf{R}}{n - p}\left(\boldsymbol{\Phi}^T\mathbf{W}\boldsymbol{\Phi}\right)^{-1} = \frac{\mathbf{R}^T\mathbf{V}^{-1}\mathbf{R}}{n - p}\left(\boldsymbol{\Phi}^T\mathbf{V}^{-1}\boldsymbol{\Phi}\right)^{-1}. \qquad (8.32)$$

Equation (8.32) looks rather complicated, but $\mathbf{R}^T\mathbf{W}\mathbf{R}$ can be calculated in a straightforward way from

$$\mathbf{R}^T\mathbf{W}\mathbf{R} = \left(\mathbf{Y} - \boldsymbol{\Phi}\widehat{\boldsymbol{\Theta}}\right)^T\mathbf{W}\left(\mathbf{Y} - \boldsymbol{\Phi}\widehat{\boldsymbol{\Theta}}\right),$$

using the measured and fitted values. In the common case where the values y_i are random variables normally distributed about f_i, then $\mathbf{R}^T\mathbf{V}^{-1}\mathbf{R}$ is the chi-squared value for the fit, and $(n-p)$ is the number of degrees of freedom n_{df}. In this case (8.32) becomes

$$\mathbf{E}^{-1} = \frac{\chi^2}{n_{df}}\left(\mathbf{\Phi}^T\mathbf{V}^{-1}\mathbf{\Phi}\right)^{-1}, \tag{8.33}$$

which is a very useful formula for making practical calculations.

Example 8.2

Calculate the errors on the best-fit parameters in Example 8.1.

Solution: These follow immediately using the matrices calculated in Example 8.1. For the primed quantities defined in Example 8.1, the error matrix is

$$\mathbf{E}'^{-1} = \left(\mathbf{\Phi}'^T\mathbf{V}'^{-1}\mathbf{\Phi}'\right)^{-1} = 10^{-2}\begin{pmatrix} 6.059 & -0.960 \\ -0.960 & 0.159 \end{pmatrix},$$

from which

$$\sigma(a) = \hat{a}\sigma(a') = 2.216 \times \sqrt{0.06059} = 0.546$$

and

$$\sigma(b) = \sigma(b') = \sqrt{0.00159} = 0.040.$$

8.1.3 Quality of the fit

To examine how well the predictions of the least-squares method fit the data we have to assume a distribution for the y_i. This will be taken to be normal about f_i, with the errors on the observations used to define the weights of the data, that is, $w = 1$ and the errors have zero mean, which is the usual situation in practice. In this case, we have seen above that the weighted sum of residuals S, of (8.9), is distributed as χ^2 with n_{df} degrees of freedom. Thus, for a fit of given order p, the expected value of S is n_{df} and a value of S close to the number of degrees of freedom indicates the function f is a good fit to the data. If the probability that the χ^2 is larger than S is small, that is, $S \gg n_{df}$, the fit should be considered unsatisfactory.

The order of the fit can be increased until this probability is as large as might be desired. To increase p beyond the point where $\chi^2 \sim n_{df}$ would result in apparently better fits to the data. However, to do so would ignore the fact that y_i are random variables and as such contain only a limited amount of information. To make the point with an extreme example, we could fit the data exactly if we selected a polynomial of order $n-1$ for f. However, for any reasonable number n of data points, one would not expect this polynomial to bear any realistic relationship to the true value of f.

What should one do if a satisfactory value of $\chi^2 \sim n_{df}$ cannot be achieved using a reasonable order p (e.g., if p is dictated by the model) that is, if $\chi^2 \gg n_{df}$? Firstly, one should examine the data to see whether there are isolated data points that contribute substantially higher-than-average values to χ^2. If this is the case, then these points should be carefully examined to see if there are any genuine reasons why they should be rejected, but as emphasized in Section 5.4 this must be done honestly, avoiding any temptation to 'massage the data', and must be defensible. In the absence of such reasons, one may have to conclude that the errors on the data have been underestimated and/or contain systematic errors. In this situation, one possibility is to scale the experimental errors by choosing a value of w so that $\chi^2 \sim n_{df}$. This will not change the values of the estimated parameters of the best fit but will increase their variances to better reflect the spread of the data. Conversely, if $\chi^2 \ll n_{df}$, the errors should be examined to see whether they have been overestimated.

Example 8.3

Calculate the χ^2 value for the fit in Example 8.2 *and comment on your result.*

Solution: The χ^2 value is found from

$$\sum_{i=1}^{7} \left[\frac{(y_i - \widehat{y}_i)}{\sigma_i} \right]^2 = 2.8,$$

and is for 5 degrees of freedom. From Table C4, $P\left[\chi_5^2 < 2.8\right]$ is approximately 0.25. This is acceptable, but since $\chi^2 < 5$, the number of degrees of freedom, it could be that the errors on the data have been somewhat overestimated, as mentioned above, and might warrant further investigation.

Another test that can be used to supplement the χ^2 test is based on the F distribution of Section 6.3. This procedure can test the significance of adding additional terms in the expansion (8.5), that is, to answer the question: Is θ_k different from zero? If S_p and S_{p-1} denote the values of S for fits of order p and $p - 1$, respectively, then from the additive property of χ^2, the quantity $(S_{p-1} - S_p)$ obeys a χ^2 distribution with one degree of freedom, and which is distributed *independently* of S_p itself. Thus, the statistic

$$F = \frac{S_{p-1} - S_p}{S_p/(n - p)}$$

obeys an F distribution with 1 and $(n-p)$ degrees of freedom. From tables of the F distribution, we can now find the probability P that the observed value F_o is greater than the expected value F_e. Thus if F_o is much larger than $F_o(1, n-p)$ we may assume $\theta_p = 0$ because otherwise, F_o should be close to F_e. It is still possible that even though $\theta_p = 0$, higher terms are nonzero, but in this case the χ^2 test would indicate that a satisfactory fit had not yet been achieved. These points will be discussed in more detail in Chapter 10, when we discuss hypothesis testing.

8.1.4 Orthogonal polynomials

The solutions for the parameters $\widehat{\Theta}$ and their error matrix $\mathbf{E}$, both require the inversion of the matrix $(\mathbf{\Phi}^T \mathbf{V}^{-1} \mathbf{\Phi})$. In the discussion so far, we have not specified the functions $\phi_k(x)$, except that they form a linearly independent set. If simple powers of x are used for $\phi_k(x)$, then the matrix is ill-conditioned[5] for even quite moderate values of k, and the degree of ill-conditioning increases as k becomes larger. This can lead to errors in $\widehat{\Theta}$ as calculated from (8.10). If a power series, or similar form, is dictated by the requirements of a particular model, the parameters of which are required to be estimated, then one can only hope to circumvent the problem by a judicious choice of method to invert the matrix. Such techniques are given in books on numerical methods. However, if all that is required is *any* form that gives an adequate representation of the data then it would clearly be advantageous to choose functions such that the matrix $(\mathbf{\Phi}^T \mathbf{V}^{-1} \mathbf{\Phi})$ is diagonal. Such functions are called *orthogonal polynomials* and their construction is briefly described here.

We will assume that the observations are uncorrelated (this is the usual situation met in practice and ensures $\mathbf{V}$ is diagonal) and denote the diagonal elements of the weight matrix $\mathbf{W} = w\mathbf{V}^{-1}$ for the data as $W(x_j)$ $(j = 1, 2, ..., n)$. Then if we fit using polynomials $\psi_k(x)$ $(k = 1, 2, ..., p)$, the matrix of the normal equations will be diagonal if

$$\sum_{j=1}^{n} W(x_j)\, \psi_r(x_j)\, \psi_s(x_j) = 0, \tag{8.34a}$$

for $r \neq s$. In this case, the least-squares estimate $\widehat{\Theta}$ from (8.10) is

$$\widehat{\theta}_k = \frac{\sum\limits_{j=1}^{n} W(x_j) y_j \psi_k(x_j)}{\sum\limits_{j=1}^{n} W(x_j) \psi_k^2(x_j)}, \quad k = 1, 2, ..., p. \tag{8.34b}$$

A valuable feature of using orthogonal polynomials is seen if we calculate the weighted sum of squared residuals at the minimum. From (8.9) this is, using p polynomials,

$$S_p = \frac{1}{w} \sum_{j=1}^{n} W(x_j) \left[y_j^2 - \sum_{k=1}^{p} \widehat{\theta}_k^2\, \psi_k^2(x_j) \right].$$

---

[5] *Ill-conditioned* refers to the situation where the magnitudes of elements of the matrix to be inverted differ greatly, so that the multiplications and subtractions involved in matrix inversion can lead to serious rounding errors in the inverted matrix.

If we now perform a new fit using $p+1$ polynomials, S_p is reduced by

$$\frac{1}{w}\widehat{\theta}_{p+1}^2 \sum_{j=1}^{n} W(x_j)\psi_{p+1}^2(x_j),$$

and the first p coefficients $\widehat{\theta}_k(k = 1, 2, ..., p)$ are unchanged.

To construct the polynomials we will assume for convenience that the values of x are normalized to lie in the interval $(-1, 1)$, and since it is desirable that none of the $\psi_k(x)$ has a large absolute value, we will arrange that the leading coefficient of $\psi_k(x)$ is 2^{k-2}. In this case it can be shown that the polynomials satisfy the following recurrence relations

$$\psi_1(x) = 1/2,$$
$$\psi_2(x) = (x - \beta_1)\psi_1(x),$$

and for $r \geq 2$,

$$\psi_{r+1}(x) = (x - \beta_r)\psi_r(x) + \gamma_{r-1}\psi_{r-1}(x).$$

To calculate the coefficients β_r and γ_r, we apply the orthogonality condition to ψ_s and ψ_{r+1}, that is,

$$\sum_{j=1}^{n} W(x_j)\,\psi_s(x_j)\,\psi_{r+1}(x_j) = 0, \quad s \neq r+1. \tag{8.35}$$

Then using the recurrence relations in (8.35) and setting first $s = r$ and then $s = r - 1$, leads to the results

$$\beta_r = \frac{\sum\limits_{j=1}^{n} W(x_j)x_j\psi_r^2(x_j)}{\sum\limits_{j=1}^{n} W(x_j)\psi_r^2(x_j)}, \quad r = 1, 2, ..., \tag{8.36a}$$

and

$$\gamma_{r-1} = -\frac{\sum\limits_{j=1}^{n} W(x_j)\psi_r^2(x_j)}{\sum\limits_{j=1}^{n} W(x_j)\psi_{r-1}^2(x_j)}, \quad r = 2, 3, ... \tag{8.36b}$$

It is straightforward to demonstrate that the first few polynomials constructed in this way are mutually orthogonal. That the rest are also orthogonal, may be shown inductively. If all polynomials of degree less than some integer r_0 are mutually orthogonal, it may be shown, by a somewhat tedious application of the recursion relations, that the polynomial of degree $r_0 + 1$ is orthogonal to all those of lesser degree.

8.1.5 Fitting a straight line

Because the least-squares method has been formulated above for any function linear in the parameters, and allows for the data to have correlated errors, the resulting formulas look a little forbidding, so it is instructive to derive explicit formulas for the simple case where the errors are uncorrelated, the situation often met in practice, and are fitted by a linear form containing just two parameters. It is worth reemphasizing that 'linear' refers to the parameters and that the fitting functions do not have to be linear, so even the two-parameter case can be far from trivial and is widely used (see Example 8.1). To make things even simpler, we shall assume the fitting function is the straight line $y = a + bx$. In this case, $p = 2$, with $\theta_1 = a$, $\theta_2 = b$, $\phi_1(x) = 1$, and $\phi_2(x) = x$, and the variances of the data values will be used to construct the weights, that is, we will set the scale factor $w = 1$. It is then straightforward, if rather tedious, to show from the general equations that for data with uncorrelated errors,

$$\widehat{a} = \frac{\overline{x^2}\,\overline{y} - \overline{x}\,\overline{xy}}{\overline{x^2} - \overline{x}^2} \quad \text{and} \quad \widehat{b} = \frac{\overline{xy} - \overline{x}\,\overline{y}}{\overline{x^2} - \overline{x}^2}, \tag{8.37a}$$

where the overbars as usual denote averages, but in this case taking account of the errors on the measurements. For example,

$$\overline{y} \equiv \frac{\displaystyle\sum_{i=1}^{n} y_i/\sigma_i^2}{\displaystyle\sum_{i=1}^{n} 1/\sigma_i^2} \longrightarrow \frac{1}{n}\sum_{i=1}^{n} y_i \quad \text{if the errors are all equal.} \tag{8.37b}$$

The denominator is the total weight and acts as a normalization factor. Thus, denoting the denominator in (8.37b) as N, (8.37a) for $\widehat{b}$ written out in full is

$$\widehat{b} = \frac{N\displaystyle\sum_{i=1}^{n} x_i y_i/\sigma_i^2 - \displaystyle\sum_{i=1}^{n} x_i/\sigma_i^2 \displaystyle\sum_{i=1}^{n} y_i/\sigma_i^2}{N\displaystyle\sum_{i=1}^{n} x_i^2/\sigma_i^2 - \left(\displaystyle\sum_{i=1}^{n} x_i/\sigma_i^2\right)^2}. \tag{8.38a}$$

A similar expression can be derived for $\widehat{a}$, but in practice, it is easier to calculate $\widehat{a}$ from

$$\widehat{a} = \overline{y} - \widehat{b}\,\overline{x} \tag{8.38b}$$

once $\widehat{b}$ has been found. The result, $y = \widehat{a} + \widehat{b}x$, can be used to interpolate for points within the data range, but where there are no measured data. In principle it can also be used to extrapolate to points outside the region where measurements exist, but care should be taken if this done, because no data have been used in these regions to constrain the parameters, and the results can rapidly become unreliable as one moves away from the fitted region.

To find the variances and covariance for the fitted parameters for the simple case of a straight-line fit we could again return to the general result (8.26). But it is simpler to use the results for $\hat{a}$ and $\hat{b}$ given in (8.37a). For example, setting $\sigma_i = \sigma$ for simplicity, $\hat{b}$ may be written

$$\hat{b} = \frac{\overline{xy} - \bar{x}\,\bar{y}}{\overline{x^2} - \bar{x}^2} = \sum_{i=1}^{n} \frac{1}{n} \frac{(x_i - \bar{x})}{(\overline{x^2} - \bar{x}^2)} y_i. \tag{8.39}$$

Using the results in Section 5.4.1 for combining errors, gives

$$\mathrm{var}\left(\hat{b}\right) = \sum_{i=1}^{n} \left[\frac{1}{n} \frac{(x_i - \bar{x})}{(\overline{x^2} - \bar{x}^2)} \right]^2 \sigma^2 = \frac{\sigma^2}{n\left(\overline{x^2} - \bar{x}^2\right)}. \tag{8.40}$$

Finally, if the errors on the data are independent but unequal, we make substitutions analogous to those in (8.37b), including setting

$$\sigma^2 \to \bar{\sigma}^2 = \frac{\sum_{i=1}^{n} \sigma_i^2/\sigma_i^2}{\sum_{i=1}^{n} 1/\sigma_i^2} = \frac{n}{\sum_{i=1}^{n} 1/\sigma_i^2}. \tag{8.41}$$

Then, writing out the result for the variance in full, gives

$$\mathrm{var}\left(\hat{b}\right) = N\left[N\sum_{i=1}^{n} x_i^2/\sigma_i^2 - \left(\sum_{i=1}^{n} x_i/\sigma_i^2 \right)^2 \right]^{-1}. \tag{8.42a}$$

In a similar way we can show that

$$\mathrm{var}(\hat{a}) = \sum_{j=1}^{n} x_j^2 / \sigma_j^2 \left[N\sum_{i=1}^{n} x_i^2/\sigma_i^2 - \left(\sum_{i=1}^{n} x_i/\sigma_i^2 \right)^2 \right]^{-1}, \tag{8.42b}$$

and

$$\mathrm{cov}\left(\hat{b},\hat{a}\right) = -\sum_{j=1}^{n} x_j / \sigma_j^2 \left[N\sum_{i=1}^{n} x_i^2/\sigma_i^2 - \left(\sum_{i=1}^{n} x_i/\sigma_i^2 \right)^2 \right]^{-1}, \tag{8.42c}$$

with a common factor appearing on the right-hand side of all three expressions. These results may also be found by applying (5.46) directly.

To find the error on the fitted value of f we can use (8.5), or (5.44), leading to

$$(\Delta f)^2 \equiv \mathrm{var} f(x) = \sum_{k=1}^{p} \sum_{l=1}^{p} \phi_k(x) E_{kl} \phi_l(x),$$ (8.43)

in which E_{kl} is the variance matrix of the parameters θ_j. For the straight-line fit $y = a + bx$, this reduces to

$$\mathrm{var}(y) = \mathrm{var}(\widehat{a}) + x^2 \mathrm{var}(\widehat{b}) + 2x \mathrm{cov}(\widehat{b}, \widehat{a}).$$ (8.44)

It is essential that the covariance term is included in (8.44). Without it, the value of $\mathrm{var}(y)$ could be seriously in error.

Example 8.4

The table below shows the values of a data set $y_i (i = 1, 2, ..., 7)$ with uncorrelated errors σ_i taken at the points x_i. Use the specific formulas for a straight-line fit $y = a + bx$ to find estimators for the parameters a and b and their error matrix. Calculate the predictions for $\widehat{y}_i$ and plot the data and the best-fit line. What is the predicted error at the point $x = 1.5$?

i	1	2	3	4	5	6	7
x_i	-3	-2	-1	0	1	2	3
y_i	0	1	2	6	6	10	12
σ_i	1	1	1	1	2	2	2

Solution: Using the notations above,

$$N = \sum_{i=1}^{7} 1/\sigma_i^2 = 4.75, \quad \sum_{i=1}^{7} x_i/\sigma_i^2 = -4.5, \quad \sum_{i=1}^{7} y_i/\sigma_i^2 = 16.0,$$

$$\sum_{i=1}^{7} x_i^2/\sigma_i^2 = 17.5, \quad \left(\sum_{i=1}^{7} x_i/\sigma_i^2 \right)^2 = 20.25, \quad \sum_{i=1}^{7} x_i y_i/\sigma_i^2 = 11.5.$$

Substituting these numbers into (8.38a) gives $\widehat{b} = 2.014$. Then from (8.38b) $\widehat{a}$ is given by $\widehat{a} = \overline{y} - \widehat{b}\overline{x} = 5.276$. To find the error matrix, substitute into Equations (8.42a–c) to find the variances and the covariance. This gives $\mathrm{var}(\widehat{a}) = 0.2783$, $\mathrm{var}\left(\widehat{b}\right) = 0.0755$ and $\mathrm{cov}(\widehat{b}, \widehat{a}) = 0.0716$, and hence the error matrix is

$$E = \begin{pmatrix} 0.2783 & 0.0716 \\ 0.0716 & 0.0755 \end{pmatrix}.$$

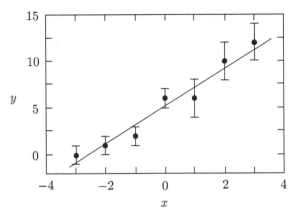

FIGURE 8.2   Least-squares fit to the data using $y = 5.276 + 2.014x$.

From the values $\widehat{a}$ and $\widehat{b}$ we can calculate the values of $\widehat{y}_i = \widehat{a} + \widehat{b}x_i$, and they are given below.

i	1	2	3	4	5	6	7
$\widehat{y}_i$	−0.77	1.25	3.26	5.28	7.29	9.30	11.32

A plot of the data and the fitted function is shown in Figure 8.2.

To calculate the predicted error at the point $x = 1.5$, we use (8.44). This gives the variance as 0.6630 and hence the error on the fitted point is 0.81.

We have previously shown that the variances of the least squares parameters are in general independent of any systematic error that may be present in the measurements. The values of the parameters *are* affected by the presence of systematic error, and it is instructive to examine this affect in the straight-line fit. For simplicity we will assume that all the systematic error consists of a common constant value ω being added to the measured value y_i. That is, $\widehat{y}_i$, defined as $y_i - \omega$, has only random error with zero mean. It is easily shown that in this case,

$$\widehat{b} = \frac{\overline{xy} - \overline{x}\,\overline{y}}{\overline{x^2} - \overline{x}^2} = \sum_{i=1}^{n} \frac{1}{n} \frac{(x_i - \overline{x})}{(\overline{x^2} - \overline{x}^2)} \widehat{y}_i. \tag{8.45}$$

That is, comparing with (8.39) the value of $\widehat{b}$ is the same as it would be if there was no systematic error.

We can use (8.38b) to determine $\widehat{a}$ since it is valid, whether systematic error is present or not. With uniform systematic error ω, $\overline{y} = \overline{\widehat{y}} + \omega$ and

$$\widehat{a} = \overline{\widehat{y}} - \widehat{b}\,\overline{x} + \omega. \tag{8.46}$$

This is in accord with common sense, because if the systematic error is the same for all data points, then they will all move in parallel the distance ω, and the slope of the fitted straight line will not change.

8.1.6 Combining experimental data

The least-squares results may be used in a simple way to combine the results of several experiments measuring the same quantities. This was considered in Example 7.6 for the simple case of repeated measurements $y_i (i = 1, 2, ..., n)$ of a single quantity y each having independent errors $\sigma_i (i = 1, 2, ..., n)$. The result was the so-called *weighted mean*,

$$\widehat{y} = \frac{\sum_{i=1}^{n} y_i / \sigma_i^2}{\sum_{i=1}^{n} 1/\sigma_i^2}, \quad \text{with} \quad \text{var}(\widehat{y}) = \frac{1}{\sum_{i=1}^{n} 1/\sigma_i^2}. \tag{8.47}$$

It also follows directly from the general solution (8.11) for the simple case where the fitted function is a constant and $\mathbf{V}$ is a diagonal matrix. Thus (8.47) are the least-squares estimators. Knowing this, we can easily generalize the result to the case where the measurements are not independent, which would occur, for example if they were based in part on the same data set. Then the expression for S of (8.9) becomes

$$S(\lambda) = \sum_{i,j=1}^{n} \left(y_i - \lambda \right) \left(\mathbf{V}^{-1} \right)_{ij} \left(y_j - \lambda \right),$$

and we seek an estimator $\widehat{\lambda}$ for the true value λ, given a set of measurements y_i of λ. As usual, this is found by setting the derivative of S with respect to λ equal to zero and gives

$$\widehat{\lambda} = \sum_{i=1}^{n} w_i y_i, \tag{8.48a}$$

where the weights are now given by

$$w_i = \sum_{j=1}^{n} \left(\mathbf{V}^{-1} \right)_{ij} \left[\sum_{k,l=1}^{n} \left(\mathbf{V}^{-1} \right)_{kl} \right]^{-1}, \tag{8.48b}$$

with the variance of $\widehat{\lambda}$ given by

$$\text{var}\left(\widehat{\lambda} \right) = \frac{1}{\sum_{i,j=1}^{n} \left(\mathbf{V}^{-1} \right)}. \tag{8.48c}$$

The formulas (8.48a), (8.48b), and (8.48c) reduce to (8.47) if the errors are uncorrelated.

Example 8.5

Three measurements of a quantity λ yield the results 3, 3.5, and 4 with a variance matrix

$$\mathbf{V} = \begin{pmatrix} 2 & 0 & 1 \\ 0 & 3 & 1 \\ 1 & 1 & 4 \end{pmatrix}.$$

Find the least-squares estimate for λ and its variance.
Solution: From the variance matrix, we have

$$\mathbf{V}^{-1} = \frac{1}{19} \begin{pmatrix} 11 & 1 & -3 \\ 1 & 7 & -2 \\ -3 & -2 & 6 \end{pmatrix}$$

and so

$$\left[\sum_{k,l=1}^{3} (\mathbf{V}^{-1})_{kl} \right]^{-1} = \frac{19}{16}.$$

Thus from (8.48b) the weights are

$$w_1 = 9/16, \quad w_2 = 6/16, \quad w_3 = 1/16$$

and from (8.48a), $\hat{\lambda} = 52/16 = 3.25$. The variance is found from (8.48c) and is $19/16$. Thus $\hat{\lambda} = 3.3 \pm 1.1$.

Combining data from different experiments must be done with care if it is to be meaningful, because the various experiments may not be compatible. A test such as that based on the Student's t distribution, or on χ^2, should first be used to establish compatibility. For example, the results in Example 8.4 yield a value $\chi^2 = 0.2$. Even if the data are compatible, averaging highly correlated data is difficult because a small error in the covariance matrix can result in a large error in the estimated value $\hat{\lambda}$ and an incorrect estimate of its variance.

The above discussion can be generalized to situations where we wish to combine data from experiments that measure combinations of quantities λ_1, λ_2, etc. An example is given in Problem 8.3.

8.2 Linear least-squares with constraints

It sometimes happens in practice that one has *some* information that can be used to refine the fit. As an example, we will generalize the discussion of Section 8.1 by considering the situation where the additional information takes the form of a set of ℓ linear constraint equations on the parameters of the form

$$\sum_{i=1}^{p} C_{li}\theta_i = Z_l,$$

or, in matrix notation

$$\mathbf{C\Theta} = \mathbf{Z}, \tag{8.49}$$

where $\ell \leq p$ so the rank of $\mathbf{C}$ is l. We must now minimize the sum of residuals S given by (8.8), subject to the constraint (8.49). This problem can be solved if we introduce an $(l \times 1)$ vector of Lagrange multipliers $\mathbf{\Lambda}$. Then the variation function to consider is

$$L = \left(\mathbf{R}^T\mathbf{V}^{-1}\mathbf{R}\right) - 2\mathbf{\Lambda}^T(\mathbf{C\Theta} - \mathbf{Z}),$$

and the minimum of S subject to (8.49) is found by setting the total differential dL with respect to the components of $\mathbf{\Theta}$ to zero. This gives

$$dL = 0 = 2\left[-\mathbf{Y}^T\mathbf{V}^{-1}\mathbf{\Phi} + \widehat{\mathbf{\Theta}}_c^T\left(\mathbf{\Phi}^T\mathbf{V}^{-1}\mathbf{\Phi}\right) - \mathbf{\Lambda}^T\mathbf{C}\right] d\mathbf{\Theta},$$

that is,

$$\mathbf{\Lambda}^T\mathbf{C} = \widehat{\mathbf{\Theta}}_c^T\left(\mathbf{\Phi}^T\mathbf{V}^{-1}\mathbf{\Phi}\right) - \mathbf{Y}^T\mathbf{V}^{-1}\mathbf{\Phi}, \tag{8.50}$$

where $\widehat{\mathbf{\Theta}}_c$ is the vector of estimates under the constraints.

We have seen previously (Equation (8.10)) that

$$\left(\mathbf{Y}^T\mathbf{V}^{-1}\mathbf{\Phi}\right) = \widehat{\mathbf{\Theta}}^T\left(\mathbf{\Phi}^T\mathbf{V}^{-1}\mathbf{\Phi}\right), \tag{8.51}$$

where $\widehat{\mathbf{\Theta}}$ is the estimate *without* the constraints. Using this relation in (8.50) gives

$$\mathbf{\Lambda}^T\mathbf{C} = \left(\widehat{\mathbf{\Theta}}_c - \widehat{\mathbf{\Theta}}\right)^T\left(\mathbf{\Phi}^T\mathbf{V}^{-1}\mathbf{\Phi}\right). \tag{8.52}$$

If, as in Section 8.1.2, we set

$$\mathbf{N} = w\left(\mathbf{\Phi}^T\mathbf{V}^{-1}\mathbf{\Phi}\right) = \left(\mathbf{\Phi}^T\mathbf{W}\mathbf{\Phi}\right). \tag{8.53}$$

Then,

$$w\mathbf{\Lambda}^T\mathbf{C}\mathbf{N}^{-1}\mathbf{C}^T = \left(\widehat{\mathbf{\Theta}}_c - \widehat{\mathbf{\Theta}}\right)^T\mathbf{C}^T = \mathbf{Z}^T - \widehat{\mathbf{\Theta}}^T\mathbf{C}^T$$

from which we obtain the result for $\mathbf{\Lambda}^T$,

$$w\mathbf{\Lambda}^T = \left(\mathbf{Z}^T - \widehat{\mathbf{\Theta}}^T\mathbf{C}^T\right)\left(\mathbf{C}\mathbf{N}^{-1}\mathbf{C}^T\right)^{-1}. \tag{8.54}$$

Substituting (8.54) into (8.52) and solving for $\widehat{\Theta}_c$ gives

$$\widehat{\Theta}_c^{\,T} = \widehat{\Theta}^{\,T} + \left(\mathbf{Z}^T - \widehat{\Theta}^{\,T}\mathbf{C}^T\right)\left(\mathbf{C}\mathbf{N}^{-1}\mathbf{C}^T\right)^{-1}\mathbf{C}\mathbf{N}^{-1}. \tag{8.55}$$

This is the solution for the least-squares estimate of Θ under the constraints, and like the unconstrained problem it only depends on the relative variances of the observations, because any scale factor in $\mathbf{V}$, and hence in $\mathbf{N}$, cancels in (8.55).

To find the variance matrix for the estimates $\widehat{\Theta}_c$ *does* require knowledge of the full variance matrix of the observations, so if we use a scale factor w as defined in (8.24), then from (8.55), using (8.27) and (8.28) and a tedious bit of matrix algebra, we find

$$\text{var}\!\left(\widehat{\Theta}_c\right) = w\left[\mathbf{N}^{-1} - \mathbf{N}^{-1}\mathbf{C}^T\left(\mathbf{C}\mathbf{N}^{-1}\mathbf{C}^T\right)^{-1}\mathbf{C}\mathbf{N}^{-1}\right], \tag{8.56}$$

and we are again left with the problem of finding an estimate for w. This may be done in a similar way to the unconstrained problem. Thus, we consider the expected value of the weighted sum of the residues under the constraints. This is

$$E[S] = E\left[(\mathbf{Y} - \mathbf{\Phi}\widehat{\Theta}_c)^T\mathbf{V}^{-1}(\mathbf{Y} - \mathbf{\Phi}\widehat{\Theta}_c)\right]$$

or equivalently,

$$E[S] = E\left[\left(\mathbf{R}^T\mathbf{V}^{-1}\mathbf{R}\right) + \left(\widehat{\Theta}_c - \widehat{\Theta}\right)^T\left(\mathbf{\Phi}^T\mathbf{V}^{-1}\mathbf{\Phi}\right)\left(\widehat{\Theta}_c - \widehat{\Theta}\right)\right], \tag{8.57}$$

where $\mathbf{R}$ is the matrix of residuals without constraints as defined in (8.9) and we have again made use of (8.51). Using the same technique previously used in Section 8.1.2, we can show that the second term has an expected value of l, the rank of the constraint matrix $\mathbf{C}$, and we have already shown that the expected value of the first term is $(n-p)$. So, an unbiased estimate of w is

$$\widehat{w} = \frac{\left(\mathbf{R}^T\mathbf{W}\mathbf{R}\right) + \left(\widehat{\Theta}_c - \widehat{\Theta}\right)^T\left(\mathbf{\Phi}^T\mathbf{W}\mathbf{\Phi}\right)\left(\widehat{\Theta}_c - \widehat{\Theta}\right)}{n - p + l}. \tag{8.58}$$

The second term may be written in a form that is independent of $\widehat{\Theta}_c$ by using (8.55) for $\left(\widehat{\Theta}_c - \widehat{\Theta}\right)$. This gives

$$\widehat{w} = \frac{\left(\mathbf{R}^T\mathbf{W}\mathbf{R}\right) + \left(\mathbf{Z} - \mathbf{C}\widehat{\Theta}\right)^T\left(\mathbf{C}\mathbf{N}^{-1}\mathbf{C}^T\right)^{-1}\left(\mathbf{Z} - \mathbf{C}\widehat{\Theta}\right)}{n - p + l}. \tag{8.59}$$

Finally, the error matrix for the parameters $\widehat{\Theta}_c$ is given by (8.56) with $\widehat{w}$ given by (8.59). An example using these results in given in Problem 8.4.

Analogous formulas to those above may be derived for situations where the constraints are directly on the least squares fit to the measurements themselves. As before, we will only consider the simple case of a set of linear constraint equations of the form

$$\mathbf{B}\widehat{\mathbf{Y}} = \mathbf{Z}$$

analogous to (8.49). Since $\widehat{\mathbf{Y}}$ is defined as $\mathbf{\Phi}\widehat{\mathbf{\Theta}}_c$, this constraint is the same as the constraint (8.49) with $\mathbf{C}$ replaced by $\mathbf{B\Phi}$. Making this substitution in (8.55) and (8.56), and for simplicity setting $w = 1$, we find

$$\widehat{\mathbf{Y}}_c^T = \widehat{\mathbf{Y}}^T + \left(\mathbf{Z}^T - \widehat{\mathbf{Y}}^T\mathbf{B}^T\right)\left(\mathbf{BVB}^T\right)^{-1}\mathbf{B}\,\mathbf{V}, \tag{8.60}$$

with an associated variance matrix

$$\mathrm{var}\left(\widehat{\mathbf{Y}}_c^T\right) = \mathrm{var}\left(\widehat{\mathbf{Y}}^T\right) - \mathbf{VB}^T\left(\mathbf{BVB}^T\right)^{-1}\mathbf{BV}, \tag{8.61}$$

in which we have written $\mathrm{var}\left(\widehat{\mathbf{Y}}^T\right)$ instead of $\mathbf{V}$ to emphasize the relationship between the constrained and unconstrained variance of the fit. The use of (8.60) and (8.61) is illustrated in the following example.

Example 8.6

Independent measurements of the three angles $y_i(i = 1,2,3)$ of a triangle yield (in degrees) the values 89 ± 1, 33 ± 2, and 64 ± 2. Find the least-squares estimate for the angles and their variance matrix, subject to the constraint that the sum of the angles is exactly 180 degrees.

Solution: We interpret the given measurements and error range as the unconstrained least squares fit and resulting standard deviation. The various matrices we will need are

$$\mathbf{V} = \begin{pmatrix} 1 & 0 & 0 \\ 0 & 4 & 0 \\ 0 & 0 & 4 \end{pmatrix},$$

$$\mathbf{B} = (1 \quad 1 \quad 1), \quad \mathbf{Y} = (89 \quad 33 \quad 64), \quad \text{and} \quad \mathbf{Z} = 180.$$

Then $\left(\mathbf{BVB}^T\right)^{-1} = 1/9$ and $\mathbf{BV} = (1 \quad 4 \quad 4)$. So, from (8.60),

$$\widehat{\mathbf{Y}}_c^T = (89 \quad 33 \quad 64) - \frac{2}{3}(1 \quad 4 \quad 4),$$

and hence

$$\widehat{y}_1 = 88\frac{1}{3}, \quad \widehat{y}_2 = 30\frac{1}{3}, \quad \text{and} \quad \widehat{y}_3 = 61\frac{1}{3}.$$

As expected, the 'excess' of 6 in the measured sum of the angles has been divided unequally, with least being subtracted from y_1 because it is more precisely determined than the other angles. The variance matrix follows from (8.61) and is

$$\text{var}\left(\widehat{\mathbf{Y}}_c^T\right) = \frac{1}{9}\begin{pmatrix} 8 & -4 & -4 \\ -4 & 20 & -16 \\ -4 & -16 & 20 \end{pmatrix},$$

so that $\widehat{y}_1 = 88.3 \pm 0.9$, $\widehat{y}_2 = 30.3 \pm 1.5$, and $\widehat{y}_3 = 61.3 \pm 1.5$. Imposing the constraint has improved the precision of the angles, as expected.

The above discussion may be extended in several ways. For example, to situations where there are constraints on both the data and the parameters to be estimated from them, or where the constraints are nonlinear, although the general formalism is considerably more complicated, and in the latter case the solution can usually only be obtained by iteration.

8.3 Nonlinear least-squares

If the fitting functions $\mathbf{F}(\boldsymbol{\Theta})$ are not linear in the parameters, then the weighted sum of residuals to be minimized is

$$S = [\mathbf{Y} - \mathbf{F}(\boldsymbol{\Theta})]^T \mathbf{W}[\mathbf{Y} - \mathbf{F}(\boldsymbol{\Theta})]. \tag{8.62}$$

Differentiating S with respect to $\boldsymbol{\Theta}$ and setting the result to zero, leads to a set of nonlinear simultaneous equations, and consequently presents a difficult problem to be solved. In practice, S is minimized directly by an iterative procedure, starting from some initial estimates for $\boldsymbol{\Theta}$, which may be suggested by the theoretical model, or in extreme situations may be little more than educated guesses. We will illustrate how such a scheme might *in principle* be applied.

The method is based on trying to convert the nonlinear problem to a series of linear ones. Let the initial estimate of $\boldsymbol{\Theta}$ be $\boldsymbol{\Theta}_0$, then if $\boldsymbol{\Theta}_0$ is close enough to the 'true' value $\boldsymbol{\Theta}$, we may expand the quantity $[\mathbf{Y} - \mathbf{F}(\boldsymbol{\Theta})]$ in a Taylor series about $\boldsymbol{\Theta}_0$ and keep only the first term. The technique relies on the truncation of the series being valid. Thus,

$$\Delta_0 = \mathbf{Y} - \mathbf{F}(\boldsymbol{\Theta}_0) \simeq \frac{\partial \mathbf{F}(\boldsymbol{\Theta}_0)}{\partial \boldsymbol{\Theta}_0}\delta_0, \tag{8.63}$$

where δ_0 is a vector of small increments of $\boldsymbol{\Theta}$. The problem of calculating δ_0 is now reduced to one of linear least squares, since both Δ_0 and the design matrix

$$\boldsymbol{\Phi}_0 = \frac{\partial \mathbf{F}(\boldsymbol{\Theta}_0)}{\partial \boldsymbol{\Theta}}$$

are obtainable. Given a solution for δ_0 from the normal equations, a new approximation

$$\mathbf{F}(\mathbf{\Theta}_1) = \mathbf{F}(\mathbf{\Theta}_0 + \delta_0)$$

may be calculated. This in turn will lead to a new design matrix

$$\mathbf{\Phi}_1 = \frac{\partial \mathbf{F}(\mathbf{\Theta}_1)}{\partial \mathbf{\Theta}}$$

and a new vector $\mathbf{\Delta}_1$ and hence, via the normal equations, to a new incremental vector δ_1. This linearization procedure may now be iterated until the changes in $\mathbf{\Theta}$ from one iteration to the next are very small. At the close of the iterations the variance matrix for the parameters is again taken to be the inverse of the matrix of the normal equations.

As we have emphasized, the above procedure is only to illustrate a possible method of finding the minimum of S. In practice several difficulties could occur, for example the initial estimates $\mathbf{\Theta}_0$ could be such as to invalidate the truncation of the Taylor series at its first term. In general, such a method is not guaranteed to converge to any value, let alone to values representing a true minimum of S.

The problem of minimizing S is an example of a more general class of problems that come under the heading of 'optimization of a function of several nonlinear variables' and in Appendix B there is a brief review of the methods that have proved successful in practice.

8.4 Other methods

Estimation using maximum likelihood, as described in Chapter 7, is a very general technique and is widely used in practical work, as is the method of least squares described above. But several other methods are also in common use and may be more suitable for certain applications. Three of them are briefly described below.

8.4.1 Minimum chi-square

Consider the case in which all the values of a population fall into k mutually exclusive categories $c_i(i = 1, 2, ..., k)$ and let p_i denote the proportion of values falling into category c_i, where

$$\sum_{i=1}^{k} p_i = 1. \tag{8.64}$$

Furthermore, in a random sample of n observations, let o_i and $e_i = np_i$ denote the *observed* and *expected* frequency in category c_i, where

$$\sum_{i=1}^{k} o_i = \sum_{i=1}^{k} e_i = n. \tag{8.65}$$

In Section 4.2.2 we considered the multinomial distribution with density function

$$f(r_1, r_2, ..., r_{k-1}) = n! \prod_{i=1}^{k} p_i^{r_i} \left(\prod_{i=1}^{k} r_i! \right)^{-1}, \tag{8.66}$$

where r_i denotes the frequency of observations in the ith category for which the true proportion of observations is $p_i (i = 1, 2, ..., k)$. We recall that the multinomial density function gives exact probabilities for any set of observed frequencies

$$r_1 = o_1, r_2 = o_2, ..., r_k = o_k. \tag{8.67}$$

Each r_i is distributed binomially and we have seen in Section 4.2.3 that the binomial distribution tends rapidly to a Poisson distribution with mean and variance both equal to np_i. The Poisson distribution in turn tends to a normal distribution as np_i increases, and is commonly considered approximately normal if the mean $\mu \geq 9$. Thus if $np_i \geq 9$, r_i is approximately normally distributed with mean and variance np_i. By converting to standard measure, it follows that the statistic

$$u_i = \frac{r_i - np_i}{(np_i)^{1/2}} \tag{8.68}$$

is approximately normally distributed with mean zero and unit variance. Furthermore,

$$\chi^2 = \sum_{i=1}^{k} u_i^2 = \sum_{i=1}^{k} \frac{(r_i - np_i)^2}{np_i} = \sum_{i-1}^{k} \frac{(o_i - e_i)^2}{e_i} \tag{8.69}$$

is distributed as χ^2 with $(k-1)$ degrees of freedom. Equation (8.69) can be used to test whether data are consistent with a specific distribution. We will return to this use of chi-squared in Chapter 11 when we discuss hypothesis testing.

A more common situation that arises in practice is where the generating density function is not completely specified, but instead, contains some unknown parameters. If the observed frequencies are used to provide estimates of the p_i, then the quantity analogous to χ^2 of (8.69) is

$$\chi'^2 = \sum_{i=1}^{k} \frac{(o_i - n\widehat{p}_i)^2}{n\widehat{p}_i}. \tag{8.70}$$

There now arise two questions; (1) what is the best way of estimating p_i and (2) what is the distribution of χ'^2? There are clearly many different methods available to estimate the p_i, but one which is widely used is to choose values which minimize χ'^2. This may in general be a difficult problem and is another example of the general class of optimization problems mentioned above, and which are briefly discussed in Appendix B. It can be shown that for a wide class of methods of estimating the p_i, including that of minimum chi-square, χ'^2 is asymptotically distributed as χ^2 with $(k-1-c)$ degrees of freedom where c is the number of independent parameters of the distribution used to estimate the p_i.

In general, if x_i is a sample of size n from a multinomial population with mean $\mu(\theta)$ and variance matrix $V(\theta)$, where θ is to be estimated, then the value $\hat{\theta}$ $(x_1, x_2, ..., x_n)$ which minimizes

$$\chi^2 = \frac{1}{n}[\bar{x} - \mu(\theta)]^T[V(\theta)]^{-1}[\bar{x} - \mu(\theta)],$$

that is, the minimum-χ^2 estimate of θ, is known to be consistent, asymptotically efficient, and asymptotically normally distributed if x is distributed like the binomial, Poisson, or normal distribution (and many others).

Example 8.7

A method for generating uniformly distributed random integers in the range 0–9 has been devised and tested by generating 1000 digits with results shown below.

Digit	0	1	2	3	4	5	6	7	8	9
Frequency	106	89	85	110	123	93	82	110	91	111

Do these results support the idea that the method of generation is suitable?

Solution: If the digits were uniformly distributed, then the expected frequencies would all be 100. So, using (8.69), we find $\chi^2 = 16.86$ and this is for 9 degrees of freedom. From Table C.4, $P[\chi^2 \geq 16.9]$ for 9 degrees of freedom is 0.05. So, although it cannot be ruled out, as this is a low probability, it raises some doubt that the method really is producing uniformly distributed integers. (Such statements will be made more precise when hypothesis testing is discussed in Chapter 11.)

The minimum chi-squared method of estimation can be used in a range of other situations, including those where the parameters are subject to constraints. An example is given in Problem 8.5.

8.4.2 Method of moments

In Section 3.2.3 we saw that two distributions with a common moment-generating function were equal. This provides a method for estimating the parameters of a distribution by estimating its moments, referred as an MM estimator. To illustrate this, let $f(x; \theta_1, \theta_2, ..., \theta_p)$ be a univariate density function with p parameters $\theta_i (i = 1, 2, ..., p)$, and let the first p algebraic moments be

$$\mu_j'(\theta_1, \theta_2, \ldots, \theta_p) = \int_{-\infty}^{\infty} x^j f(x; \theta_1, \theta_2, \ldots, \theta_p)\, dx, \quad j = 1, 2, \ldots, p. \tag{8.71}$$

Let x_n be a random sample of size n drawn from the density f. The first p sample algebraic moments are given by

$$m_j' = \frac{1}{n} \sum_{i=1}^{n} x_i^j. \tag{8.72}$$

The estimators $\widehat{\theta}_i$ of the parameters θ_i are obtained from the solutions of the p equations

$$m_j' = \mu_j', \quad j = 1, 2, \ldots, p. \tag{8.73}$$

Example 8.8

Use the method of moments to find the MM estimators for the mean and variance of a normal distribution.
Solution: We have previously seen (using Equation (4.6)) that for a normal distribution,

$$\mu_1' = \mu; \quad \mu_2' = \sigma^2 + \mu^2.$$

The sample moments are

$$m_1' = \frac{1}{n} \sum_{i=1}^{n} x_i; \quad m_2' = \frac{1}{n} \sum_{i=1}^{n} x_i^2.$$

Applying (8.73) then gives

$$\widehat{\mu} = \frac{1}{n} \sum_{i=1}^{n} x_i = \overline{x},$$

and

$$\widehat{\sigma}^2 + \widehat{\mu}^2 = \frac{1}{n} \sum_{i=1}^{n} x_i^2,$$

that is,

$$\widehat{\sigma}^2 = \frac{1}{n} \left[\sum_{i=1}^{n} x_i^2 - n\overline{x}^2 \right] = \frac{1}{n} \sum_{i=1}^{n} (x_i - \overline{x})^2.$$

Thus, the MM estimators are, for this example, the same as those obtained by the maximum likelihood method, as derived in Example 7.5 and this is commonly the case for other examples.

In some applications where the population density function is not completely known it may be advantageous to use linear combinations of moments. Consider, for example, a density function $f(x; \theta_1, \theta_2, ..., \theta_p)$, which is unknown but may be expanded in the form

$$f(x; \theta_1, \theta_2, ..., \theta_p) = \sum_{j=1}^{p} \theta_j P_j(x), \tag{8.74}$$

where $P_j(x)$ is a set of orthogonal polynomials normalized such that

$$\int P_i(x) P_j(x) \, dx = \begin{cases} \phi_j, & i = j \\ 0, & i \neq j \end{cases}. \tag{8.75}$$

The population moments deduced from (8.74) are

$$\mu_i' = \int \sum_{j=1}^{p} \theta_j P_j(x) \, x^i \, dx. \tag{8.76}$$

However, we may also consider the linear combination of moments given by

$$\Omega_i = \int \sum_{j=1}^{p} \theta_j P_j(x) P_i(x) \, dx, \tag{8.77}$$

which by (8.75) is

$$\Omega_i = \theta_i \phi_i. \tag{8.78}$$

The equivalent sample moments are

$$m_i = \frac{1}{n} \sum_{j=1}^{n} P_i(x_j), \tag{8.79}$$

and so, by equating the two, we have

$$\widehat{\theta}_i = \frac{1}{n\phi_i} \sum_{j=1}^{n} P_i(x_j). \tag{8.80}$$

This method is useful, for example, for finding the distribution coefficients a_j in the expansion of a cross-section in particle scattering problems where this is expressed in terms of the scattering angle θ. In this case, the differential angular cross-section $d\sigma/d\cos\theta$ is written as

$$\frac{d\sigma}{d\cos\theta} = \sum_j a_j P_j(\cos\theta), \tag{8.81}$$

where P_j are Legendre polynomials and the coefficients are

$$\widehat{a}_j = \left(\frac{2j+1}{2n}\right) \sum_{i=1}^{n} P_j(x_i).$$

An example is given in Problem 8.10.

The modifications necessary to the above simple account for it to be applicable for situations involving binned data are like those that have been discussed for the maximum likelihood and least-squares methods, and so we will not discuss these further. Under quite general conditions, it can be shown that estimators obtained by the method of moments are consistent, but not in general most efficient.

Finally, although MM estimators usually require less computations than ML estimators, and often give similar results, as in Example 8.8, they can sometimes give results that are unreasonable, or even not physical, and so must be rejected. This is illustrated in the following simple example.

Example 8.9

A sample of four measurements 60, 20, 15, 5 is taken of a random variable x distributed as a uniform distribution in the interval $[0, m]$. Use the data to find the MM estimator for m.

Solution: The first moment about zero of the population is given by

$$\frac{1}{m} \int_0^m x\mathrm{d}x = \frac{m}{2},$$

and the first moment of a sample is $\bar{x}$. Hence, $\widehat{m} = 2\bar{x}$. However, the mean of the sample is $\bar{x} = 25$ and hence $\widehat{m} = 50$, which is unreasonable, since we know from the sample that one value in the sample exceeds 50, whereas the true value of m must exceed all the data values.

8.4.3 Bayes' estimators

In Section 2.5.2 we discussed the Bayesian interpretation of probability. There are several advantages of the Bayesian viewpoint. Foremost of these is that it can incorporate prior information about the parameter to be estimated. However, we saw from Bayes' theorem that to maximize the posterior probability requires knowledge of prior probabilities, and in general these are not completely known. Nevertheless, cases do occur where partial information is available, and in these circumstances, it would clearly be advantageous to include it in the estimation procedure if possible. The objection to the Bayesian approach is that one must choose a prior *pdf*, and as this is necessarily subjective, different choices can lead to different outcomes. The Bayesian answer to this objection is that it is a fact of life that different people will have different views about data and so it is entirely reasonable that different interpretations should exist. There is no definite answer to this question, but it can make it difficult to compare different inferences drawn from comparable data sets.

We will consider the case where the prior information about the parameter is such that the parameter itself can be *formally* regarded as a random variable with a prior density $f_{prior}(\theta)$, as in the maximum likelihood method. There has been much theoretical work done on the question of how to choose a prior density, but all suggestions have problems. Empirically, the form for $f_{prior}(\theta)$ could be obtained, for example, by plotting all previous estimates of θ. This will very often be found to be an approximately Gaussian form, and from the results, estimates of the mean and variance of the associated normal distribution could be made. In these cases, where both the usual variable and the parameter can be regarded as random variables, the corresponding *pdf* will be denoted as $f_R(x\,;\theta)$.

In Bayesian estimation, the emphasis is not on satisfying the requirements of 'good' point estimators, as discussed in Section 5.1.2, but rather on minimizing 'information loss', expressed through a so-called *loss function* $l(\widehat{\theta}\,;\theta)$. Expressed loosely, the latter gives the loss of information incurred by using the estimate $\widehat{\theta}$ instead of the true value θ. In practice it is difficult to know what form to assume for the loss function, but a simple, common-sense, assumption is the loss of information is a minimum when the estimate is the true value, and that minimum is zero. The first nonzero term of the Taylor series expansion of $\widehat{\theta}$ about the true value θ is then the quadratic term, which suggests the form

$$l(\widehat{\theta};\theta) = (\widehat{\theta} - \theta)^2, \tag{8.82}$$

and is often used. (A loss function that is bounded by zero, as in (8.82), is an example of a more general function used in decision theory, called a *risk function*.) The other quantities we need follow directly from work of previous chapters. Thus

$$f_R(x\,;\theta) = j(x_1, x_2, ..., x_n; \theta) = f(x_1, x_2, ..., x_n|\theta)f_{prior}(\theta) \tag{8.83}$$

is the joint density of $x_1, x_2, ..., x_n$ and θ; and

$$m(x_1, x_2, ..., x_n) = \int_{-\infty}^{\infty} j(x_1, x_2, ..., x_n, \theta)d\theta \tag{8.84}$$

is the marginal distribution of the x's. From Equation (3.23) it then follows that the conditional distribution of θ given $x_1, x_2, ..., x_n$ is

$$c(\theta|x_1, x_2, ..., x_n) = \frac{j(x_1, x_2, ..., x_n; \theta)}{m(x_1, x_2, ..., x_n)} = \frac{f(x_1, x_2, ..., x_n|\theta)f_{prior}(\theta)}{m(x_1, x_2, ..., x_n)}. \tag{8.85}$$

This is the posterior density $f_{post}(\theta|x_1, x_2, ..., x_n)$. We can now define a Bayes' estimator.

Let $x_1, x_2, ..., x_n$ be a random sample of size n drawn from a density $f_R(x;\theta)$. Also, let $f_{prior}(\theta)$ be the prior density of θ, and $f(x_1, x_2, ..., x_n|\theta)$ be the conditional density of the set

x_i given θ. Furthermore, let $f_{\text{post}}(\theta|\mathbf{x}))$ be the posterior density of θ given the set x_i, and let $l\left(\widehat{\theta}, \theta\right)$ be the loss function. Then the *Bayes' estimator* of θ is that function defined by

$$\widehat{\theta} = d(x_1, x_2, \ldots, x_n)$$

which minimizes the quantity

$$B\left(\widehat{\theta}; x_1, x_2, \ldots, x_n\right) = \int_{-\infty}^{\infty} l(\widehat{\theta}, \theta) f_{\text{post}}(\theta|x_1, x_2, \ldots, x_n) \, d\theta. \tag{8.86}$$

The disadvantage in using (8.86) is the necessity of assuming a form for both $f_{\text{prior}}(\theta)$ and $l\left(\widehat{\theta}; \theta\right)$. The following example illustrates the use of the method.

Example 8.10

Let $x_1, x_2, \ldots, x_n$ be an independent random sample of size n drawn from a normal density $f_R(x; \theta, a)$ with unknown mean θ and unit variance $a^2 = 1$. If θ is assumed to be normally distributed with known mean μ and unit variance $b^2 = 1$, find the Bayes' estimator for θ, using a loss function of the form $l(\widehat{\theta}, \theta) = (\widehat{\theta} - \theta)^2$.

Solution: From the above, setting $a = 1$,

$$f_R(x; \theta, a) = (2\pi)^{-1/2} \exp\left[-\frac{1}{2}(x - \theta)^2\right],$$

and hence

$$f(x_1, x_2, \ldots, x_n|\theta) = \frac{1}{(2\pi)^{n/2}} \exp\left[-\frac{1}{2}\left(\sum_{i=1}^{n} x_i^2 - 2\theta \sum_{i=1}^{n} x_i + n\theta^2\right)\right].$$

Also, setting $b = 1$,

$$f_{\text{prior}}(\theta) = (2\pi)^{-1/2} \exp\left[-(\theta - \mu)^2/2\right],$$

so that from (8.83)

$$j(x_1, x_2, \ldots, x_n, \theta) = \frac{1}{(2\pi)^{(n+1)/2}} \exp\left[-\frac{1}{2}\left(\sum_{i=1}^{n} x_i^2 + \mu^2\right)\right] \exp\left[-\frac{1}{2}(n+1)\theta^2 + (n\bar{x} + \mu)\theta\right],$$

and from (8.84)

$$m(x_1, x_2, \ldots, x_n) = (2\pi)^{-(n+1)/2} \exp\left[-\frac{1}{2}\left(\sum x_i^2 + \mu^2\right)\right] \int_{-\infty}^{\infty} \exp\left[\theta(n\bar{x} + \mu) - \frac{1}{2}(n+1)\theta^2\right] d\theta$$

$$= \frac{1}{(n+1)^{1/2}(2\pi)^{n/2}} \exp\left[-\frac{1}{2}\left(\sum x_i^2 + \mu^2\right) + \frac{1}{2}\frac{(n\bar{x} + \mu)^2}{n+1}\right].$$

Then using these in (8.85) gives

$$f_{\text{post}}(\theta|x_1, x_2, ..., x_n) = \left(\frac{n+1}{2\pi}\right)^{1/2} \exp\left\{ -\frac{(n+1)}{2}\left[\theta - \frac{n\bar{x}+\mu}{n+1}\right]^2 \right\},$$

and using $l\left(\widehat{\theta}, \theta\right) = \left(\widehat{\theta} - \theta\right)^2$ and the above expression for $f_{\text{post}}(\theta|x_1, x_2, ..., x_n)$ in (8.86), we find, after some algebra,

$$B\left(\widehat{\theta}; x_1, x_2, ..., x_n\right) = \left(\frac{n+1}{2\pi}\right)^{1/2} \int_{-\infty}^{\infty} \left(\widehat{\theta} - \theta\right)^2 \exp\left\{ -\frac{(n+1)}{2}\left[\theta - \frac{n\bar{x}+\mu}{n+1}\right]^2 \right\} d\theta$$

$$= \widehat{\theta}^2 - \frac{2\widehat{\theta}(\bar{x}n + \mu)}{n+1} + \frac{1}{n+1} + \left(\frac{\bar{x}n+\mu}{n+1}\right)^2.$$

Finally, to minimize B we set

$$\frac{\partial B}{\partial \widehat{\theta}}\left(\widehat{\theta}; x_1, x_2, ..., x_n\right) = 0,$$

giving

$$\widehat{\theta} = \frac{\mu + n\bar{x}}{n+1},$$

which is the Bayes' estimator for θ. It can be seen that $\widehat{\theta}$ is the weighted average of the sample mean $\bar{x}$ and the prior mean μ.

If we extend the case studied in Example 8.10 to the situation where the variances a and b are not equal, then a useful general result is as follows, which is given without proof, but may be obtained by repeating the steps in Example 8.10. If $\bar{x}$ is the mean of a random sample of size n from a normal population with known variance a^2, and the prior distribution of the population mean is a normal distribution with mean μ and variance b^2, then the posterior distribution of the population mean is also a normal distribution and the Bayes' estimators for the mean and variance are

$$\mu_1 = \frac{a^2\mu + nb^2\bar{x}}{a^2 + nb^2} \quad \text{and} \quad \sigma_1^2 = \frac{a^2b^2}{a^2 + nb^2}. \tag{8.87}$$

If the prior was uniform, the posterior density is also normal, although in this case with $\mu_1 = \bar{x}$ and $\sigma_1^2 = a^2/n$, which are the limits of (8.87) as $n \to \infty$. For *large* samples, Equations (8.87) hold for an independent random sample of size n drawn from *any* distribution with a finite variance. This is the Bayesian statement of the central limit theorem.

Under very general conditions it can be shown that Bayes' estimators, independent of the assumed prior distribution $f_{\text{prior}}(\theta)$, are efficient, consistent, and a function of sufficient estimators.

It is useful to consider the relation between Bayes' estimators and those obtained from the maximum likelihood method. Using Bayes' theorem, the posterior *pdf* of (8.85) may be written in terms of the likelihood (which, as mentioned earlier, is *not* a *pdf*) as

$$f_{\text{post}}(\theta|\mathbf{x}) = \frac{L(\mathbf{x}|\theta)f_{\text{prior}}(\theta)}{\int L(\mathbf{x}|\theta')f_{\text{prior}}(\theta')d\theta'}. \tag{8.88}$$

In the absence of any prior information, it is common to take $f_{\text{prior}}(\theta)$ to be a constant and in this case the posterior *pdf* is proportional to the likelihood and the two methods are very similar. However, a uniform prior has potential problems. Firstly, if the parameter can take on *any* values, $f_{\text{prior}}(\theta)$ cannot be normalized, although in practice this is not usually a difficulty because in the denominator it appears multiplied by the likelihood function. But a second problem is that one could take the prior to be uniform in a function of θ, rather than the parameter itself and this would lead to a different posterior *pdf* and hence a different estimate. Thus, Bayes' estimators with a uniform prior do not have the useful invariance property that ML estimators have. In practice, the distinction between different methods of estimation lessens as the sample size increases (because of the central limit theorem) and Bayes' estimators depend less on the assumed prior density.

Problems 8

8.1 Figure 8.3 shows some data fitted with polynomials of order 1, 2, and 3. Assuming the data are normally distributed, the χ^2 values for the fits are 13.9, 12.0, and 5.1, respectively. Comment on these results.

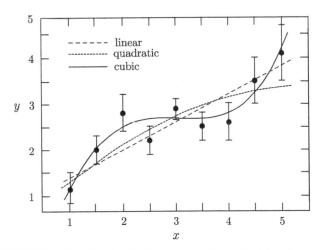

FIGURE 8.3 Data fitted with linear, quadratic, and cubic polynomials.

8.2 The table below shows the values of a quantity y, assumed to be normally distributed, and their associated errors σ, measured at six values of x.

i	1	2	3	4	5	6
x_i	1	2	3	4	5	6
y_i	2.5	3.0	6.0	9.0	10.5	10.5
σ_i	1	1	1	1	1	1

By successively fitting polynomials of increasing order, deduce the lowest order polynomial that gives an acceptable fit to the data and justify your answer. Find the coefficients of the polynomials corresponding to the best fit and their errors and plot the resulting best-fit curve.

8.3 An experiment determines two parameters λ_1 and λ_2, and finds values $y_1^{(1)} = 1.0$ and $y_2^{(1)} = -1.0$ with a variance matrix

$$\mathbf{V}^{(1)} = \begin{pmatrix} 2.0 & -1.0 \\ -1.0 & 1.5 \end{pmatrix} \times 10^{-2}.$$

A second experiment finds a new value of λ_2 to be $y_2^{(2)} = -1.1$ with a variance 10^{-2}. Find the least-squares estimates for λ_1 and λ_2 and their associated error matrix.

8.4 Rework Problem 8.3, but now with the constraint $\lambda_1 + \lambda_2 = 0$.

8.5 Measurements are made of the lengths $x_i(i = 1,2,3)$ of the sides of a right-angled triangle and the values $h_i(i = 1,2,3)$ found. If these are assumed to be normally distributed with equal variances σ^2, find the minimum chi-squared estimates for $x_i(i = 1,2,3)$.

8.6 Use the method of moments to find the estimator for the parameter λ in the exponential density

$$f(x; \lambda) = \lambda \exp(-\lambda x).$$

8.7 Use the methods of moments to find an estimator for the parameters α and β in the two-parameter distribution:

$$f(x; \alpha, \beta) = \alpha \exp[-\alpha(x - \beta)], \quad \alpha, \beta > 0; x > 0$$

in terms of the first two sample moments.

8.8 Let $r_1, r_2, \ldots, r_n$ be an independent random sample of size n drawn from a binomial density $f_R(r; p, n)$ with unknown parameter p. If p is assumed to be uniformly distributed in the interval $(0, 1)$, find the Bayes' estimator for p, using a loss function of the form $l(\hat{p}, p) = (\hat{p} - p)^2$. Compare your solution with that obtained by using the maximum likelihood method (Problem 7.6). Note the integral:

$$\int_0^1 x^n (1 - x)^m dx = \frac{n! m!}{(n + m + 1)!}.$$

8.9 Two determinations are made of the parameters of a straight line $y = ax + b$. The first is $a_1 = 4, b_1 = 12$ and the second is $a_2 = 3, b_2 = 14$. The associated variance matrices are

$$\mathbf{V}_1 = \begin{pmatrix} 1 & -1 \\ -1 & 2 \end{pmatrix} \quad \text{and} \quad \mathbf{V}_2 = \begin{pmatrix} 1 & -1 \\ -1 & 3 \end{pmatrix}.$$

Find the best estimate for a and b and the associated error matrix.

8.10 A beam of n particles is scattered from a fixed target through an angle θ. The angular distribution of the scattered particles has the form

$$\frac{dn}{d\cos\theta} - a + b\cos^2\theta,$$

where a and b are parameters. Find the MM estimator for the ratio b/a.

CHAPTER

9

Interval estimation

Overview

In Chapters 7 and 8, point estimation was discussed, that is, the estimation of the value of a parameter. In practice, point estimation alone is not enough. It is also necessary to supply a statement about the error on the estimate. In those chapters this was done by calculating the variance on the estimator and taking its square root, the standard deviation, as a 'standard error' to define error bars. In practice, because of the central limit theorem, most density functions lead to a normal form for the sampling density of the estimate in the case of large samples. In cases where this is not true, we could still use the standard deviation as a measure of uncertainty, but in these situations, it is more usual to consider a generalization called *interval estimation*, based on the concept of a *confidence interval*, which are intervals constructed in such a way that a predetermined percentage of such intervals will contain the true value of the parameter. This chapter will describe these ideas and their application, including the problematic case where an estimate leads to a value of a parameter that is close to its physical boundary.

9.1 Confidence intervals: Basic ideas

We have already encountered the idea of a confidence interval in Chapter 1, although it was not the term used there. In Section 1.5 we noted that the distribution of the observations on a random variable x for large samples often, indeed usually, had a density $n(x)$ of approximately normal form about the sample mean $\bar{x}$ with variance σ^2. In that case we could find values

$$C = \int_{x_L}^{x_U} n(x)\mathrm{d}x$$

for any values x_L and x_U. The quantity C is called the *confidence coefficient* and is usually written $C = (1 - 2\alpha)$. The reason for using the quantity $(1 - 2\alpha)$ will become clear later. We also refer to $100C\% = 100(1 - 2\alpha)\%$ as the *confidence level*. The confidence coefficient corresponds to a random interval (x_L, x_U), called the *confidence interval* $x_L \leq x \leq x_U$, which depends only on the observed data. For example, from tables of the normal density we know that $C = 0.683$, that is, a confidence level of 68.3%, for a confidence interval

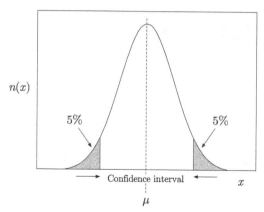

$n(x)$

5%   5%

Confidence interval x

μ

FIGURE 9.1   Central confidence interval corresponding to a 90% confidence level for a normal distribution.

$\mu - \sigma \leq x \leq \mu + \sigma$. In general, if the confidence coefficient is $C = (1 - 2\alpha)$, then $100(1 - 2\alpha)\%$ of the corresponding confidence intervals computed will include the true value of the parameter being estimated. Figure 9.1 shows an example of a confidence interval for a 90% confidence level of a normal distribution. In this case the shaded areas both contain $\frac{1}{2}(100 - 90)\% = 5\%$ of the area of the distribution. In general, if the probabilities above and below the interval are equal, the interval is called a *central interval*.

Confidence intervals are not uniquely defined by the value of the confidence level. In addition to the choice used in Figure 9.1, where the probabilities lead to a central interval, we could, for example, have chosen a symmetric interval about the mean, so that $(x_U - \mu)$ and $(\mu - x_L)$ were equal, or values of x_L and x_U that minimize $(x_U - x_L)$, although in practice the construction of confidence intervals that are shortest for a given confidence coefficient is difficult, or may not even be possible. For symmetric distributions like the normal, all three choices produce the same confidence intervals, but this is not true in general for asymmetric probability densities. The usual choice is the central interval.

Suppose we are interested in estimating a single parameter θ from an experiment consisting of n observations of a random variable x drawn from a probability density $f(x; \theta)$. The sample $x_1, x_2, \ldots, x_n$ is used to construct an estimator $\hat{\theta}(x_1, x_2, \ldots, x_n)$ for θ, by one of the methods discussed in earlier chapters or any other useful method. If $\hat{\theta}_e$ is the value of the estimator observed in the experiment, and $\hat{\sigma}_e$ is the estimate of its standard deviation, then the measurement would be given as $\hat{\theta} = \hat{\theta}_e \pm \hat{\sigma}_e$. The interpretation of this is that if repeated estimates are made, all based on n observations of the random variable x, they will be distributed according to some sampling distribution $g(\hat{\theta}; \theta)$ centred around the true value θ and with a true standard deviation σ_θ, that are estimated to be $\hat{\theta}_e$ and $\hat{\sigma}_e$. For most practical cases $g(\hat{\theta}; \theta)$ will be approximately normal for large samples. In this chapter, we take an alternate approach and find intervals about the estimator $\hat{\theta}$ such that we may make probabilistic statements concerning the probability of the true value θ being within the intervals.

A method that is applicable in many cases is the following. One finds, if possible, a function of the sample data and the parameter to be estimated, say u, which has a distribution independent of the parameter. Then a probability statement of the form

$$P[u_1 \leq u \leq u_2] = p$$

is constructed and converted into a probability statement about the parameter to be estimated. It is not always possible to find such a function, and when this occurs, more general methods (to be described in Section 9.2) must be used. For now, we illustrate the method using a distribution independent of the parameter by an example.

Example 9.1

A sample of size 100 is drawn from a population with unit variance but unknown mean μ. If $\hat{\mu}$ is estimated from the sample to be $\hat{\mu}_e = 1.0$, find a random interval for a confidence coefficient of 0.95.

Solution: Because the sample size is large, the quantity

$$u = \left(\frac{\hat{\mu}_e - \mu}{\sigma/\sqrt{n}} \right) = 10(\hat{\mu}_e - \mu),$$

is, in general, normally distributed with mean zero and unit variance, and so has a density function

$$f(u) = \frac{1}{\sqrt{2\pi}} \exp\left(-\frac{u^2}{2} \right),$$

which is independent of μ. The probability that u lies between any two arbitrary values u_1 and u_2 is thus

$$P[u_1 \leq u \leq u_2] = \int_{u_1}^{u_2} f(t)dt.$$

Then, from Table C.1 we can find values $u_1 = -u_2 = -1.96$ such that

$$P[-1.96 \leq u \leq 1.96] = \int_{-1.96}^{1.96} f(t)dt = 0.95.$$

Transforming back to the variable μ, this becomes

$$P[\hat{\mu}_e - 0.196 \leq \mu \leq \hat{\mu}_e + 0.196] = 0.95,$$

and since $\hat{\mu}$ is estimated from the sample to be $\hat{\mu}_e = 1.0$, we have

$$P[0.804 \leq \mu \leq 1.196] = 0.95.$$

This is the required confidence interval. Note that it is *not* correct to interpret this result as the probability is 0.95 that the true mean falls in the interval 0.804–1.196. The correct interpretation is as follows: If samples of size 100 were repeatedly drawn from the population, and if random intervals were computed as above for each sample, then 95% of those intervals would be expected to contain the true mean.

For obvious reasons, the intervals discussed above are called *two-tailed confidence intervals*. *One-tailed confidence intervals* are also commonly used. In these cases the confidence coefficients are defined by

$$C_U = P[x < x_U] = \int_{-\infty}^{x_U} f(x)\,\mathrm{d}x$$

if one is only interested in the upper limit of the variable, or

$$C_L = P[x > x_L] = \int_{x_L}^{\infty} f(x)\,\mathrm{d}x$$

if one is only interested in its lower limit. It is worth emphasizing that a central interval corresponding to a confidence level C is not the same as a one-tailed limit corresponding to the same value of C. For example, for a normal distribution, the upper limit of a 90% two-tailed central confidence interval has 95% of the distribution below it and 5% above, whereas for a one-tailed confidence interval, a 90% upper limit has 90% of the distribution below it and 10% above.

Example 9.2

Out of 1000 decays of an unstable particle, 9 are observed to be of type E. What can be said about the upper limit for the probability of a decay of this type?

Solution: The Poisson distribution is applicable here and we note that the probability p of an event of type E occurring is approximately $9/1000$. The Poisson parameter λ is then $1000 \times p = 9$ and we have $\mu = \sigma^2 = 9$. However, we also know that for $\mu \geq 9$ the Poisson distribution is well approximated by a normal distribution. Thus the quantity

$$u = (x - \mu)/\sigma = (x - 9)/3$$

is a standard normal variate. So, for example, from Table C.1,

$$P[u \leq 1.645] = 0.95,$$

and hence $x \leq 13.9$ and 14 is approximately the 95% confidence upper limit on the number of events of type E per 1000 events. Hence the upper limit for the probability of this type of decay is $P \leq 0.014$ with 95% confidence.

The concept of interval estimation for a single parameter may be extended in a straightforward way to include simultaneous estimation of intervals for several parameters. Thus, a $100(1 - 2\alpha)\%$ *confidence region* is a region constructed from the sample such that, for repeatedly drawn samples, $100(1-2\alpha)\%$ of the regions would be expected to contain the set of parameters under estimation.

It should be remarked immediately that confidence intervals and regions are essentially arbitrary, because they depend on what function of the observations is chosen to be an estimator. This is easily illustrated by reference to the normal distribution of Example 9.1. If we use the sample mean as an estimator of the population mean, then for a confidence coefficient of 0.95,

$$P\left[\bar{x}-\frac{1.96\sigma}{\sqrt{n}} \leq \mu \leq \bar{x}+\frac{1.96\sigma}{\sqrt{n}}\right] = 0.95 \qquad (9.1)$$

and the length of the interval is $2 \times 1.96\sigma/\sqrt{n}$. However, we could also use any given single observation to be an estimator, in which case the confidence interval would be $\sqrt{n}$ times as long. An important property of ML estimators is that, for large samples, they provide confidence intervals and regions that on average are smaller than intervals and regions determined by any other method of estimation of the parameters.

9.2 Confidence intervals: General method

The method used in Section 9.1 requires the existence of functions of the sample and parameters that are distributed independently of the parameters. This is its disadvantage, for in many cases such functions do not exist. However, for these cases there exists a more general method that we now describe.

Let $g\left(\widehat{\theta};\theta\right)$ be the sampling *pdf* of $\widehat{\theta}$, the estimator for samples of size n drawn from a population density $f(x;\theta)$ containing a parameter θ. Figure 9.2 shows a plot of $g\left(\widehat{\theta};\theta\right)$ as a function of $\widehat{\theta}$ for a given value of the true parameter θ. Also shown are two shaded regions that give the values of $\widehat{\theta}$ for which

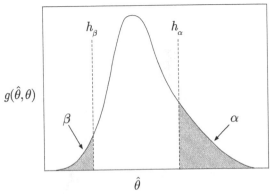

FIGURE 9.2 The density function of $g\left(\widehat{\theta};\theta\right)$ for a given value of the true parameter θ.

$$P\left[\widehat{\theta} \geq h_\alpha(\theta)\right] = \int_{h_\alpha(\theta)}^{\infty} g\left(\widehat{\theta}; \theta\right) d\widehat{\theta} = 1 - G(h_\alpha; \theta) = \alpha \qquad (9.2)$$

and

$$P\left[\widehat{\theta} \leq h_\beta(\theta)\right] = \int_{-\infty}^{h_\beta(\theta)} g\left(\widehat{\theta}; \theta\right) d\widehat{\theta} = G(h_\beta; \theta) = \beta, \qquad (9.3)$$

where G is the distribution function corresponding to the density $g\left(\widehat{\theta}; \theta\right)$. Thus, for a fixed value of θ, a $100(1 - \alpha - \beta)\%$ confidence interval for $\widehat{\theta}$ is

$$P\left[h_\alpha(\theta) \leq \widehat{\theta} \leq h_\beta(\theta)\right] = \int_{h_\alpha(\theta)}^{h_\beta(\theta)} g\left(\widehat{\theta}, \theta\right) d\widehat{\theta} = 1 - \alpha - \beta. \qquad (9.4)$$

Equations (9.2) and (9.3) determine the functions $h_\alpha(\theta)$ and $h_\beta(\theta)$. If the equations $\widehat{\theta} = h_\alpha(\theta)$ and $\widehat{\theta} = h_\beta(\theta)$ are plotted as a function of the true parameter θ, a diagram such as that shown in Figure 9.3 would result. The region between the two curves is called the *confidence belt*. A vertical line through any value of θ, say $\overline{\theta}$, intersects $h_\alpha(\theta)$ and $h_\beta(\theta)$ at the values $\widehat{\theta} = h_\alpha(\overline{\theta})$ and $\widehat{\theta} = h_\beta(\overline{\theta})$, which determine the $100(1 - \alpha - \beta)\%$ confidence interval for $\overline{\theta}$. Thus (9.4) gives the probability for the estimator to be within the belt, *regardless of the value of θ.*

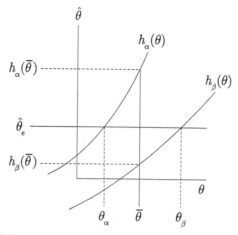

FIGURE 9.3   General method for constructing a confidence interval.

A horizontal line through some experimental value of $\widehat{\theta} = \widehat{\theta}_e$, corresponding to an estimate based on a sample of size n, cuts the curves at values $\theta_\alpha\left(\widehat{\theta}_e\right)$ and $\theta_\beta\left(\widehat{\theta}_e\right)$, where θ_α and θ_β are the values of the inverse functions $h_\alpha^{-1}\left(\widehat{\theta}\right)$ and $h_\beta^{-1}\left(\widehat{\theta}\right)$, respectively, evaluated at $\widehat{\theta}_e$.

By definition, given the experimental value $\widehat{\theta}_e$ of the estimator, θ_α is the value of θ for which the probability of an estimator being larger is α and θ_β is the value of θ which the probability of an estimator being smaller is β. That is, for any sample of size n, the probability that the resulting estimator will lie between θ_α and θ_β is $1 - \alpha - \beta$.

Thus, to construct a confidence interval for θ, an estimate $\widehat{\theta}_e$ from a sample of size n is first calculated. Then a horizontal line through $\widehat{\theta}_e$ to cut the curves at values $\theta_\alpha\left(\widehat{\theta}_e\right)$ and $\theta_\beta\left(\widehat{\theta}_e\right)$ is drawn, as shown in Figure 9.3, so that, by construction, the required confidence limit is

$$P\left[\theta_\alpha\left(\widehat{\theta}_e\right) \leq \theta \leq \theta_\beta\left(\widehat{\theta}_e\right)\right] = 1 - \alpha - \beta. \tag{9.5}$$

A confidence interval is often expressed by asymmetric error bars in the same way as the use of a standard deviation. Thus the result of the measurement would be written $\theta = \widehat{\theta}_e{}^{+d}_{-c}$, where $c = \widehat{\theta}_e - \theta_\alpha$ and $d = \theta_\beta - \widehat{\theta}_e$. If we are only interested in one-sided confidence intervals, then θ_α represents a lower limit on θ, that is, $P[\theta_a \leq \theta] = 1 - \alpha$, and similarly θ_β represents an upper limit with $P[\theta_\beta \geq \theta] = \beta$.

To find the curves $h_\alpha(\theta)$ and $h_\beta(\theta)$ may be a lengthy procedure. However, in some cases the needed values θ_α and θ_β may be obtained without knowing these curves. From (9.2) and (9.3), $a = \theta_\alpha$ and $b = \theta_b$ are solutions of the equations

$$\alpha = \int_{\widehat{\theta}_e}^{\infty} g\left(\widehat{\theta}, a\right) d\widehat{\theta} = 1 - G\left(\widehat{\theta}_e; a\right), \tag{9.6a}$$

and

$$\beta = \int_{-\infty}^{\widehat{\theta}_e} g\left(\widehat{\theta}, b\right) d\widehat{\theta} = G\left(\widehat{\theta}_e; b\right). \tag{9.6b}$$

So, if these equations can be solved (possibly numerically), the confidence interval results directly.

The general method given above can be extended to the case of confidence regions for the p parameters of the population $f(x; \theta_1, \theta_2, ..., \theta_p)$, that is, that region R in the parameter space such that

$$P\left[\widehat{\theta}_1, \widehat{\theta}_2, ..., \widehat{\theta}_p \text{ are contained in } R\right]$$

$$= \int_R ... \int g\left(\widehat{\theta}_1, \widehat{\theta}_1, ..., \widehat{\theta}_p; \theta_1, \theta_2, ..., \theta_p\right) \prod_{i=1}^{p} d\widehat{\theta}_i \qquad (9.7)$$

$$= 1 - \alpha - \beta.$$

This can be done assuming that the sampling distribution of the estimators is a multivariate normal distribution with a given covariance matrix. This will not be pursued further here except to say that the confidence region for two variables is approximately an ellipse, and for n variates is an n-dimensional ellipsoid.

Finally, we note that the method cannot be used to obtain confidence regions for a subset r of the p parameters in the density $f(x; \theta_1, \theta_2, ..., \theta_p)$, except for the case of large samples. This is discussed in Section 9.5 below.

9.3 Normal distribution

Because the normal distribution is very widely used in physical sciences, we will obtain specific confidence intervals for its parameters.

9.3.1 Confidence intervals for the mean

From (9.1), it is clear that a confidence interval for the mean μ cannot be calculated unless the variance σ^2 is known and so we will initially assume that this is the case. We will also assume, as usual, that the distribution of $\bar{x}$, the sample mean, is approximately normal with mean μ and standard deviation σ, that is, its sampling distribution function is

$$G(\bar{x}; \mu, \sigma) = \frac{1}{\sqrt{2\pi}\sigma} \int_{-\infty}^{\bar{x}} \exp\left[-\frac{1}{2}\left(\frac{\bar{x}' - \mu}{\sigma}\right)^2\right] d\bar{x}'.$$

Of course, if the distribution from which the sample is selected is normal, then as shown in Section 4.1.2, the distribution of $\bar{x}$ is exactly normal.

A confidence interval $[a, b]$ may be constructed if equations (9.6) can be solved. These are

$$\alpha = 1 - N(\bar{x}; a, \sigma) \text{ and } \beta = N(\bar{x}; b, \sigma),$$

where N is the standardized form of the normal distribution function. The solutions for a and b are

$$a = \bar{x} - \sigma N^{-1}(1 - \alpha) \qquad (9.8a)$$

and

$$b = \bar{x} + \sigma N^{-1}(1 - \beta), \tag{9.8b}$$

where N^{-1} is the inverse function of N, that is, the quantile of the standardized normal distribution function, and we have taken $N^{-1}(\beta) = -N^{-1}(1-\beta)$ for symmetry. The relationship between the inverse function and the confidence level is illustrated in Figure 9.4.

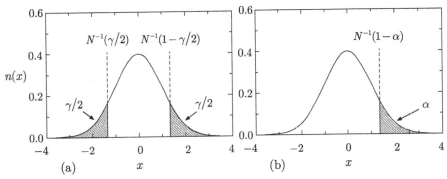

FIGURE 9.4 The standardized normal density $n(x)$ and the relationship between the inverse function N^{-1} and the confidence level for (a) a two-tailed central confidence level and (b) a one-tailed confidence level.

If we consider a central confidence interval so that $\alpha = \beta = \gamma/2$, a common choice for the interval is to use values such that $N^{-1}(1 -\gamma/2) = 1, 2, \ldots$. Similarly, for a one-sided interval we could choose $N^{-1}(1-\alpha) = 1, 2, \ldots$ Tables of the inverse function N^{-1} are published, but in practice only a few values are commonly used. The resulting confidence levels for these values are shown in Table 9.1(a). Alternatively, we could choose a convenient number for the confidence level itself and find the corresponding values of N^{-1}. Again, the commonly used values are shown in Table 9.1(b). The conventional 68.3% central confidence interval has $\alpha = \beta = \gamma/2$ with $N^{-1}(1 -\gamma/2) = 1$ and corresponds to 'one σ' errors bars.

TABLE 9.1(A)   Values of the confidence level for different values of the inverse of the standardized normal distribution N^{-1}: (A) for a central confidence interval with confidence level $(1 -\gamma)$ (see Figure 9.4(a)); and (B) a one-tailed confidence interval with confidence level $(1 -\alpha)$ (see Figure 9.4(b)).

(A) Central two-tailed		(B) One-tailed	
$N^{-1}(1 - \gamma/2)$	$1 - \gamma$	$N^{-1}(1 - \alpha)$	$1 - \alpha$
1	0.6826	1	0.8413
2	0.9546	2	0.9773
3	0.9974	3	0.9987

TABLE 9.1(B)   Values of the inverse of the standardized normal distribution N^{-1} for different values of the confidence level: (A) for a central confidence interval and $N^{-1}(1 - \gamma/2)$ (see Figure 9.4(a)); and (B) a one-tailed confidence interval with $N^{-1}(1 - \alpha)$ (see Figure 9.4(b)).

(A) Central two-tailed		(B) One-tailed	
$1 - \gamma$	$N^{-1}(1 - \gamma/2)$	$1 - \alpha$	$N^{-1}(1 - \alpha)$
0.90	1.645	0.90	1.282
0.95	1.960	0.95	1.645
0.99	2.575	0.99	2.327

If σ^2 is not known, then for large samples we could use an estimate $\hat{\sigma}^2$ for this quantity without significant loss of precision, but for small samples this procedure is not satisfactory. The solution is to use the quantity

$$t = \frac{\bar{x} - \mu}{(s^2/n)^{1/2}} = (\bar{x} - \mu)\left[\frac{1}{n(n-1)} \sum_{i=1}^{n} (x_i - \bar{x})^2\right]^{-1/2}, \tag{9.9}$$

which we have seen in Section 6.2 has a Student's t distribution with $(n-1)$ degrees of freedom, and only involves μ. Thus, with $f(t; n-1)$ the density function of the t distribution, we can find a number t_α such that

$$P[-t_\alpha \leq t \leq t_\alpha] = \int_{-t_\alpha}^{t_\alpha} f(t; n-1)dt = (1 - 2\alpha). \tag{9.10}$$

As in Example 9.1, we may now transform the inequality in (9.10) to give

$$P[\bar{x} - T_\alpha \leq \mu \leq \bar{x} + T_\alpha] = (1 - 2\alpha), \tag{9.11}$$

where

$$T_\alpha = t_\alpha \left[\frac{1}{n(n-1)} \sum_{i=1}^{n} (x_i - \bar{x})^2\right]^{1/2}.$$

The width of the interval is then $2T_\alpha$. The number t_α is called the $100\alpha\%$ *level of* t, and gives the point that cuts off $100\alpha\%$ of the area under the curve $f(t)$ on the upper tail.

Example 9.3

Using a sample size of 60, find: (a) a 95% central confidence interval for the mean of a normal distribution with unknown variance, given that the sample mean and sample variance are $\bar{x} = 5$ and $s^2 = 6$, respectively; and (b) an exact 95% central confidence interval using the Student's t distribution for the same statistics. Repeat the calculations for a sample size of 8 and comment on your results.

Solution: (a) For $n = 60$ we can use the normal approximation and the interval $[a, b]$ is then given by equations (9.8). Using Table 9.1(b), a 95% confidence interval is

$$\left[\left(\bar{x} - \frac{1.96 \times s}{\sqrt{n}}\right), \left(\bar{x} + \frac{1.96 \times s}{\sqrt{n}}\right)\right] = [4.38, \, 5.62]$$

and has a length 1.24. (b) For an exact confidence level we use the t distribution and (9.11). Then, using Table C.5, a 95% confidence interval is

$$\left[\left(\bar{x} - \frac{2.00 \times s}{\sqrt{n}}\right), \left(\bar{x} + \frac{2.00 \times s}{\sqrt{n}}\right)\right] = [4.36, \, 5.64]$$

which has a length 1.28, that is, slightly larger than the interval using the normal distribution, as expected. Repeating the calculation using a sample size of 8, the confidence intervals are [3.30, 6.70], with a length 3.40, in the normal approximation, and [3.00, 7.05] with a length 4.05, using the exact form from the t distribution. These differences are greater for $n = 8$ than for $n = 60$, because in the former, the small sample size means that the normal approximation is less accurate.

9.3.2 Confidence intervals for the variance

To find confidence intervals for the variance we use the χ^2 distribution. We know that the quantity

$$\chi^2 = \frac{1}{\sigma^2} \sum_{i=1}^{n} (x_i - \bar{x})^2 \tag{9.12}$$

has a χ^2 distribution with $(n-1)$ degrees of freedom, and so we can use it to find numbers χ_1^2 and χ_2^2 such that

$$P[\chi_1^2 \leq \chi^2 \leq \chi_2^2] = \int_{\chi_1^2}^{\chi_2^2} f(\chi^2; n-1) d\chi^2 = 1 - 2\alpha,$$

or, equivalently,

9. Interval estimation

$$P\left[\frac{1}{\chi_2^2}\sum_{i=1}^{n}(x_i - \bar{x})^2 \le \sigma^2 \le \frac{1}{\chi_1^2}\sum_{i=1}^{n}(x_i - \bar{x})^2\right] = 1 - 2\alpha. \tag{9.13}$$

Since the χ^2 distribution is not symmetric, the shortest confidence interval cannot be simply obtained for a given α. However, provided the number of degrees of freedom is not too small, a good approximation is to choose χ_1^2 and χ_2^2 such that $100\alpha\%$ of the area of $f(\chi^2)$ is cut off from each tail, that is, such that

$$\int_{\chi_1^2}^{\infty} f(\chi^2; n-1)d\chi^2 = 1 - \alpha,$$

and

$$\int_{\chi_2^2}^{\infty} f(\chi^2; n-1)d\chi^2 = \alpha.$$

Such numbers can easily be obtained from tables of the χ^2 distribution function.

Example 9.4

The following random sample was drawn from a normal distribution with variance σ^2:

$$10 \quad 11 \quad 13 \quad 13 \quad 12 \quad 13 \quad 10 \quad 14 \quad 12 \quad 12$$

Find an approximate 99% central confidence interval for σ^2.
Solution: This is found using (9.13). Firstly, we find the sample mean $\bar{x} = 12$, and hence

$$\sum_{i=1}^{10}(x_i - \bar{x})^2 = 16.$$

For an approximate central confidence interval, we need to find values of χ_1^2 and χ_2^2 such that equal areas are cut off from the upper and lower tails of the chi-squared distribution function. So for a 99% confidence level,

$$\int_{0}^{\chi_1^2} f(x, n-1)dx = 0.005 \text{ and } \int_{\chi_2^2}^{\infty} f(x, n-1)dx = 0.995,$$

where $f(x, n-1)$ is the chi-squared density for $n-1$ degrees of freedom. Using Table C.4, for $n = 9$, gives $\chi_1^2 = 1.73$ and $\chi_2^2 = 23.6$. Hence from (9.13) the interval is [0.68, 9.25], with width 8.57.

9.3.3 Confidence interval for a ratio of variances

In this section we will use the properties of the F-distribution, discussed in Section 6.3, to derive the confidence interval for a ratio R of the two variances σ_1^2 and σ_2^2 of two normal distributions. The random variable F involves this ratio in its definition, which may be written

$$F(n,m) = \frac{(s_1^2/\sigma_1^2)}{(s_2^2/\sigma_2^2)} = \frac{\sigma_2^2 s_1^2}{\sigma_1^2 s_2^2}, \tag{9.14}$$

where $n = n_1 - 1$ and $m = n_2 - 1$ are the degrees of freedom associated with the calculation of the sample variances s_1^2 and s_2^2, respectively. The confidence interval is then constructed by finding the percentage points $F_{1-\alpha}(n,m)$ and $F_\alpha(n,m)$ such that

$$P[F_{1-\alpha}(n,m) \leq F \leq F_\alpha(n,m)] = 1 - 2\alpha, \tag{9.15}$$

which may be manipulated, using (6.28) to give

$$P\left[\frac{s_2^2}{s_1^2}\frac{1}{F_\alpha(m,n)} \leq \frac{\sigma_2^2}{\sigma_1^2} \leq \frac{s_2^2}{s_1^2}F_\alpha(n,m)\right] = 1 - 2\alpha. \tag{9.16}$$

Example 9.5

Two random samples of sizes $n_1 = 12$ and $n_2 = 10$ are taken from a normal distribution and produced sample variances $s_1^2 = 3.1$ and $s_2^2 = 2.8$. Find a 90% confidence interval for the ratio s_2^2/s_1^2.

Solution: Using (9.16), the 90% confidence interval is given by

$$(2.8/3.1)F_{0.95}(11,9) \leq \frac{\sigma_2^2}{\sigma_1^2} \leq (2.8/3.1)F_{0.05}(11,9)$$

which may be evaluated using, from Table C.6,

$$F_{0.95}(11,9) = 1/F_{0.05}(9,11) = 1/2.90 = 0.34 \text{ and } F_{0.05}(11,9) = 2.90.$$

The confidence interval is thus (0.34, 2.9).

9.3.4 Confidence regions for the mean and variance

In constructing a confidence region for the mean and variance simultaneously we cannot use the region bounded by the limits of the confidence intervals obtained separately for μ and σ^2 (a rectangle in the (μ, σ^2) plane), because the quantities t of (9.9) and the sum of squares in (9.12) are not independently distributed, and hence the joint probability that the two intervals contain the true parameter values is not equal to the product of the separate probabilities. However, the distributions of $\bar{x}$ and $\sum(x_i - \bar{x})^2$ *are* independent and may be used to construct the required confidence region. Thus, for a $100(1-2\alpha)\%$ confidence region we may find numbers $a_i(i = 1, 4)$ such that

$$P\left[-a_1 \leq \left(\frac{\bar{x} - \mu}{\sigma/\sqrt{n}}\right) \leq a_2\right] = (1 - 2\alpha)^{1/2}, \tag{9.17a}$$

and

$$P\left[a_3 \leq \left(\frac{\sum\limits_{i=1}^{n}(x_i - \bar{x})^2}{\sigma^2}\right) \leq a_4\right] = (1 - 2\alpha)^{1/2}. \tag{9.17b}$$

The joint probability is then $(1 - 2\alpha)$ by virtue of the independence of the variables. The region defined by equations (9.17) will not, in general, be the smallest possible or a rectangle, but will not differ much from the minimum (which is roughly elliptical) unless the sample size is very small.

9.4 Poisson distribution

Another important distribution commonly met in physical science is the Poisson, that we discussed in Section 4.2.3. Recall that the probability of observing k events is given by the Poisson density (4.47),

$$f(k; \lambda) = \frac{\lambda^k}{k!}\exp(-\lambda), \lambda > 0, \ k = 0, 1, 2, \ldots \tag{9.18}$$

and λ is the mean of the distribution, that is, $\lambda = E[k]$. The aim is to construct a confidence interval for a single measurement $\widehat{\lambda}_e = k_e$. For values of $k_e \geq 9$ we can use the normal approximation to the Poisson, as we did in Example 9.2, but for smaller values we must use the exact form of the distribution. The general technique given in Section 9.2 is not directly applicable because for a discrete parameter the functions h_α and h_β that define the confidence corridor do not exist for all values of the parameter. For example, in the present case we would need to find values of h_α and h_β satisfying the conditions

$$P\left[\widehat{\lambda} \geq h_\alpha(\lambda)\right] = \alpha \ \text{ and } \ P\left[\widehat{\lambda} \geq h_\beta(\lambda)\right] = \beta$$

for all values of λ. But if α and β have fixed values, then because $\widehat{\lambda}_e$ only takes on the discrete values k_e, these inequalities hold only for particular values of λ. However, we can still construct a confidence interval $[a,b]$ by using equations (9.6). For discrete variables these become

$$\alpha = P\left[\widehat{\lambda} \geq \widehat{\lambda}_e; a\right] \ \text{ and } \ \beta = P\left[\widehat{\lambda} \leq \widehat{\lambda}_e; b\right] \tag{9.19a}$$

and for the case of a Poisson variable, using (9.18), they take the forms

$$\alpha = \sum_{k=k_e}^{\infty} f(k;a) = 1 - \sum_{k=0}^{k_e-1} f(k;a) = 1 - \sum_{k=0}^{k_e-1} \frac{a^k}{k!} e^{-a} \qquad (9.19b)$$

and

$$\beta = \sum_{k=0}^{k_e} f(k;b) = \sum_{k=0}^{k_e} \frac{b^k}{k!} e^{-b}. \qquad (9.19c)$$

For a given estimate $\widehat{\lambda} = k_e$, these equations may be solved numerically by iteration to yield values for a and b. Some values of the upper and lower limits obtained for a range of values of k_e are given in Table 9.2. Note that a lower limit is not obtainable if $k_e = 0$.

TABLE 9.2   Lower and upper limits for a Poisson variable for various observed values k_e.

k_e	Lower limit a			Upper limit b		
	$\alpha = 0.1$	$\alpha = 0.05$	$\alpha = 0.01$	$\beta = 0.1$	$\beta = 0.05$	$\beta = 0.01$
0	—	—	—	2.30	3.00	4.61
1	0.11	0.05	0.01	3.89	4.74	6.64
2	0.53	0.36	0.15	5.32	6.30	8.41
3	1.10	0.82	0.44	6.68	7.75	10.04
4	1.74	1.37	0.82	7.99	9.15	11.60
5	2.43	1.97	1.28	9.27	10.51	13.11
6	3.15	2.61	1.79	10.53	11.84	14.57
7	3.89	3.29	2.33	11.77	13.15	16.00
8	4.66	3.98	2.91	12.99	14.43	17.40
9	5.43	4.70	3.51	14.21	15.71	18.78
10	6.22	5.43	4.13	15.41	16.96	20.14

The interpretation of equation (9.19a) is that if $\lambda = a$, the probability of observing a value greater than or equal to the one observed is α. Likewise, if $\lambda = b$, the probability of observing a value less than or equal to the one observed is β. The confidence intervals for the mean are

$$P[\lambda \geq a] = \alpha, \quad P[\lambda \leq b] = \beta$$

and

$$P[a \leq \lambda \leq b] \geq 1 - \alpha - \beta.$$

An important case is when $k_e = 0$, that is, no events are observed. In this case (9.19c) becomes $\beta = \exp(-b)$, or $b = -\ln\beta$ and we have $P[\lambda \le -\ln\beta] = \beta$.

Example 9.6

How does the probability calculated in Example 9.2 change if no events were observed?

Solution: With less than about nine events, the normal approximation used in Example 9.2 is not appropriate and we have to use the Poisson distribution. If we still work at a confidence level of 95%, so that $\beta = 0.05$, the upper limit is obtained from (9.19c) is $b = -\ln(0.05) \approx 3$, as shown in Table 9.2. Thus, if the number of occurrences of a rare event follows a Poisson distribution with mean λ and no such event is observed, the 95% upper limit for the mean is 3. That is, if the true mean were 3, then the probability of observing zero events is 5%. So if no events were seen, the probability of the occurrence of a type E event is $P \le 0.003$ with 95% confidence.

9.5 Large samples

In Chapter 7 we have seen that the large-sample distribution of the ML estimator $\widehat{\theta}$ of a parameter θ in the density function $f(x; \theta)$ is approximately normal about θ as mean. In this situation approximate confidence intervals may be simply constructed. The method is, by analogy with Example 9.1, to convert an inequality of the form

$$P\left[-u_\alpha \le \frac{\widehat{\theta} - \theta}{\left(\mathrm{var}\widehat{\theta}\right)^{1/2}} \le u_\alpha\right] \simeq 1 - 2\alpha \tag{9.20}$$

for the distribution of $\widehat{\theta}$ expressed in standard measure, to an inequality for θ itself. Recall that α is defined by

$$\frac{1}{\sqrt{2\pi}} \int_{-u_\alpha}^{u_\alpha} \exp\left(-\frac{u^2}{2}\right) du = 1 - 2\alpha. \tag{9.21}$$

This will be illustrated by the following examples.

Example 9.7

Find an approximate 95% confidence interval for p, the parameter of the binomial distribution.

Solution: If we apply Equation (7.14) to a single experiment of n trials with r events using the binomial distribution of Equation (4.35) we find

$$\mathrm{var}\left(\widehat{\theta}\right) \equiv \widehat{\theta}^2 = \frac{p(1-p)}{n}. \tag{9.22}$$

An approximate $(1 - 2\alpha)$ confidence interval is then obtained from (9.20) by considering the statement

$$P\left[-u_\alpha \le \frac{\widehat{p}-p}{[\widehat{p}(1-\widehat{p})/n]^{1/2}} \le u_\alpha\right] \simeq 1 - 2\alpha, \tag{9.23}$$

which may be written as

$$P\left[\widehat{p}-u_\alpha\left\{\frac{\widehat{p}(1-\widehat{p})}{n}\right\}^{1/2} \le p \le \widehat{p}+u_\alpha\left\{\frac{\widehat{p}(1-\widehat{p})}{n}\right\}^{1/2}\right] \simeq 1 - 2\alpha. \tag{9.24}$$

So, using Table 9.1b, a 95% confidence interval for p is defined by

$$P\left[\widehat{p}-1.96\left(\frac{\widehat{p}(1-\widehat{p})}{n}\right)^{1/2} \le p \le \widehat{p}+1.96\left(\frac{\widehat{p}(1-\widehat{p})}{n}\right)^{1/2}\right] \simeq 0.95.$$

Example 9.8

Extend the work of Section 9.3.1 to the case of two normal populations $N(\mu_1,\sigma_1^2)$ and $N(\mu_2,\sigma_2^2)$ to derive a confidence interval for the difference $(\mu_1-\mu_2)$, where the variances are known.

Solution: Firstly, we note that the point estimator of $(\mu_1-\mu_2)$ is the difference of the sample means $\bar{x} = \bar{x}_1 - \bar{x}_2$. Thus, to obtain an estimate of $(\mu_1-\mu_2)$, we take random samples of size n_1 and n_2, respectively, from the two distributions. Then $\bar{x}$ is normally distributed with mean $\bar{\mu} = \mu_1 - \mu_2$ and variance

$$\widehat{\sigma}^2(\bar{x}) = \sigma_1^2/n_1 + \sigma_2^2/n_2.$$

So a $(1-2\alpha)\%$ confidence interval for the standardized variable $z = (\bar{x}-\bar{\mu})/\widehat{\sigma}(\bar{x})$ is $-z_\alpha < z < z_\alpha$, which after rearranging is

$$\bar{x} - z_\alpha\widehat{\sigma}(\bar{x}) < (\mu_1 - \mu_2) < \bar{x} + z_\alpha\widehat{\sigma}(\bar{x}).$$

As shown in Chapter 6, when the variances are unknown but equal, the z statistic is replaced by a Student's t statistic of $n_1 + n_2 - 2$ degrees of freedom and the pooled sample variance given in (6.21) replaces $\widehat{\sigma}^2(\bar{x})$. If the variances are not equal, or nearly equal, the t statistic does not apply and the problem is much more difficult. This case is examined in detail in volume 2 of Kendall and Stuart, cited in the Bibliography.

Example 9.9

If a national physics examination is taken by a sample of 40 women and 80 men, and produces average marks of 60 and 70, with standard deviations of 8 and 10, respectively, use the result of Example 9.8 to construct a 95% confidence interval for the difference in the marks for all men and women eligible to take the examination.

Solution: To apply this to the data, we set (m = men, w = women). Then

$$n_m = 80, \ n_w = 40, \ \bar{x} = 10, \ \sigma_m^2/n_m = 1.25, \ \sigma_w^2/n_w = 1.60,$$

giving a 95% confidence level $6.69 < \mu_m - \mu_w < 13.31$.

When the variances σ_1^2 and σ_2^2 of two normal distributions are unknown, a confidence interval for their ratio σ_1^2/σ_2^2 can be constructed using the method described in Section 9.3.3. If samples of size n_1 and n_2, with sample variances s_1^2 and s_2^2, respectively, are taken from the two distributions, then the statistic $F = s_1^2\sigma_2^2/s_2^2\sigma_1^2$, where the numerator has $(n_1 - 1)$ degrees of freedom and the denominator has $(n_2 - 1)$ degrees of freedom, has the $F(n_1 - 1, n_2 - 1)$ distribution. Thus a $(1 - 2\alpha)\%$ confidence interval for the ratio σ_1^2/σ_2^2 is

$$\frac{s_2^2}{s_1^2}F_{1-\alpha}(n_1 - 1, n_2 - 1) \leq \frac{\sigma_2^2}{\sigma_1^2} \leq \frac{s_2^2}{s_1^2}F_\alpha(n_1 - 1, n_2 - 1). \tag{9.25}$$

The above method may be extended to confidence regions. In terms of the matrix M_{ij}, defined in Equation (7.34), we know that

$$\chi^2 = \sum_{i=1}^{p} \sum_{j=1}^{p} \left(\hat{\theta}_i - \theta_i\right) M_{ij} \left(\hat{\theta}_j - \theta_j\right), \tag{9.26}$$

is approximately distributed as χ^2 with p degrees of freedom. So, just as we used (9.15) for the normal distribution, we can use the α percentage points of the χ^2 distribution to set up a confidence region for the parameters θ_i. It is an ellipsoid with center at $(\theta_1, \theta_2, ..., \theta_p)$.

At the end of Section 9.2 we remarked that it was not possible, in general, to obtain a confidence region for a subset of the p parameters for samples of arbitrary size. However, for large samples this *is* possible. If we wish to construct a region for a subset of r parameters with $(r < p)$, then the elements of the matrix M'_{ij} analogous to M_{ij} above are given by

$$M'_{ij} = (V')^{-1}_{ij}, \tag{9.27}$$

and the quadratic form

$$\chi'^2 = \sum_{i=1}^{r} \sum_{j=1}^{r} \left(\hat{\theta}_i - \theta_i\right) M'_{ij} \left(\hat{\theta}_j - \theta_j\right) \tag{9.28}$$

is then approximately distributed as χ^2 with r degrees of freedom and will define an ellipsoid in the $\theta_i(1, 2, ..., r)$ space.

9.6 Confidence intervals near boundaries

In the discussion of point estimation in Chapters 7 and 8, it was implicitly assumed that an individual measurement could take on any value.[1] However, this assumption is not always true. An example often cited is that of the mass of a body, which cannot be negative. If the mass of a body is obtained by a direct measurement, for example by weighing it, then this condition is automatically satisfied. But a direct measurement may not always be possible. For example, in the case of a subatomic particle, it is usual to measure its energy E and momentum p. The mass m is then found from the general kinematical relation $m^2c^4 = E^2 - p^2c^2$, where c is the speed of light. But in this situation E and p are both random variables with associated uncertainties, so that even though both will be positive, the resulting experimental value of the squared mass, being the difference of two terms, can be, and sometimes is, found to be negative. In general, a measurement of any quantity θ that is known to be positive for physical reasons, when found from the differences of random variables, can result in a negative value for a specific measurement of its estimator $\widehat{\theta}$. In these circumstances, the construction and interpretation of confidence intervals must be treated with care if one is not to end up by making misleading probability statements. Similar difficulties occur whenever a parameter has a physical boundary, but the method of estimation allows its estimator to take values in unphysical regions. An example will illustrate this.

Consider the simple case where an estimator $\widehat{\theta}$ of a parameter θ, known for physical reasons to be nonnegative, is given in terms of two independent random variables x and y by $\widehat{\theta} = x - y$. If x and y are both normally distributed with means μ_x and μ_y and variances σ_x^2 and σ_y^2, respectively, we know from the work of previous chapters that $\widehat{\theta}$ is also normally distributed, with mean $\theta = \mu_x - \mu_y$ and variance $\sigma_\theta^2 = \sigma_x^2 + \sigma_y^2$. If now an experiment gives a value $\widehat{\theta}_e$ for $\widehat{\theta}$, then the upper limit for θ at a confidence level of $(1 - \beta)$ is obtained from (9.8b) and is

$$\theta_{up} = \widehat{\theta}_e + \sigma_\theta N^{-1}(1 - \beta). \tag{9.29a}$$

For example, if $\widehat{\theta}_e$ is measured to be -3.0 with $\sigma_\theta = 1.5$, where the latter is either known or estimated from the data, and we use a 95% confidence interval, with $N^{-1}(0.95) = 1.645$ (see Table 9.1b) then from (9.29a), $\theta_{up} = -0.532$ The interval $[-\infty, -0.532]$ therefore will contain, by construction, the true value θ with a probability of 95%, regardless of the actual value of θ. Although this may look odd at first sight, there is nothing intrinsically wrong with this. If the true value of θ were zero, half of such estimates would be expected to be negative. But the upper limit is also in the unphysical region. Again, we would expect this for 5% of similar experiments if θ really were zero. There is nothing incorrect with the procedure; we have simply encountered an experiment that does not lie within the interval constructed by applying the frequency definition of probability. So, unless there are other compelling reasons for doing so, the data should certainly not be discarded as being 'wrong'. Nevertheless,

---

[1] The constraints discussed in the context of the least-squares method were on combinations of data, or the parameters used in the fitting procedure.

since we know that θ cannot be negative, the measurement has not added to our prior knowledge in any significant way, so the question arises as to whether this single estimate can be better used.

Unfortunately, there is no unique answer to this question. The first possibility is to do nothing except to report the measurement. Other experiments will produce different values of $\widehat{\theta}_e$ and by combining them (e.g., by combining the likelihood functions for each experiment, as mentioned in Chapter 7, or by using the least-squares method as in Section 8.1.6), a more precise overall estimate $\widehat{\theta}$ may be found. A second possibility is to increase the confidence limit until the upper limit enters the physical region. Using the example above, if the confidence limit is increased to 99%, with $N^{-1}(0.99) = 2.327$ from Table 9.1b, then from (9.29a), $\theta_{up} = 0.490$. Although this is comfortably greater than zero, it could be smaller than the precision of the experiment as measured by σ_θ, and in this example it is because we took $\sigma_\theta = 1.5$. An extreme example of the difficulty that this strategy can lead to is if a confidence level is deliberately chosen so that θ_{up} is only just positive. Thus, if we choose a confidence level of 97.725%, we would quote the value $\theta_{up} = 1.5 \times 10^{-5}$ at this confidence level, which is clearly absurd. A third possibility is to move a negative value of $\widehat{\theta}_e$ to zero before using (9.29a) to calculate θ_{up}, so that

$$\theta_{up} = \max\left(\widehat{\theta}_e, 0\right) + \sigma_\theta N^{-1}(1 - \beta). \tag{9.29b}$$

For the example above, this gives $\theta_{up} = 2.468$, which is both in the physical region and compatible with the precision of the experiment. The drawback with this method is that we can no longer interpret the computed interval as a range that will include the true value with a probability $(1 - \beta)$. The actual probability will always be greater, because the value of θ_{up} from (9.29b) will always be greater than the value calculated from (9.29a) when $\widehat{\theta}_e$ is negative.

9.7 Bayesian confidence intervals

The strategies outlined in Section 9.6 to handle the problem of confidence intervals when the estimated value of a parameter is close to a boundary all use the frequency interpretation of probability. A final possibility is to incorporate our prior knowledge, including the fact that the parameter cannot have certain values, by using subjective probability, leading to so-called Bayesian intervals. Although such intervals may look similar to the confidence intervals discussed previously, because they are based on posterior probability densities their interpretation is very different, and for this reason they are usually called *probability intervals* or *credible intervals*, to distinguish them from confidence intervals constructed using the frequency interpretation of probability. In the latter, x is a random variable and gives rise to a random interval that has a specific probability of containing the fixed, but unknown, value of the parameter θ. In the Bayesian approach, the parameter is random in the sense that our belief about its value is expressed in terms of probabilities, and the interval can be thought of as fixed once this information is available. Only if the prior distribution of the unknown parameter is chosen to be a uniform distribution are the two intervals equivalent.

The starting point for constructing credible intervals is Bayes' theorem, which we first introduced for discrete variables in equation (2.15) and generalized to the case of continuous variables in equation (3.22c). In the present context it is more convenient to write Bayes' theorem in terms of the likelihood function, as was done in Section 8.4.3. Thus for a single variable θ, if $L(\mathbf{x}; \theta)$ is the likelihood function of the set of n variables $\mathbf{x}(x_1, x_2, ..., x_n)$ for a given value of θ, that is,

$$L(\mathbf{x}|\theta) = \prod_{i=1}^{n} f(x_i; \theta),$$

where $f(x; \theta)$ is the density function of the variables $\mathbf{x}$, then rewriting (8.88), the posterior probability density $f_{\text{post}}(\theta|\mathbf{x})$ is given by

$$f_{\text{post}}(\mathbf{x}|\theta) = \frac{L(\mathbf{x}|\theta)f_{\text{prior}}(\theta)}{\int L(\mathbf{x}|\theta')f_{\text{prior}}(\theta')d\theta'}, \qquad (9.30)$$

where $f_{\text{prior}}(\theta)$ is the prior probability density for θ. The density $f_{\text{post}}(\theta|\mathbf{x})$ replaces the distributions (usually the normal) assumed in previous sections and can be used to construct an interval $[a, b]$ such that for any given probabilities α and β,

$$\alpha = \int_{-\infty}^{a} f_{\text{post}}(\theta|\mathbf{x})d\theta \quad \text{and} \quad \beta = \int_{b}^{\infty} f_{\text{post}}(\theta|\mathbf{x})d\theta.$$

Thus $\alpha = \beta$ gives a central interval with a predetermined probability $(1 - \alpha - \beta)$; alternatively, one could choose $f_{\text{post}}(a|\mathbf{x}) = f_{\text{post}}(b|\mathbf{x})$, which leads to the shortest interval and is what is usually used in practice. The advantage of the subjective approach over the frequency approach is that in principle prior knowledge can be incorporated via the density $f_{\text{prior}}(\theta)$. The credible interval then contains a fraction of one's total belief about the parameter, in the sense that one would be prepared to bet, with well-defined odds that depend on α and β, that the true value of θ lies in the interval. The qualifier 'in principle' is necessary because as in all applications of subjective probability the problem arises in choosing a form for $f_{\text{prior}}(\theta)$. An example of the construction of a credible interval is given in Problem 9.7.

For the case discussed in Section 9.6, where $\theta > 0$, we can certainly set $f_{\text{prior}}(\theta) = 0$ for $\theta \leq 0$. Then using (9.26), the upper limit is given by

$$1 - \beta = \int_{-\infty}^{\theta_{\text{up}}} f_{\text{post}}(\theta|\mathbf{x})d\theta$$

$$= \int_{0}^{\theta_{\text{up}}} L(\theta|\mathbf{x})f_{\text{prior}}(\theta)d\theta \left[\int_{0}^{\infty} L(\theta|\mathbf{x})f_{\text{prior}}(\theta)d\theta\right]^{-1}.$$

But for $\theta > 0$, it is not so clear what to do. If we invoke Bayes' postulate, we would choose

$$f_{prior}(\theta) = \begin{cases} 0 & \theta \le 0 \\ 1 & \theta > 0 \end{cases}.$$ (9.31)

While this has the advantage of simplicity, it also has a serious problem. Continuing with our example of identifying θ with a mass, usually one would have some knowledge of at least its order of magnitude from physical principles. If the body were an atom, for example, we would expect to find values of the order of 10^{-22} g, and it would be unrealistic to assume that the probabilities of obtaining *any* positive value were all equal. Other forms have been suggested for $f_{prior}(\theta)$, but none are without their difficulties. Moreover, Bayes' postulate applied to different *functions* of θ result in different credible intervals, that is, the method is not invariant with respect to a nonlinear transformation of the parameter. Despite these limitations, in practice the simple form (9.31) is often used.

Problems 9

9.1 Potential voters in an election are asked whether they will vote for candidate X. If 700 out of a sample of 2000 indicate that they would, find a 95% confidence interval for the fraction of voters who intend to vote for candidate X.

9.2 A signal of initial strength S is sent between two points. Between the two points it is subject to random noise N, known to be normally distributed with mean zero and variance 5, so that the received signal R has strength $(S + N)$. If the signal is sent 10 times with the results:

6   7   11   15   12   8   9   14   5   13

construct (a) a 95% central confidence interval and (b) 95% upper and lower bounds, for S.

9.3 Rework Problem 9.2(a) for the case where the variance of the noise is unknown.

9.4 In the study of very many decays of an elementary particle, eight events of a rare decay mode are observed. If the expected value of events due to other random processes is 2, what are the 90% confidence limits for the actual number of decays?

9.5 Electrical components are manufactured consecutively and their effective lifetimes τ are found to be given by an exponential distribution with a common parameter λ. Construct a 90% confidence interval for the population lifetime, given that the sample mean for the first 15 components is 200 hours. Use the fact that the random variable

$2\lambda n\bar{\tau}$, where $\bar{\tau}$ is the sample mean lifetime of n components, has the χ^2 distribution with $2n$ degrees of freedom.

9.6 A national physics examination is taken by a sample of 40 women and 80 men, and produces average marks of 60 and 70, with known standard deviations of 8 and 10, respectively. Following Example 9.8, construct a 95% confidence interval for the difference in the marks for all men and women eligible to take the examination.

9.7 Electrical components are manufactured with lifetimes that are approximately normally distributed with a standard deviation of 15. Prior experience suggests that the lifetime is a normal random variable with a mean of 800 hours and a standard deviation of 10 hours. If a random sample of 25 components has an average lifetime of 780 hours, use the results given in Section 9.7 to find a 95% credible interval for the mean and compare it with a 95% confidence interval.

9.8 All school leavers applying for a place at a university take a standard national English language test. At a particular school, the test was taken by 16 boys and 13 girls. The average mark of the boys was 75 with a standard deviation of 8, and the corresponding numbers for the girls were 80 and 6. Assuming a normal distribution, find a 95% confidence interval for the ratio σ_b^2/σ_g^2, where $\sigma_{b,g}^2$ are, respectively, the variances of all the boys and girls who took the test nationally.

9.9 Solve Problem 9.6 in the case the samples are 10 women and 15 men and the given standard deviations are those of the samples, not the populations. The population variances are equal but unknown.

9.10 The heights of 20 students were measured to the nearest centimeter with the following results:

163	9176	166	169	170	166	163	173	172	160
182	175	166	175	168	170	174	164	173	177

Assuming the heights are distributed normally, find the simultaneous 0.95% confidence region for the mean and the variance of the students' heights.

9.11 Two chemicals, A and B, combine in equal parts in a manufacturing process to form a liquid product AB. Small quantities of impurities in the process combine with a small fraction of the reactants to produce a substance C which binds to the equipment surfaces and is difficult to detect. The substance C can damage the equipment if enough of it builds up over time, so it is important to determine how much of C is likely to be produced in each manufacturing run. 10 batches of each of A and B are reacted to form 10 batches of AB. The differences between sum of the masses of A and B and the mass of AB are measured for each batch and found to be, in micrograms:

$$-0.974 \quad 1.714 \quad -0.421 \quad 0.191 \quad 0.375 \quad -0.291 \quad -0.918 \quad 1.099 \quad 0.811 \quad -1.503$$

Find the symmetric 95% confidence interval for the mean mass of the substance C. What is the confidence interval if the negative values are set to zero or ignored?

Hypothesis testing I: Parameters

Overview

Another important branch of statistics as applied to physical science is testing hypotheses. This chapter starts with a general discussion of statistical hypotheses and illustrates the general technique of hypothesis testing using a simple example. This is then extended to consider more general cases, which are addressed by the use of the method of likelihood ratios. There is a section on specific tests for the parameters of the normal distribution and a brief discussion of how to test hypotheses concerning the parameters of the binomial and Poisson distributions. This is followed by a short discussion of the technique called 'analysis of variance', which is an extension of the earlier methods for comparing two means to the case of many means. The chapter ends with a brief section on the Bayesian approach to hypothesis testing.

In earlier chapters we have discussed one of the two main branches of statistical inference as applied to physical science: estimation. We now turn to the other main branch: hypothesis testing. This is a large topic and so for convenience it has been split into two parts. In this chapter we consider situations where the parent distribution is known, usually a normal distribution, either exactly or approximately, and the aim is to test hypotheses about its parameters, for example whether they have certain values, rather than to estimate the values of the parameters. In Chapter 11, we consider the second part of the topic, which concerns other types of hypotheses, such as whether a sample of data is compatible with an assumed distribution, or whether a sample really is random. We also discuss in Chapter 11 hypotheses that can be applied to situations that are occasionally met in physical science where the data are nonnumeric.

The topic of hypotheses about parameters was touched upon in previous chapters, particularly Chapter 6 where we discussed the use of the χ^2, t, and F distributions, and much of the preliminary work has been done in Chapter 9 where confidence intervals were constructed. The aim of the present chapter is to bring together and extend those ideas to discuss hypothesis testing on parameters in a systematic way.

We approach hypothesis testing for parameters using a precise definition in terms of likelihood ratios. Explanation of the application of the general idea of hypothesis testing is carried out through examples, starting with simple hypotheses, and continuing through complicated composite hypotheses. Occasionally hypotheses concerning the question of whether more than two samples are drawn from the same population arise in the physical sciences so the

Probability and Statistics for Physical Sciences, Second Edition
https://doi.org/10.1016/B978-0-443-18969-2.00010-6

method of analysis of variance is presented in this chapter for relatively simple situations. Hypothesis testing for binomial and Poisson distributions are also encountered in the physical sciences, although not as commonly as for the normal distribution. Application of the ideas explained in detail for normal distributions is briefly discussed for these two distributions.

While most of the discussion will use the frequency approach to probability, as used in earlier chapters, there is also a brief discussion of the Bayesian approach. This is very similar from a calculational viewpoint to the frequency approach but uses the appropriate posterior probability distributions as defined in Chapter 9, and hence, by analogy with Bayesian confidence intervals, the interpretation is different.

10.1 Statistical hypotheses

Consider a set of random variables defining a sample space S of n dimensions. If we denote a general point in the sample space by E, then if R is a region in S, any hypothesis concerning the probability that E falls within R, that is, $P[E \in R]$, is called a *statistical hypothesis*. Furthermore, if the hypothesis determines $P[E \in R]$ completely, then it is called *simple*, otherwise it is called *composite*. For example, when testing the significance of the mean of a sample, it is a statistical hypothesis that the parent population is normal. Furthermore, if the parent population is postulated to have mean μ and variance σ^2, then the hypothesis is simple, because the density function is then completely determined.

The hypothesis under test is called the *null* hypothesis and denoted H_0. The general procedure for testing H_0 is as follows. Assuming the hypothesis to be true, we can find a region S_c in the sample space S such that the probability of E falling in S_c is any preassigned value α, called the *significance level*. The region $S_0 = (S - S_c)$ is called the *region of acceptance*, and S_c is called the *critical region*, or *region of rejection*. If the observed event E falls in S_c we reject H_0, otherwise we do not reject it. The general idea is to choose S_c so that α is small. Thus, the probability of the observed E falling in S_c is small when H_0 is true. In this case, when E does fall in S_c, the evidence against H_0 being true is strong. It is usually said that when E falls in S_0 that H_0 is accepted. This is not precisely correct. It is more precise to say that H_0 is not rejected because the evidence against it is not strong, or equivalently, the data are *not inconsistent* with the hypothesis, in the sense that the observed value falling in S_0 would be expected $100(1 - \alpha)\%$ of the time if the null hypothesis were true. In practice, as we shall discuss below, the critical region is determined by a statistic, the nature of which depends upon the hypothesis to be tested.

Just as there are many confidence intervals for a given confidence level, so there are many possible acceptance regions for a given hypothesis at a given significance level α. For all of them the hypothesis will be rejected, although true, in some cases. Such 'false negatives' are called *type I errors* and their probability, denoted by $P[\text{I}]$, is equal to the significance level of the test. The value of α is arbitrary, and the choice of a suitable value depends on how important the consequence of rejecting H_0 is. Thus, if its rejection could have serious consequences, such as a substantial loss, either of money, time, etc., or even lives in extreme cases, then one would tend to be conservative and choose a value $\alpha = 0.05, 0.01$, or even smaller. Values commonly used in physical science are 0.1 and 0.05. It is also possible that even though the hypothesis is false, we fail to reject it. These 'false positives' are called *type II errors*. We are led to the following definition of *error probabilities*.

Consider a parameter θ and two hypotheses, the null hypothesis $H_0 : \theta \in R_0$ and the alternative $H_a : \theta \in R_a$, where R_0 and R_a are two mutually exclusive and exhaustive regions of the parameter space. Further, let S_0 and S_a be the acceptance and critical regions of the sample space S associated with the event $E \equiv (x_1, x_2, ..., x_n)$, assuming H_0 to be true. Then the probability of a type I error is

$$P[\text{I}] = P[E \in S_a | H_0 : \theta \in R_0] = \alpha, \tag{10.1}$$

and, if H_0 is false, but H_a is true, the probability of a type II error is

$$P[\text{II}] = P[E \in S_0 | H_a : \theta \in R_a] = \beta, \tag{10.2}$$

where we follow the usual convention in the statistics literature by denoting $P[\text{II}]$ as β. The two types of error are inversely related; for a fixed sample size, one can only be reduced at the expense of increasing the other. A useful quantity related to β, called the *power*, can be used to compare the relative merits of two tests. It is defined as

$$power = P[E \in S_a | H_a : \theta \in R_a] \tag{10.3}$$

and so is the probability of rejecting the hypothesis when it is false. Clearly an acceptable test should result in a power that is large. From (10.2) and (10.3) it follows that

$$power = 1 - P[\text{II}] = 1 - \beta. \tag{10.4}$$

To see how these definitions are used in practice we will look at a simple example concerning a normal population with unknown mean μ and known variance σ^2. We will test the null hypothesis $H_0 : \mu = \mu_0$ against the alternative $H_a : \mu \neq \mu_0$, where μ_0 is a constant, using a sample of size n. Since the arithmetic mean is an unbiased estimator for μ, it is reasonable to accept the null hypothesis if $\bar{x}$ is not too different from μ_0, so the critical region for the test is determined by the condition

$$P[|\bar{x} - \mu_0| > c : H_0] = \alpha,$$

where H_0 indicates that the probability is calculated under the assumption that H_0 is true, that is, that the mean is μ_0. If H_0 is true, then $\bar{x}$ is normally distributed with mean μ_0 and variance σ^2/n, so that the quantity $z = (\bar{x} - \mu_0)/(\sigma/\sqrt{n})$ has a standard normal distribution, and thus

$$P[z > c\sqrt{n}/\sigma] = \alpha/2.$$

However, we know that

$$P[z > z_{\alpha/2}] = \alpha/2,$$

and so $c = z_{\alpha/2}\sigma/\sqrt{n}$. Thus, at the significance level α, we conclude that

$$\frac{\sqrt{n}}{\sigma}|\bar{x} - \mu_0| > z_{\alpha/2} \Rightarrow H_0 \text{ is rejected} \tag{10.5a}$$

and

$$\frac{\sqrt{n}}{\sigma}|\bar{x} - \mu_0| \leq z_{\alpha/2} \Rightarrow H_0 \text{ is accepted.} \tag{10.5b}$$

This is shown in Figure 10.1 and for obvious reasons this is called a *two-tailed test*.

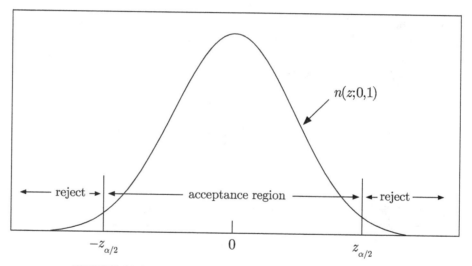

FIGURE 10.1   Two-tailed test for $H_0 : \mu = \mu_0$ against $H_a : \mu \neq \mu_0$.

The decision of whether to accept a null hypothesis using a test with a fixed value of α often depends on very small changes in the value of the test statistic. It is therefore useful to have a way of reporting the result of a significance test that does not depend on having to choose a value of the significance level beforehand. This may be done by calculating the so-called *p-value* of the test. It may be defined in several ways. For example: the smallest level of significance which would lead to rejection of the null hypothesis using the observed sample; or the probability of observing a value of the test statistic that contradicts the null hypothesis at least as much as that computed from the sample. Thus H_0 will be accepted (rejected) if the significance level α is less than (greater than or equal to) the *p*-value.

Example 10.1

A manufacturer aims to produce power packs with an output O of 10 watts and a standard deviation of 1 watt. To control the quality of its output, 20 randomly chosen power packs are tested and found to have an average value of 9.5 watts. Test the hypothesis $H_0 : O = 10$ against the alternative $H_a : O \neq 10$ at a 5% significance level. What is the lowest significance level at which H_0 would be accepted?

Solution: If we assume we can use the normal distribution, then the test statistic is

$$z = \frac{|\bar{x} - O|}{\sigma/\sqrt{n}} = 2.236,$$

if H_0 is true. This value is to be compared with $z_{0.025}$ because we are using a two-tailed test. From Table C.1, $z_{0.025} = 1.96$ and since $z > z_{0.025}$, the null hypothesis must be rejected at the 5% significance level, and the critical region from (10.5a) is

$$\bar{x} < 9.56 \quad \text{and} \quad \bar{x} > 10.44.$$

Notice that even though the lower of these values is very close to the observed mean, we are forced to reject H_0 because we have set the value of α before the test. The p-value for the sample is given by

$$p = P[|z| > 2.236] = 2P[z > 2.236] = 0.0254,$$

using Table C.1. Thus for $\alpha > 0.0254$, H_0 would still be rejected, but for values lower than this it would not be rejected.

We also consider type II errors, because if the probability of these is large, the test may not be useful. In principle this is not an easy question to address. In the present example, each value of $\mu \neq \mu_0$ leads to a different sampling distribution, and hence a different value of P[II]. However, in practice what can be done is to find a curve, that for a given value of α, gives the probability of a type II error for any value of $(\mu - \mu_0)$. This is called the *operating characteristic (OC) curve*. From the definition (10.2), and using the fact that

$$z = \frac{\bar{x} - \mu}{\sigma/\sqrt{n}}$$

is approximately distributed as standard normal distribution, it is straightforward to show that the OC curve is given by

$$P[\text{II}, \mu] = \beta(\mu) = N\left(\frac{\mu_0 - \mu}{\sigma/\sqrt{n}} + z_{\alpha/2}\right) - N\left(\frac{\mu_0 - \mu}{\sigma/\sqrt{n}} - z_{\alpha/2}\right) \tag{10.6}$$

and is symmetric about the point where $(\mu - \mu_0)$ equals zero.

Figure 10.2(a) shows a plot of (10.6) as a function of μ for the data of Example 10.1. For this example, the maximum of the OC curve is at $\mu = 10$, where $(1 - \alpha) = 0.95$. The curve shows that the probability of making a type II error is larger when the true mean is close to the value μ_0, and decreases as the difference becomes greater, but if $\mu \approx \mu_0$, a type II error will presumably have smaller consequences. The associated power curve $1 - P[\text{II}; \mu]$ is given in Figure 10.2(b). This shows that the power of the test increases to unity as the true mean gets further away from the H_0 value, which is a statement of the fact that it is easier to reject a hypothesis as it gets further from the truth. Decreasing α with a fixed sample size, reduces the power of the test, whereas increasing the sample size produces a power curve with a sharper minimum and will increase the power of the test, except at $\mu = \mu_0$. Once α and the sample size n are chosen, the size of the type II error is determined, but we can also use the OC curve to calculate the sample size for a fixed α that gives a specific value for the type II error. From (10.6), this is equivalent to

$$P[\text{II}; \mu_n] = N\left(\frac{\mu_0 - \mu_n}{\sigma/\sqrt{n}} + z_{\alpha/2}\right) - N\left(\frac{\mu_0 - \mu_n}{\sigma/\sqrt{n}} - z_{\alpha/2}\right),$$

where $P[\text{II}; \mu_n]$ is the probability of accepting H_0 when the true mean is μ_n. Given $P[\text{II}; \mu_n]$ and α, a numerical solution for n may be obtained from this relation.

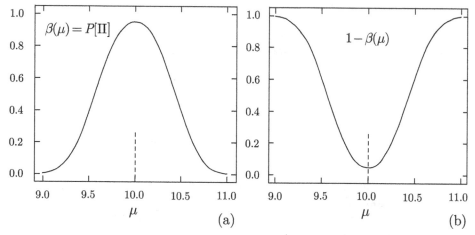

FIGURE 10.2   (a) Operating characteristic curve as a function of μ and (b) power curve, both for the data for Example 10.1.

10.2 General hypotheses: Likelihood ratios

Example 10.1 exhibits all the steps necessary for testing a hypothesis. In this section we turn to tests derived from a consideration of likelihood ratios, starting with the simplest of all possible situations, that of a simple hypothesis and one simple alternative. This case is not very useful in practice, but serves as an introduction to the method.

10.2.1 Simple hypothesis: One simple alternative

The likelihood ratio λ for a sample $x_i (i = 1, 2, ..., n)$ of size n having a density $f(x, \theta)$ is defined by

$$\lambda \equiv \frac{\prod_{i=1}^{n} f(x_i; \theta_0)}{\prod_{i=1}^{n} f(x_i; \theta_a)} = \frac{L(\theta_0)}{L(\theta_a)}. \tag{10.7a}$$

Then, for a fixed $k > 0$, the *likelihood ratio test* for deciding between a simple null hypothesis $H_0 : \theta = \theta_0$ and the simple alternative $H_a : \theta = \theta_a$ is

$$
\begin{aligned}
&\text{for } \lambda > k, \qquad H_0 \text{ is accepted} \\
&\text{for } \lambda < k, \qquad H_0 \text{ is rejected} \\
&\text{for } \lambda = k, \qquad \text{either action is taken.}
\end{aligned}
\qquad (10.7b)
$$

The inequality $\lambda > k$ determines the acceptance and critical region S_0 and S_a, respectively, as illustrated by the following example.

Example 10.2

If x is a random sample of size one drawn from a normal distribution with mean and variance both equal to 1, find the acceptance and critical regions for testing the null hypothesis $H_0 : \mu = -1$ against the alternative $H_a : \mu = 0$ for $k = e^{1/2}$.

Solution: The normal density with unit variance is

$$
f(x; \theta) = \frac{1}{\sqrt{2\pi}} \exp\left[-\frac{(x - \mu)^2}{2} \right],
$$

and from (10.7a), the likelihood ratio is

$$
\lambda = \exp\left[-\frac{1}{2}(x + 1)^2 \right] \exp\left[\frac{1}{2}x^2 \right] = \exp\left[-\frac{1}{2}(2x + 1) \right].
$$

So, with $k = e^{1/2}$, the inequality $\lambda > k$ implies $e^{-x} > e$, which is true for $-\infty < x < -1$ and is the acceptance region for H_0. Likewise, $\infty > x \geq -1$ is the critical region.

For a fixed sample size, the method of testing described in Section 10.1 concentrates on controlling only type I errors, and type II errors are calculated *a posteriori*. A better test would be one that for a null hypothesis $H_0 : \theta \in R_0$, with an alternative $H_a : \theta \in R_a$, gives

$$
P[\mathrm{I}] \leq \alpha, \text{ for } \theta \in R_0,
$$

and maximizes the power

$$
power = 1 - \beta(\theta), \text{ for } \theta \in R_a.
$$

For the case of a simple null hypothesis and a simple alternative, such a test exists and is defined as follows.

The critical region R_k, which for a fixed significance level α, maximizes the power of the test of the null hypothesis $H_0 : \theta = \theta_0$ against the alternative $H_a : \theta \neq 0_a$, where $x_1, x_2, \ldots x_n$ is a sample of size n from a density $f(x; \theta)$, is that region for which the likelihood ratio

$$
\lambda = \frac{L(\theta_0)}{L(\theta_a)} < k, \qquad (10.8a)
$$

for a fixed number k, and

$$\int_{R_k} \cdots \int \prod_{i=1}^{n} f(x_i; \theta_0) dx_i = \alpha. \tag{10.8b}$$

This result is known as the *Neyman-Pearson lemma*. The proof is as follows. The object is to find the region R that maximizes the power

$$1 - \beta = \int_R L(\theta_a) d\mathbf{x},$$

subject to the condition imposed by Equation (10.8b), that is,

$$\int_R L(\theta_0) d\mathbf{x} = \alpha.$$

Consider the region R_k defined to be that where the likelihood ratio is bounded by k:

$$\lambda = \frac{L(\theta_0)}{L(\theta_a)} < k.$$

In R_k it follows that

$$\int_{R_k} L(\theta_a) d\mathbf{x} > \frac{1}{k} \int_{R_k} L(\theta_0) d\mathbf{x}. \tag{10.9a}$$

Now for any region R outside R_k,

$$\lambda = \frac{L(\theta_0)}{L(\theta_a)} > k,$$

and hence

$$\frac{1}{k} \int_R L(\theta_0) d\mathbf{x} > \int_R L(\theta_a) d\mathbf{x}. \tag{10.9b}$$

Combining the two inequalities (10.9a) and (10.9b) gives

$$\int_{R_k} L(\theta_a) d\mathbf{x} > \int_R L(\theta_a) d\mathbf{x}, \tag{10.9c}$$

which is true for any R, except R_k. Thus R_k is the required critical region. Once λ, or equivalently k, is chosen, the values of type I and type II errors, and hence the power, are determined.

The use of the Neyman-Pearson lemma is illustrated by Example 10.3 involving the normal distribution.

Example 10.3

If θ is the mean of a normal population with unit variance, test the null hypothesis $H_0 : \theta = 2$ against the alternative $H_a : \theta = 0$, given a sample of size n.

Solution: Using the normal density, the likelihood ratio is

$$\lambda = \exp\left[-\frac{1}{2}\sum_{i=1}^{n}(x_i - 2)^2\right]\left\{\exp\left[-\frac{1}{2}\sum_{i=1}^{n}x_i^2\right]\right\}^{-1}$$

$$= \exp\left[\sum_{i=1}^{n}(2x_i - 2)\right] = \exp[2n\bar{x} - 2n],$$

and thus, from (10.7b), H_0 is accepted if $\lambda > k$, that is, if

$$\bar{x} > c = \frac{\ln k}{2n} + 1, \tag{10.10}$$

and rejected if $\bar{x} < c$.

The error probabilities for Example 10.3 are given by the shaded areas in Figure 10.3. To find the point for which $P[\mathrm{I}]$ is a given value for a fixed value of n we note that when $\theta = 2$

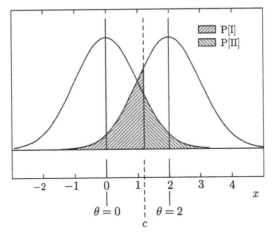

FIGURE 10.3 Error probabilities for Example 10.3.

$$P[\mathrm{I}] = P[\bar{x} < c | \theta = 2] = \alpha,$$

so, for $\alpha = 0.05$ and $n = 4$ (say), using Table C.1 gives

$$c = \theta - \frac{1.645}{\sqrt{4}} = 1.1775,$$

and for this value of c,

$$P[\mathrm{II}] = P[\bar{x} > 1.1775 | \theta = 0] = 0.009 = \beta.$$

It is also possible to find the sample size necessary to control the values of $P[\text{I}]$ and $P[\text{II}]$. Thus, for example, if we want the error probabilities to be $P[\text{I}] = 0.03$ and $P[\text{II}] = 0.01$, then using approximate numbers from Table C.1, we require that

$$c = \theta_0 - \frac{1.88}{\sqrt{n}} = 2 - \frac{1.88}{\sqrt{n}}, \text{ and } c = \theta_a + \frac{2.33}{\sqrt{n}} = \frac{2.33}{\sqrt{n}},$$

simultaneously. This gives $n = 4.43$, and so a sample size of 5 would suffice.

10.2.2 Composite hypotheses

The case considered in Section 10.2.1 is only useful for illustrative purposes. More realistic situations usually involve composite hypotheses. The first that we will consider is when the null hypothesis is simple, and the alternative is composite but may be regarded as an aggregate of simple hypotheses. If the alternative is H_a, then for each of the simple hypotheses in H_a, say H_a', we may construct, for a given α, a region R for testing H_0 against H_a'. However, R will vary from one hypothesis H_a' to the next and we are therefore faced with the problem of determining the best critical region for the totality of hypotheses H_a'. Such a region is called the *uniformly most powerful (UMP)* and a UMP test is defined as follows.

A test of the null hypothesis $H_0 : \theta \in R_0$ against the alternative $H_a : \theta \in R_a$ is a *UMP test* at the significance level α if the critical region of the test is such that

$$P[\text{I}] \le \alpha \text{ for all } \theta \in R_0$$

and

$$1 - \beta(\theta) \text{ is a maximum for each } \theta \in R_a.$$

The following simple example will illustrate how such a UMP test may be constructed.

Example 10.4

Test the null hypothesis $H_0 : \mu = \mu_0$ against the alternative $H_a : \mu > \mu_0$, for a normal distribution with unit variance, using a sample of size n.

Solution: The hypothesis H_a may be regarded as an aggregate of hypotheses H_a' of the form $H_a' : \mu = \mu_a$ where $\mu_a > \mu_0$. The likelihood ratio for testing H_0 against H_a' is, from the definition (10.7a) and as derived in Example 10.3,

$$\lambda = \exp\left\{-\frac{1}{2}\left[2n\bar{x}(\mu_a - \mu_0) + n(\mu_0^2 - \mu_a^2)\right]\right\}.$$

The Neyman-Pearson lemma may now be applied for a given k, and gives the critical region

$$\bar{x} > c = \frac{-\ln k}{n(\mu_a - \mu_0)} + \frac{1}{2}(\mu_0 + \mu_a).$$

Thus, the critical region is of the form $\bar{x} > c$ regardless of the value of μ_a, provided $\mu_a > \mu_0$. Therefore, to accept H_0 if $\bar{x} > c$ tests H_0 against $H_a : \mu > \mu_0$. The number c may be found from

$$P[\mathrm{I}] = \alpha = \sqrt{\frac{n}{2\pi}} \int_c^\infty \exp\left[-\frac{n}{2}(\bar{x} - \mu_0)^2 \right] \mathrm{d}\bar{x}.$$

The integral is evaluated by substituting $u = \sqrt{n}(\bar{x} - \mu_0)$ and using Table C.1. For example, choosing $\alpha = 0.025$ gives $c = \mu_0 + 1.96/\sqrt{n}$.

A more complicated situation that can occur is testing one composite hypothesis against another, for example testing the null hypothesis $H_0 : \theta_1 < \theta < \theta_2$ against $H_a : \theta < \theta_1, \theta > \theta_2$. In such cases a UMP test does not exist, and other tests must be devised with power not too inferior to the maximum power tests. A useful method is to construct a test having desirable large-sample properties and hope that it is still reasonable for small samples. One such test is the *generalized likelihood ratio* described next.

Let $x_1, x_2, ..., x_n$ be a sample of size n from a population density $f(x; \theta_1, \theta_2, ..., \theta_p)$ where S is the parameter space. Let the null hypothesis be $H_0 : (\theta_1, \theta_2, ..., \theta_p) \in R_0$, and the alternative be $H_a : (\theta_1, \theta_2, ..., \theta_p) \in (S - R_0)$. Then, if the likelihood of the sample is denoted by $L(S)$ and its maximum value with respect to the parameters in the region S and R_0 denoted by $L\left(\widehat{S}\right)$ and $L\left(\widehat{R}_0\right)$ respectively, the *generalized likelihood ratio* is given by

$$\lambda = \frac{L\left(\widehat{R}_0\right)}{L\left(\widehat{S}\right)}, \tag{10.11}$$

and $0 < \lambda < 1$. Furthermore, if $P[\mathrm{I}] = \alpha$, then the critical region for the generalized likelihood ratio test is $0 < \lambda < A$, where

$$\int_0^A g(\lambda|H_0)\mathrm{d}\lambda = \alpha,$$

and $g(\lambda|H_0)$ is the density of λ when the null hypothesis H_0 is true. Again, we will illustrate the method by a simple example.

Example 10.5

Use the generalized likelihood ratio to test the null hypothesis $H_0 : \mu = 2$ against $H_\alpha : \mu \neq 2$, for a normal density with unit variance.

Solution: In this example the region R_0 is a single point $\mu = 2$, and $(S - R_0)$ is the remaining real axis. The likelihood is

$$L = \left(\frac{1}{2\pi}\right)^{n/2} \exp\left[-\frac{1}{2}\sum_{i=1}^{n}(x_i - \mu)^2\right]$$

$$= \left(\frac{1}{2\pi}\right)^{n/2} \exp\left[-\frac{1}{2}\sum_{i=1}^{n}(x_i - \bar{x})^2 - \frac{n}{2}(\bar{x} - \mu)^2\right],$$

and the maximum value of $L(S)$ is obtained when $\mu = \bar{x}$, that is,

$$L\left(\widehat{S}\right) = \left(\frac{1}{2\pi}\right)^{n/2} \exp\left[-\frac{1}{2}\sum_{i=1}^{n}(x_i - \bar{x})^2\right]. \tag{10.12}$$

Similarly,

$$L\left(\widehat{R}_0\right) = \left(\frac{1}{2\pi}\right)^{n/2} \exp\left[-\frac{1}{2}\sum_{i=1}^{n}(x_i - \bar{x})^2 - \frac{n}{2}(\bar{x} - 2)^2\right], \tag{10.13}$$

and the generalized likelihood ratio is thus

$$\lambda = \exp\left[-\frac{n}{2}(\bar{x} - 2)^2\right]. \tag{10.14}$$

If we use $\alpha = 0.025$, the critical region for the test is given by $0 < \lambda < A$, where

$$\int_0^A g(\lambda|H_0)d\lambda = 0.025.$$

Now if H_0 is true, $\bar{x}$ is normally distributed with mean 2 and variance $1/n$. Then $n(\bar{x} - 2)^2$ is distributed as chi-square with one degree of freedom. Taking the logarithm of (10.14), it follows that $(-2 \ln \lambda)$ is also distributed as chi-square with one degree of freedom. Setting $\chi^2 = -2 \ln \lambda$ and using Table C.4 gives

$$0.025 = \int_0^A g(\lambda|H_0)d\lambda = \int_{-2\ln A}^{\infty} f(\chi^2; 1)d\chi^2 = \int_{5.02}^{\infty} f(\chi^2; 1)d\chi^2.$$

Thus, the critical region is defined by $-2 \ln \lambda > 5.02$, that is, $n(\bar{x} - 2)^2 > 5.02$, or

$$\bar{x} > 2 + \frac{2.24}{\sqrt{n}}; \quad \bar{x} < 2 - \frac{2.24}{\sqrt{n}}. \tag{10.15}$$

The generalized likelihood ratio test has useful large-sample properties. These can be stated as follows. Let $x_1, x_2, ..., x_n$ be a random sample of size n drawn from a density $f(x; \theta_1, \theta_2, ..., \theta_p)$, and let the null hypothesis be

$$H_0: \; \theta_i = \bar{\theta}_i, \; i = 1, 2, ..., k \leq p,$$

with the alternative

$$H_a: \theta_i \neq \bar{\theta}_i.$$

Then, when H_0 is true, $-2 \ln \lambda$ is approximately distributed as chi-square with k degrees of freedom, if n is large. The proof will not be given here, but it follows from the result demonstrated in Chapter 7 that maximum likelihood estimators are asymptotically normally distributed for large n. This results in a multinormal for the probability density of the parameters and a chi-square distribution for $-2 \ln \lambda$ with k squared normal terms rather than the single term in (10.14).

To use this result to test the null hypothesis H_0 with $P[I] = \alpha$, we need only compute $-2 \ln \lambda$ from the sample, and compare it with the α level of the chi-square distribution. If $-2 \ln \lambda$ exceeds the α level H_0 is rejected, if not H_0 is accepted.

10.3 Normal distribution

Because of the great importance of the normal distribution, in this section we shall give some more details concerning tests involving this distribution.

10.3.1 Basic ideas

In Section 10.1 we discussed a two-tailed test of the hypothesis $H_0 : \mu = \mu_0$ against the alternative $H_a : \mu \neq \mu_0$ for the case where the population variance is known $n(x : \mu, \sigma^2)$. To recap the result obtained there: The null hypothesis is rejected if the quantity $W = \sqrt{n}(\bar{x} - \mu_0)/\sigma$ is greater than W_γ in modulus, that is, if $|W| > W_\gamma$, where

$$P[|W| \geq W_\gamma] = \left(\frac{1}{2\pi}\right)^{1/2} \left\{ \int_{-\infty}^{-W_\gamma} e^{-t^2/2} dt + \int_{W_\gamma}^{\infty} e^{-t^2/2} dt \right\}$$

$$= 2[1 - N(W_\gamma; 0, 1)] = 2\gamma.$$

If the alternative hypothesis is $H_a : \mu > \mu_0$, then $P[I]$ is the area under only *one* of the tails of the distribution, and the significance level of the test is thus

$$P[I] = \alpha = \gamma.$$

Such a test is called a *one-tailed test*. We also showed how to find the probability of a type II error and from this the power of the test. If for definiteness the alternative is taken to be $H_a : \mu = \mu_a$, then the power is given by

$$1 - \beta = 1 - P\left[-W_{\alpha/2} - \left(\frac{\mu_a - \mu_0}{\sigma/\sqrt{n}}\right) \leq W \leq W_{\alpha/2} + \left(\frac{\mu_a - \mu_0}{\sigma/\sqrt{n}}\right) \right].$$

If $\mu - \mu_0$ is small,

$$P[II] \simeq 1 - P[I],$$

and hence

$$1 - \beta \simeq \alpha.$$

Thus, the power of the test will be very low. This situation can only be improved by making $(\mu_a - \mu_0)$ large, or by having n large. This is in accord with the common-sense view that it is difficult to distinguish between two close alternatives without a large quantity of data. The situation is illustrated in Figure 10.4, which shows the power $1 - \beta$ as a function of the parameter $\Delta = \sqrt{n}(\mu_a - \mu_0)/\sigma$ for two sample values of α, the significance level. This is a generalization of the specific case shown in Figure 10.2(b).

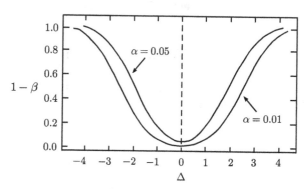

FIGURE 10.4   The power of a test comparing two means for a normal population with known variance.

We are now in a position to review the general procedure followed to test a hypothesis:

1. State the null hypothesis H_0, and its alternative H_a;
2. Specify $P[\text{I}]$ and $P[\text{II}]$, the probabilities for errors of types I and II, respectively, and compute the necessary sample size n.[1] In practice, $P[\text{I}] = \alpha$ and n are commonly given. However, since even a relatively small $P[\text{II}]$ is usually of importance, a check should always be made to ensure that the values of α and n used lead to a suitable $P[\text{II}]$.
3. Choose a test statistic and determine the critical region for the test.
4. Accept or reject the null hypothesis H_0 depending on whether the value obtained for the sample statistic falls outside or inside the critical region.

A graphical interpretation of the above scheme is shown in Figure 10.5. The curve $f_0(\theta|H_0)$ is the density function of the test statistic θ if H_0 is true and $f_a(\theta|H_a)$ is its density function if H_a is true. The hypothesis H_0 is rejected if $\theta > \theta_\alpha$, and H_a is rejected if $\theta < \theta_\alpha$. The probabilities of the errors of types I and II are also shown. It is worth repeating that failure to reject a hypothesis does not necessarily mean that the hypothesis is true. However, if we can reject the hypothesis based on the test, then we *can* say that there is experimental evidence against it.

---

[1] Tables for this purpose applying to some of the tests we will consider are given in O.L. Davies *Design and Analysis of Industrial Experiments, Research Vol 1*, Oliver and Boyd Ltd (1948).

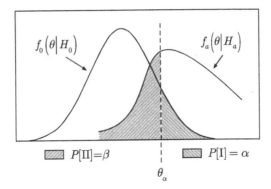

$$P[\text{II}] = \beta \qquad\qquad P[\text{I}] = \alpha$$

FIGURE 10.5   Graphical representation of a general hypothesis test.

10.3.2 Specific tests

We shall now turn to more practical cases where one of the parameters of the distribution is unknown, and use the general procedure given above to establish some commonly used tests, deriving them from the likelihood ratio.

(a) Test of whether the mean is different from some specified value.

The null hypothesis in this case is

$$H_0: \mu = \mu_0, \quad 0 < \sigma^2 < \infty,$$

and the alternative is

$$H_0: \mu \neq \mu_0, \quad 0 < \sigma^2 < \infty.$$

Since the variance is unknown, the parameter space is

$$S = \left\{ -\infty < \mu < \infty; \quad 0 < \sigma^2 < \infty \right\},$$

and the acceptance region associated with the null hypothesis is

$$R_0 = \left\{ \mu = \mu_0 : 0 < \sigma^2 < \infty \right\}.$$

In this case the null hypothesis is not simple because it does not specify the value of σ^2. For large samples we could use an estimate s^2 for σ^2 and then take over the results of Section 10.3.1. However, for small samples this procedure could lead to errors and so we must devise a test where σ^2 is not explicitly used. Such a test is based on the use of the Student's t distribution. Its derivation is as follows.

The likelihood function for a sample of size n drawn from the population is given by

$$L = \frac{1}{(2\pi)^{n/2}} \frac{1}{\sigma^n} \exp\left[-\frac{1}{2} \sum_{i=1}^{n} \left(\frac{x_i - \mu}{\sigma} \right)^2 \right], \tag{10.16}$$

and we have seen in Chapter 7 that the maximum likelihood estimators of μ and σ^2 are

$$\hat{\mu} = \frac{1}{n} \sum_{i=1}^{n} x_i \quad \text{and} \quad \sigma^2 = \frac{1}{n} \sum_{i=1}^{n} (x_i - \bar{x})^2. \tag{10.17}$$

Using (10.17) in (10.16) gives

$$L(\hat{S}) = \left[\frac{n}{2\pi \sum (x_i - \bar{x})^2} \right]^{n/2} e^{-n/2}. \tag{10.18}$$

To maximize L in R_0 we set $\mu = \mu_0$, giving

$$L' = \frac{1}{(2\pi)^{n/2}} \frac{1}{\sigma^n} \exp \left[-\frac{1}{2} \sum_{i=1}^{n} \left(\frac{x_i - \mu_0}{\sigma} \right)^2 \right].$$

Then the value of σ^2 that maximizes L' is

$$\hat{\sigma}^2 = \frac{1}{n} \sum_{i=1}^{n} (x_i - \mu_0)^2,$$

and hence

$$L(\hat{R}_0) = \left[\frac{n}{2\pi \sum (x_i - \mu_0)^2} \right]^{n/2} e^{-n/2}. \tag{10.19}$$

So, from (10.18) and (10.19), the generalized likelihood ratio is

$$\lambda = \left[\frac{\sum (x_i - \bar{x})^2}{\sum (x_i - \mu_0)^2} \right]^{n/2}. \tag{10.20}$$

We must now find the distribution of λ if H_0 is true. Rewriting (10.20) gives

$$\lambda = \left[1 + \frac{n(\bar{x} - \mu_0)^2}{\sum (x_i - \bar{x})^2} \right]^{-n/2} = \left(1 + \frac{t^2}{n - 1} \right)^{-n/2}, \tag{10.21}$$

where

$$t = \left[\frac{n(n - 1)}{\sum (x_i - \bar{x})^2} \right]^{1/2} (\bar{x} - \mu_0)$$

is distributed as the t-distribution with $(n - 1)$ degrees of freedom. From (10.21), a critical region of the form $0 < \lambda < A$ is equivalent to the region $t^2 > F(A)$, where $F(A)$ is the inverse of

(10.21) giving t^2 as a function of λ, evaluated at $\lambda = A$. Thus, a significance level α corresponds to the pair of intervals

$$t < -t_{\alpha/2} \quad \text{and} \quad t > t_{\alpha/2},$$

where

$$\int_{t_{\alpha/2}}^{\infty} f(t; n-1) dt = \alpha/2, \tag{10.22}$$

and $f(t; n-1)$ is the t-distribution with $(n-1)$ degrees of freedom. If t lies between $-t_{\alpha/2}$ and $t_{\alpha/2}$, then H_0 is accepted, otherwise it is rejected. This is a typical example of a two-tailed test and is exactly equivalent to constructing a $100(1-2\alpha)\%$ confidence interval for μ, and accepting H_0 if μ lies within it. The test is summarized as follows:

Observations	n values of x
Significance level	α
Null hypothesis	$H_0 : \mu = \mu_0, \quad 0 < \sigma^2 < \infty$
Alternative hypothesis	$H_0 : \mu \neq \mu_0, \quad 0 < \sigma^2 < \infty$
Test statistic	$t - \left[\dfrac{n(n-1)}{\sum (x_i - \overline{x})^2} \right]^{1/2} (\overline{x} - \mu_0) - \dfrac{(\overline{x} - \mu_0)}{s/\sqrt{n}}$ (10.23)

Decision criterion. The test statistics obeys a t-distribution with $(n-1)$ degrees of freedom if the null hypothesis is true, so if the observed value of t lies between $-t_{\alpha/2}$ and $t_{\alpha/2}$, where the latter is defined by (10.23), the null hypothesis is accepted, otherwise it is rejected.

This may be generalized in an obvious way to test the null hypothesis against the alternatives $H_\alpha : \mu > \mu_0$ and $H_\alpha : \mu < \mu_0$. The test statistic is the same, but the critical regions are now $t > t_\alpha$ and $t < -t_\alpha$, respectively.

The above procedure has controlled type I errors by specifying the significance level. We must now consider the power of the test. This is no longer a simple problem because if H_0 is not true then the statistic t no longer has a Student's t distribution. If the alternative hypothesis is

$$H_a : \mu = \mu_a, \quad 0 < \sigma^2 < \infty$$

and H_a is true, it can be shown that t obeys a *noncentral t-distribution*. Unfortunately, this distribution, apart from being exceedingly complex, contains the population variance σ^2 in the *noncentrality parameter δ.* An estimate of the power of the test may be obtained by replacing σ^2 by the sample variance s^2 in the noncentral distribution and then using tables of the distribution.

Noncentral distributions arise typically if we wish to consider the power of a test. They are generally functions of a noncentrality parameter that itself is a function of the alternative hypothesis and a population parameter that is unknown.

Example 10.6

An experiment finds the mean lifetime of a rare nucleus to be $\tau = 125$ μs. A second experiment finds 20 examples of the same decay with lifetimes given below (in μs).

$$120 \quad 113 \quad 120 \quad 140 \quad 136 \quad 117 \quad 140 \quad 136 \quad 110 \quad 119$$
$$123 \quad 119 \quad 118 \quad 120 \quad 134 \quad 121 \quad 137 \quad 129 \quad 137 \quad 140$$

Test whether the data from the second experiment are compatible with those from the first experiment at the 10% level.

Solution: Formally, we are testing the null hypothesis $H_0 : \tau = 125$ with the alternative $H_a : \tau \neq 125$, using test (a) above. Firstly, using the data given, we find the mean to be $\bar{x} = 126.45$ and the standard deviation $\sigma = 10.02$, and using these the calculated value of t is 0.647. Next, we find $t_{\alpha/2}$ from (10.23) for $\alpha = 0.10$ and 19 degrees of freedom. Using Table C.5, this gives $t_{\alpha/2} = 1.73$. As this value is greater than the observed value of t, the null hypothesis is accepted, that is, the two experiments are compatible at this significance level.

Another use of Student's t distribution is contained in the following test, which we will state without proof.

(b) *Test of whether the means of two populations having the same, but unknown, variance differ.*

Observations	m values of x_1, n values of x_2
Significance level	α
Null hypothesis	$H_0 : \mu_1 = \mu_2, \; -\infty < \sigma_{1,2}^2 < \infty \; (\sigma_1^2 = \sigma_2^2)$
Alternative hypothesis	$H_a : \mu_1 \neq \mu_2, \; -\infty < \sigma_{1,2}^2 < \infty \; (\sigma_1^2 = \sigma_2^2)$
Test statistic	

$$t = \frac{(\bar{x}_1 - \bar{x}_2)}{s_p\sqrt{(1/m) + (1/n)}}, \tag{10.24a}$$

where the pooled sample variance s_p^2 is given by

$$s_p^2 = \frac{(m-1)s_1^2 + (n-1)s_2^2}{m+n-2}. \tag{10.24b}$$

Decision criterion. The test statistic obeys a t-distribution with $(m+n-2)$ degrees of freedom if the null hypothesis is true, so if the observed value of t lies in the range $t_{-\alpha/2} < t < t_{\alpha/2}$, where

$$\int_{t_{\alpha/2}}^{\infty} f(t; m+n-2)\mathrm{d}t = \alpha/2, \qquad (10.25)$$

the null hypothesis is accepted, otherwise it is rejected.

Again, this test may be extended to the cases $\mu_1 > \mu_2$ and $\mu_1 < \mu_2$, with critical regions $t > t_\alpha$ and $t < -t_\alpha$, respectively. It may also be extended in a straightforward way to test the null hypothesis $H_0 : \mu_1 - \mu_2 = d$, where d is a specified constant, and to cases where the two variances are both unknown and unequal.

We will now consider two tests associated with the variance of a normal population, and by analogy with the discussion of tests involving the mean, we shall start with a test of whether the variance is equal to some specific value.

(c) Test of whether the variance is equal to some specific value

The null hypothesis in this case is

$$H_0 : \sigma^2 = \sigma_0^2, \qquad -\infty < \mu < \infty,$$

and the alternative is

$$H_\alpha : \sigma^2 \neq \sigma_0^2, \qquad -\infty < \mu < \infty.$$

Since the mean is unknown, the parameter space is

$$S = \{ -\infty < \mu < \infty; 0 < \sigma^2 < \infty \},$$

and the acceptance region associated with the null hypothesis is

$$R_0 = \{ -\infty < \mu < \infty; 0 < \sigma^2 = \sigma_0^2 \}.$$

The test will involve the use of the χ^2 distribution and will again be derived by the method of likelihood ratios.

As before, the likelihood function for a sample of size n drawn from the population is given by

$$L = \left(\frac{1}{2\pi}\right)^{n/2} \frac{1}{\sigma^n} \exp\left[-\frac{1}{2}\sum_{i=1}^{n} \left(\frac{x_i - \mu}{\sigma}\right)^2 \right]$$

and in the acceptance region R_0

$$L = \left(\frac{1}{2\pi}\right)^{n/2} \frac{1}{\sigma_0^n} \exp\left[-\frac{1}{2}\sum_{i=1}^{n}\left(\frac{x_i - \mu}{\sigma_0}\right)^2\right].$$

This expression is a maximum when the summation is a minimum, that is, when $\bar{x} = \mu$. Thus

$$L(\widehat{R}_0) = \left(\frac{1}{2\pi}\right)^{n/2} \frac{1}{\sigma_0^n} \exp\left[-\frac{1}{2}\sum_{i=1}^{n}\left(\frac{x_i - \bar{x}}{\sigma_0}\right)^2\right]$$

$$= \left(\frac{1}{2\pi}\right)^{n/2} \frac{1}{\sigma_0^n} \exp\left[-\frac{(n-1)s^2}{2\sigma_0^2}\right],$$

where s^2 is the sample variance. To maximize L in S we must solve the maximum likelihood equations. The solutions have been given in (10.18) and hence

$$L(\widehat{S}) = \left(\frac{1}{2\pi}\right)^{n/2} \frac{1}{s^n}\left(\frac{n}{n-1}\right)^{n/2} \exp\left(-\frac{n}{2}\right)$$

We may now form the generalized likelihood ratio

$$\lambda = \frac{L(\widehat{R}_0)}{L(\widehat{S})} = \left[\frac{(n-1)s^2}{n\sigma_0^2}\right]^{n/2} \exp\left[\frac{n}{2} - \frac{(n-1)s^2}{2\sigma_0^2}\right].$$

It is easy to see that λ has a single maximum and there are two values of s^2/σ_0^2 for any $\lambda = k < 1$. The critical region of the form $\lambda < k$ is equivalent to the region

$$\frac{s^2}{\sigma_2^2} < k_1 \text{ and } \frac{s^2}{\sigma_0^2} > k_2$$

and the region of acceptance is

$$k_1 < \frac{s^2}{\sigma_0^2} < k_2,$$

where k_1 and k_2 are constants depending on n and α, the significance level of the test. If H_0 is true then $(n-1)s^2/\sigma_0^2$ obeys a χ^2 distribution with $(n-1)$ degrees of freedom and so, in principle, the required values of k_1 and k_2 could be found. A good approximation is to choose values of k_1 and k_2 using equal right and left tails of the chi-squared distribution. Thus, we are led to the following test procedure.

Observations	n values of x
Significance level	α
Null hypothesis	$H_0 : \sigma^2 = \sigma_0^2$
Alternative hypothesis	$H_a : \sigma^2 \neq \sigma_0^2$
Test statistic	

$$\chi^2 = \sum_{i=1}^{n} \left(\frac{x_i - \bar{x}}{\sigma_0} \right)^2 = \frac{s^2}{\sigma_0^2}(n-1) \tag{10.26}$$

Decision criterion. The test statistic obeys a χ^2 distribution with $(n-1)$ degrees of freedom if the null hypothesis is true, so if the observed value of χ^2 lies in the interval $\chi^2_{1-\alpha/2} < \chi^2 < \chi^2_{\alpha/2}$, where

$$\int_{\chi^2_{0/2}}^{\infty} f(\chi^2; n-1) d\chi^2 = \alpha/2, \tag{10.27}$$

the null hypothesis is accepted, otherwise it is rejected.

As previously, we must now examine the question: What is the probability of a type II error in this test? If the alternative hypothesis is

$$H_a : \sigma^2 = \sigma_a^2,$$

and if H_a is true, then the quantity

$$\chi_a^2 = \frac{s^2}{\sigma_a^2}(n-1) = \chi^2 \frac{\sigma_0^2}{\sigma_a^2}$$

will be distributed as χ^2 with $(n-1)$ degrees of freedom. In this formula, χ^2 is the statistic given by (10.26). Thus, from the definition of the power function, we have

$$1 - \beta = 1 - P \left[\chi^2_{\alpha/2} \geq \frac{s^2}{\sigma_0^2}(n-1) \geq \chi^2_{1-\frac{\alpha}{2}} \right],$$

and therefore,

$$1 - \beta = 1 - P \left[\chi^2_{\alpha/2} \frac{\sigma_0^2}{\sigma_a^2} \geq (n-1) \frac{s^2}{\sigma_a^2} \geq \chi^2_{1-\frac{\alpha}{2}} \frac{\sigma_0^2}{\sigma_a^2} \right]. \tag{10.28}$$

Having fixed the significance level α and the values of σ_0 and σ_a, we can read off from tables the probability that a chi-square variate with $(n-1)$ degrees of freedom lies between the two limits in the square brackets.

Again, this test may be simply adapted to deal with the hypotheses $H_a : \sigma^2 > \sigma_0^2$ and $H_a :$ $\sigma^2 < \sigma_0^2$. The critical regions are $\chi^2 > \chi_\alpha^2$ and $\chi^2 < \chi_{1-\alpha}^2$, respectively.

Example 10.7

Steel rods of notional standard lengths are produced by a machine whose specifications state that regardless of the length of the rods, their standard deviation will not exceed 2 (in centimeter units). During commissioning, a check was made on a random sample of 20 rods, whose lengths (in centimeters) were found to be:

105	104	103	98	100	102	103	97	99	106
105	102	99	100	98	97	102	101	100	99

Test at the 10% level whether the machine is performing according to its specification.

Solution: Here we are testing the null hypothesis $H_0 : \sigma^2 \leq \sigma_0^2$ against the alternative $H_a : \sigma^2 > \sigma_0^2$, where $\sigma_0^2 = 4$. That is, we are using a one-tailed test, which is an adaptation of test (c) above. From the data, $\bar{x} = 101$ and $s^2 = 7.474$, so the test statistic is

$$\chi^2 = s^2(n-1)/\sigma_0^2 = 35.5.$$

Now from Table C.4 we find that for $\alpha = 0.10$ and 19 degrees of freedom, $\chi_{0.1}^2 = 27.2$. As this value is less than χ^2, we conclude that the machine is not performing according to its specification.

The final test concerns the equality of the variances of two normal populations, which we quote without proof as it follows easily from the discussion in Section 9.3.3.

(d) *Test of whether the variances of two populations with different but unknown means differ.*

Observations	n values of x_1, m values of x_2
Significance level	α
Null hypothesis	$H_0 : \sigma_1^2 = \sigma_2^2$
Alternative hypothesis	$H_a : \sigma_1^2 \neq \sigma_2^2$
Test statistic	

$$F = \frac{s_1^2}{s_2^2} = \frac{(m-1)\sum_i (x_{1i} - \bar{x}_1)^2}{(n-1)\sum_i (x_{2i} - \bar{x}_2)^2} \tag{10.29}$$

Decision criterion. The test statistic obeys the F distribution with $(n-1)$ and $(m-1)$ degrees of freedom if the null hypothesis is true, so if the observed value of F lies in the interval

$$[F_{\alpha/2}(m-1, n-1)]^{-1} < F < F_{\alpha/2}(n-1, m-1)$$

where

$$\int_{F_{\alpha/2}}^{\infty} f(F; n-1, m-1)\mathrm{d}F = \alpha/2, \tag{10.30}$$

the null hypothesis is accepted, otherwise it is rejected.

To calculate the power of the test, we note that $P[\text{II}]$ depends on the value of σ^2 / σ_2^2. If the true value of this ratio is δ then, since $(n-1)s_1^2/\sigma_1^2$ for a sample from a normal population is distributed as χ_{n-1}^2, we find s_1^2/s_2^2 is distributed as

$$\frac{\sigma_1^2}{\sigma_2^2}F(n-1, m-1) = \delta F(n-1, m-1).$$

Thus,

$$1 - \beta = 1 - P\left[F_{1-\frac{\alpha}{2}}(n-1, m-1) \leq \frac{s_1^2}{s_2^2} \leq F_{\alpha/2}(n-1, m-1)\right],$$

is equivalent to

$$1 - \beta = 1 - P\left[\frac{F_{1-\frac{\alpha}{2}}(n-1, m-1)}{\delta} \leq F \leq \frac{F_{\alpha/2}(n-1, m-1)}{\delta}\right]. \tag{10.31}$$

For any given value of δ, these limits may be found from tables of the F distribution. It can be shown by consulting these tables that the power of the F test is rather small unless the ratio of variances is large, a result that is in accordance with common sense. As before, this test may be adapted to test the null hypotheses $\sigma_1^2 > \sigma_2^2$ and $\sigma_1^2 < \sigma_2^2$. The critical regions are $F > F_\alpha$ and $F < F_{1-\alpha}$, respectively.

A summary of some of the tests mentioned above is given in Table 10.1.

10.4 Other distributions

The ideas discussed in previous sections can be applied to other distributions and we will look briefly at two important examples, the binomial and Poisson.

Consider a situation where there are only two outcomes of a trial, for example either satisfactory or unsatisfactory, with the latter outcome having a probability p. In this situation, the number of unacceptable outcomes in n independent random trials is given by the binomial distribution. As an example, we will test the null hypothesis $H_0 : p \leq p_0$ against the alternative $H_a : p > p_0$, where p_0 is some specified value. If the observed number of unacceptable outcomes is x, then from Equation (4.34),

$$P[x \geq k] = \sum_{r=k}^{n} \binom{n}{r} p^r (1-p)^{n-r}.$$

TABLE 10.1 Summary of hypothesis tests on a normal distribution.

H_0	Test statistic	H_a	Critical region
$\mu = \mu_0$ σ^2 unknown	$t = \frac{\bar{x}-\mu_0}{s/\sqrt{n}}$ distributed as t with $(n-1)$ degrees of freedom	$\mu \neq \mu_0$ $\mu > \mu_0$ $\mu < \mu_0$	$t > t_{\alpha/2}$ and $t < -t_{\alpha/2}$ $t > t_\alpha$ $t < -t_\alpha$
$\mu_1 = \mu_2$ $\sigma_1^2 = \sigma_2^2$ unknown	$t = \frac{(\bar{x}_1 - \bar{x}_2)}{s_p\sqrt{(1/m)+(1/n)}}$, where there are m, n values of x_1, x_2 $s_p^2 = \frac{(m-1)s_1^2+(n-1)s_2^2}{n+m-2}$ is distributed as t with $(n+m-2)$ degrees of freedom	$\mu_1 \neq \mu_2$ $\mu_1 > \mu_2$ $\mu_1 < \mu_2$	$t > t_{\alpha/2}$ and $t < -t_{\alpha/2}$ $t > t_\alpha$ $t < -t_\alpha$
$\sigma^2 = \sigma_0^2$ μ unknown	$\chi^2 = s^2(n-1)/\sigma_0^2$ distributed as χ^2 with $(n-1)$ degrees of freedom	$\sigma^2 \neq \sigma_0^2$ $\sigma^2 > \sigma_0^2$ $\sigma^2 < \sigma_0^2$	$\chi^2 < \chi^2_{1-\alpha/2}$ and $\chi^2 > \chi^2_{\alpha/2}$ $\chi^2 > \chi^2_\alpha$ $\chi^2 < \chi^2_{1-\alpha}$
$\sigma_1^2 = \sigma_2^2$ $\mu_1 \neq \mu_2$ unknown	$F = s_1^2/s_2^2$ distributed as F with $(n-1)$ and $(m-1)$ degrees of freedom in the numerator and denominator, respectively	$\sigma_1^2 \neq \sigma_2^2$ $\sigma_1^2 > \sigma_2^2$ $\sigma_1^2 < \sigma_2^2$	$F < F_{1-\alpha/2}$ and $F > F_{\alpha/2}$ $F > F_\alpha$ $F < F_{1-\alpha}$

But the probability of k unacceptable outcomes is smaller if the probability p is smaller, so it follows that when H_0 is true, that is when $p \leq p_0$,

$$P[x \geq k] \leq \sum_{r=k}^{n} \binom{n}{r} p_0^r (1-p_0)^{n-r} \equiv B(k, p_0).$$

In practice, if we calculate $B(k, p_0)$, from the number of unacceptable outcomes measured, we accept the null hypothesis if $B(k, p_0) > \alpha$, otherwise reject it.

Example 10.8

In the catalogue of a supplier of capacitors it is stated that no more than 1% of its products are defective. A buyer checks this claim by testing a random sample of 100 capacitors at the 5% level and finds that 3 are defective. Does the buyer have a claim against the supplier?

Solution: We calculate the probability that in a random sample of 100 capacitors there would be at least 3 that are defective if $p_0 = 0.01$. From the binomial distribution with $p = p_0$,

$$P[x \geq 3] = 1 - P[x < 3]$$

$$= 1 - \sum_{r=0}^{2} \binom{100}{r} (0.01)^r (0.99)^{100-r}$$

This may be evaluated exactly by direct calculation. It is tempting to use the normal approximation to the binomial for large n, but in this case, $np_0 = 1$, which is too small for the approximation to be valid. By the normal approximation, the value of $P[x \geq 3]$ would correspond approximately to

$$P\left[z \geq \frac{3.0 - \mu}{\sigma}\right],$$

where $\mu = np_0 = 1$ and $\sigma = [np_0(1 - p_0)]^{1/2} = 0.995$. The direct calculation gives 0.0794 compared to the normal approximation of 0.023, which must be ignored. As the direct calculation is greater than 0.05 the hypothesis cannot be rejected, and no claim can be made against the supplier.

In the case of a random Poisson variable k distributed with parameter λ, as an example, we could consider testing the null hypothesis $H_0 : \lambda = \lambda_0$ against the alternative $H_a : \lambda \neq \lambda_0$. By analogy with the procedure followed for the binomial distribution, for a given significance level we would calculate the probability that the observed value of k is greater than or less than the value predicted if the null hypothesis is true and compare this with the significance level. An example is given in Problem 10.9.

Just as for the normal distribution, these tests may be extended to other related cases, for example to test the equality of the parameters p_1 and p_2 in two Bernoulli populations, or the equality of the parameters of two Poisson distributions, but we will not pursue this further.

Finally, can we test a null hypothesis, such as $H_0 : \mu = \mu_0$, against a suitable alternative when the population distribution is unknown? The answer is yes, but the methods are generally less powerful than those available when the distribution is known. Hypothesis testing in the former situation is an example of *nonparametric statistics* and is discussed in Chapter 11.

10.5 Analysis of variance

In Section 10.3 we discussed how to test the hypothesis that the means of two normal distributions with the same, but unknown, variance differ. It is natural to consider how to extend that discussion to the case of several means. The technique for doing this is called *analysis of variance*, usually abbreviated to ANOVA. It can be used, for example, to test the consistency of a series of measurements carried out under different conditions, or whether different manufacturers are producing a particular component to the same standard. ANOVA is an important technique in biological and social sciences, as well as in industrial process development, but used much less in the physical sciences, and so the discussion here will be brief.

Consider the case of m groups of measurements each leading to an average value $\mu_i (i = 1, 2, \ldots, m)$. We wish to test the null hypothesis

$$H_0 : \mu_1 = \mu_2 = \cdots = \mu_m \tag{10.32a}$$

against the alternative

$$H_a : \mu_i \neq \mu_j \text{ for some } i \neq j, \tag{10.32b}$$

at some significance level. Thus, we are testing whether the various measurements all come from the same population and hence have the same variance. ANOVA is a method for splitting the total variation in the data into independent components that measure different sources of variation. A comparison is then made of these independent estimates of the common variance σ^2. In the simplest case there are two independent estimates. The first of these, s_W^2, is obtained from measurements *within* groups, and their corresponding means. It is a valid estimator of σ^2 *independent* of whether H_0 is true or not. The second estimator, s_B^2, is obtained from the measurements *between* groups, and their corresponding means. It is *only* valid when H_0 is true, and moreover if H_0 is false, it will tend to exceed σ^2. The test is therefore to compare the values of these two estimators and to reject H_0 if the ratio s_B^2/s_W^2 is sufficiently large. This is done using the F distribution.

The above is an example of a *one-way analysis* because the observations are classified under a single classification, for example the manufacturing process in Example 10.8. ANOVA may be extended to deal with multiway classifications. In our example, a second classification might be the country where the components were made. This situation is referred to as *multiway analysis*, but we will discuss only the simpler one-way analysis. In addition, we will assume that all samples are of the same size. This condition is preferable in practical work, but can be relaxed if necessary, at the expense of somewhat more complicated equations.

The procedure starts with random samples each of size n selected from m populations. Each sample is assumed to be normally and independently distributed with means $\mu_1 = \mu_2 = \cdots = \mu_m$ and a common, but unknown, variance σ^2. If x_{ij} is the jth sample value from the ith population, the sample mean of the ith population is

$$\bar{x}_i = \frac{1}{n}\sum_{j=1}^{n} x_{ij}, \qquad i = 1, 2, \ldots, m \tag{10.33a}$$

and the mean of the whole sample is

$$\bar{x} = \frac{1}{mn}\sum_{i=1}^{m}\sum_{j=1}^{n} x_{ij} = \frac{1}{m}\sum_{i=1}^{m}\bar{x}_i \tag{10.33b}$$

The estimate of the variance from the entire sample is given by

$$s^2 = SST/(mn - 1), \tag{10.34}$$

where SST is the *total sum of squares* given by

$$SST = \sum_{i=1}^{m}\sum_{j=1}^{n}(x_{ij} - \bar{x})^2. \tag{10.35}$$

By writing the right-hand side of this expression as

$$\sum_{i=1}^{m}\sum_{j=1}^{n}(x_{ij} - \bar{x}_i + \bar{x}_i - \bar{x})^2,$$

and noting that by virtue of (10.33a),

$$\sum_{i=1}^{m}\sum_{j=1}^{n}(x_{ij}-\bar{x}_i)(\bar{x}_i-\bar{x})=0,$$

the quantity SST may be written as the sum of two terms,

$$SST = SSB + SSW,\tag{10.36}$$

where

$$SSB = n\sum_{i=1}^{m}(\bar{x}_i-\bar{x})^2\tag{10.37}$$

is the sum of squares between the groups, and

$$SSW = \sum_{i=1}^{m}\sum_{j=1}^{n}(x_{ij}-\bar{x}_i)^2\tag{10.38}$$

is the *sum of squares within the groups*. Alternative forms for (10.35), (10.37), and (10.38) that are more convenient for calculations are as follows. SST may be written as:

$$SST = \sum_{i=1}^{m}\sum_{j=1}^{n}x_{ij}^2 - nm\bar{x}^2,\tag{10.39a}$$

which is easily seen by using (10.33b) to rearrange SSB as:

$$SSB = n\sum_{i=1}^{m}\bar{x}_i^2 - nm\bar{x}^2\tag{10.39b}$$

and, using (10.36),

$$SSW = \sum_{i=1}^{m}\sum_{j=1}^{n}x_{ij}^2 - n\sum_{i=1}^{m}\bar{x}_i^2.\tag{10.39c}$$

The two sums SSW and SSB form the basis of the two independent estimates of the variance we require. From previous work on the chi-squared distribution, in particular its additive property, we know that if H_0 is true, the quantity SSW/σ^2 is distributed as χ^2 with $m(n-1)$ degrees of freedom and the quantity SSB/σ^2 is distributed as χ^2 with $(m-1)$ degrees of freedom. The test statistic is then $F = s_B^2/s_W^2$ where $s_W^2 = SSW/m(n-1)$ and $s_B^2 = SSB/(m-1)$. The test statistic F is distributed as $F(m-1, m(n-1))$ and is unbiased if H_0 is true. The quantities s_B^2 and s_W^2 are called the *between groups mean square* and *within*

TABLE 10.2 Analysis of variance for a one-way classification.

Source of variation	Sums of squares	Degrees of freedom	Mean square	Test statistic
Between groups	SSB	$m - 1$	$s_B^2 = SSB/(m - 1)$	$F = s_B^2/s_W^2$
Within groups	SSW	$m(n - 1)$	$s_W^2 = SSW/[m(n - 1)]$	
Total	SST	$nm - 1$	$s^2 = SST/[(nm - 1)]$	

groups mean square, respectively. $s_T^2 = SST/(nm - 1)$ is called the *total mean square* and is distributed as χ^2 with $(nm - 1)$ degrees of freedom.

The variance of SSW is the same whether H_0 is true or not, but the variance of SSB if H_0 is not true may be calculated and is greater than it would be if H_0 were true. Thus since s_B^2 overestimates σ^2 when H_0 is false, the test is a one-tailed one, with the critical region such that H_0 is rejected at a significance level α when $F > F_a[(m - 1), m(n - 1)]$. The test procedure is summarized in Table 10.2.

Example 10.9

Three students each use five identical balances to weigh the same number of objects to find their average weight, with the results as shown below.

S1	22	17	15	20	16
S2	29	23	22	20	28
S3	29	31	27	30	32

Test, at a 5% significance level, the hypothesis that the average weights obtained do not depend on which student made the measurements.

Solution: This can be solved using a simple spreadsheet. Intermediate values are:

$$\bar{x}_1 = 18.0, \bar{x}_2 = 24.4, \bar{x}_3 = 29.8 \text{ and } \bar{x} = 24.07.$$

Then from Equations (10.39)

$$SSB = 348.9, SSW = 110.0$$

and so

$$F = s_B^2/s_W^2 = 19.03.$$

From Table C.5, $F_{0.05}(2, 12) = 3.89$ and since $F \gg 3.89$, the hypothesis must be rejected at this significance level.

10.6 Bayesian hypothesis testing

In the frequentist view, we use a sample of data $\mathbf{x} = (x_1, x_2, ..., x_n)$ from one or more random variables to decide whether a parameter (or set of parameters) θ of the random variables satisfies the null hypothesis $H_0 : \theta \in \Theta_0$ or the alternate hypothesis $H_a : \theta \in \Theta_a$, where the set $\Theta = \Theta_0 \cup \Theta_a$ contains all possible values of θ and the sets Θ_0 and Θ_a are disjoint.

In the Bayesian view, the parameters θ are assigned probabilities that reflect our subjective believe about their values. In one sense this view is simpler than the frequentist view in that the goal is to calculate the posterior probability of H_0 given the data $\mathbf{x}$, and use the result to decide whether or not to accept H_0. However, the posterior probabilities cannot be calculated without an estimate of the prior probabilities, and the result depends on the subjective choice of the prior probabilities as well as the measured data. It is often noted that the purpose of Bayesian hypothesis testing is to make a decision, and all decisions are influenced by many subjective factors in addition to the measured data, regardless of whether one follows the Bayesian or frequentist views of probability. In this section we briefly summarize the fundamentals of Bayesian hypothesis testing, following the book by Lee quoted in the Bibliography, and illustrate their use in a simple example.

Given the data $\mathbf{x}$, the posterior probabilities p_0 and p_a, of the null and alternate hypotheses, respectively, are

$$p_0 = P[\theta \in \Theta_0 | \mathbf{x}] \text{ and } p_a = P[\theta \in \Theta_a | \mathbf{x}].$$

Since Θ_a is the complement of Θ_0 in Θ, we know that $p_a = 1 - p_0$. The prior probabilities π_0 and π_a also sum to one:

$$\pi_0 = P[\theta \in \Theta_0], \pi_a = P[\theta \in \Theta_a], \text{and } \pi_a = 1 - \pi_0.$$

The likelihood ratio plays a role in Bayesian hypothesis testing through the notion of *odds*. The prior odds of H_0 against H_a is the ratio π_0/π_a and the posterior odds is the ratio p_0/p_α. The *Bayes factor B*, in favor of H_0 over H_a is the ratio:

$$B = \frac{p_0/p_a}{\pi_0/\pi_a} = \frac{p_0 \pi_a}{p_a \pi_0}. \tag{10.40}$$

If we know the Bayes factor, we additionally need only the prior probability π_0 to calculate the posterior probability p_0. It is easy to show that

$$p_0 = \frac{1}{1 + \dfrac{1 - \pi_0}{\pi_0 B}}. \tag{10.41}$$

For simple hypotheses, B is the likelihood ratio of H_0 against H_a, given the data $\mathbf{x}$. This may be seen by noting that if H_0 and H_a are simple, then Θ_0 has one element θ_0 and Θ_a has one element θ_a. Then

$$p_0 = P[\theta_0|\mathbf{x}] = \frac{P[\mathbf{x}|\theta_0]P[\theta_0]}{P[\mathbf{x}]} = \frac{P[\mathbf{x}|\theta_0]\pi_0}{P[\mathbf{x}]},$$

where we follow the Bayesian view and equate $P[\theta_0]$ with π_0. In the same way,

$$p_a = \frac{P[\mathbf{x}|\theta_a]\pi_a}{P[\mathbf{x}]}$$

and the formula for B is simply

$$B = \frac{P[\mathbf{x}|\theta_0]}{P[\mathbf{x}|\theta_a]}, \tag{10.42}$$

which is exactly the likelihood ratio of H_0 against H_a. A large value of B indicates H_0 should be accepted and a small value indicates H_0 should be rejected.

For composite hypotheses, we cannot interpret the Bayes factor as the likelihood ratio because we cannot simply equate the prior probability to the probability of a simple hypothesis. To interpret the Bayes factor in the general case, we firstly let $f(\theta)$ be the prior probability density of θ and $p_0(\theta)$ be the conditional prior probability density of θ, given that H_0 is true. Then

$$f(\theta) = p_0(\theta)\pi_0$$

and

$$p_0 = P[\theta \in \Theta_0|\mathbf{x}] = \int_{\theta \in \Theta_0} f(\theta|\mathbf{x})\mathrm{d}\theta.$$

Secondly, using Bayes' theorem (2.16), with an integral over θ in the denominator instead of a sum, we have

$$f(\theta|\mathbf{x}) = \frac{f(\theta)P[\mathbf{x}|\theta]}{\int_{\theta \in \Theta_0 \cup \Theta_a} f(\theta|\mathbf{x})\mathrm{d}\theta} = \pi_0 \frac{p_0(\theta)P[\mathbf{x}|\theta]}{\int_{\theta \in \Theta \cup \Theta_a} f(\theta|\mathbf{x})\mathrm{d}\theta}.$$

Thus,

$$p_0 = \pi_0 \frac{\int_{\theta \in \Theta_0} p_0(\theta)P[\mathbf{x}|\theta]\mathrm{d}\theta}{\int_{\theta \in \Theta_0 \cup \Theta_a} f(\theta|\mathbf{x})\mathrm{d}\theta}.$$

In the same way, we can show that

$$p_a = \pi_a \frac{\int\limits_{\theta \in \Theta_a} \rho_a(\theta) P[\mathbf{x}|\theta] d\theta}{\int\limits_{\theta \in \Theta_0 \cup \Theta_a} f(\theta|\mathbf{x}) d\theta},$$

where $\rho_a(\theta)$ is the conditional prior probability density of θ, given that H_a is true. The Bayes factor is then

$$B = \frac{p_0 \pi_a}{p_a \pi_0} = \frac{\int\limits_{\theta \in \Theta_0} \rho_0(\theta) P[\mathbf{x}|\theta] d\theta}{\int\limits_{\theta \in \Theta_a} \rho_a(\theta) P[\mathbf{x}|\theta] d\theta}. \tag{10.43}$$

In the general case, we see that the Bayes factor is the ratio of likelihoods, weighted by the conditional prior probability densities $\rho_0(\theta)$ and $\rho_a(\theta)$. The selection of the prior probability densities is as important to the calculation of the value of B as is the sample data and the likelihood $P[\mathbf{x}|\theta]$. If the choice of the prior densities is arranged so that it has little effect on the value of B or the evidence for the prior densities is strong, then it may make sense to use the Bayes factor to decide whether or not to accept the null hypothesis. Otherwise, the decision to accept or reject the null hypothesis must be accompanied by a clear explanation of the prior densities so that the reader may judge whether the decision is appropriate.

Example 10.10

Consider the situation described in Problem 9.11 where a manufacturing process combines two chemicals, A and B, in equal parts to form a liquid product AB. The small amount of an impurity C produced by the reaction is determined as the difference in mass x between the reactants A+B and the product AB. The results in micrograms for 10 experiments are found to be:

x_i: −0.974 1.714 −0.421 0.191 0.375 −0.291 −0.918 1.099 0.811 −1.503

Given that the mean θ cannot be negative and prior work indicates $\theta \leq 1.0$ microgram, use the Bayesian approach to test $H_0 : 0 \leq \theta \leq 0.5$ against $H_a : 0.5 \leq \theta \leq 1$. Assume the mass is approximately normally distributed.

Solution: We calculate the Bayes factor B using (10.43). The probability $P[\mathbf{x}|\theta]$ is the likelihood of the experiments resulting in the measurements x_i for a given value of θ, after dividing by an appropriate scale factor. Since the scale factor is the same in the numerator and denominator of (10.43), we can ignore it and use

$$P[\mathbf{x}|\theta] = \prod_{i-1}^{N} n(x_i, \theta, \sigma),$$

where $N = 10$, the number of measurements, and n is the normal probability density with mean θ and variance σ^2. Thus

$$P[\mathbf{x}|\theta] = (2\pi\sigma^2)^{-N/2} \exp\left(-\sum_{i=1}^{N} \frac{(x_i - \theta)^2}{2\sigma^2} \right).$$

Since the variance is unknown, we estimate σ^2 as the sample variance, which in this case is 1.03. We know that $0 \leq \theta \leq 1$, but we do not know the probability distribution of θ in this range. Lacking any other information, it is reasonable to take the prior probability density as uniform:

$$f(\theta) = \begin{cases} 1 & \text{for } 0 \leq \theta \leq 1 \\ 0 & \text{for } \theta < 0 \text{ or } \theta > 1 \end{cases}.$$

Since we have no knowledge of whether H_0 or H_a is true, we set the prior probabilities equal, that is $\pi_0 = \pi_a = 0.5$. The prior conditional probabilities ρ_0 and ρ_a are then also equal to 0.5 and we have from (10.43)

$$B = \frac{\int_0^{0.5} \exp\left(-\frac{1}{2\sigma^2} \sum_{i-1}^N (x_i - \theta)^2 \right) d\theta}{\int_{0.5}^1 \exp\left(-\frac{1}{2\sigma^2} \sum_{i-1}^N (x_i - \theta)^2 \right) d\theta}.$$

After some algebraic manipulation, we find

$$B = \frac{K(0, 0.5)}{K(0.5, 1)},$$

where

$$K(L_1, L_2) = \int_{L_1}^{L_2} \exp\left(-\frac{N}{2\sigma^2}(\bar{x} - \theta)^2 \right) d\theta$$

and $\bar{x}$ is the sample mean. In the present case $\bar{x} = 8.3 \times 10^{-3}$. The integral can be solved in terms of the error function $\text{erf}(z)$, defined as

$$\text{erf}(z) = \frac{2}{\sqrt{\pi}} \int_0^z \exp(-u^2) du.$$

With the change of variables $u = \sqrt{N/2\sigma^2}(\theta - \bar{x})$ we find

$$K(L_1, L_2) = \sqrt{\frac{\pi\sigma^2}{2N}} \left[\text{erf}\left(\sqrt{\frac{N}{2\sigma^2}}(L_2 - \bar{x}) \right) - \text{erf}\left(\sqrt{\frac{N}{2\sigma^2}}(L_1 - \bar{x}) \right) \right].$$

Using a table of the error function or computing with any commonly available programing language, we find for the present example

$$B = \frac{0.36}{0.050} = 7.2.$$

Finally, we determine the posterior probability p_0 of H_0 from (10.41). Since $\pi_0 = 0.5$, we have

$$p_0 = \frac{1}{1 + \dfrac{1 - 0.5}{0.5B}} = \frac{1}{1 + B} = 0.89$$

We conclude that the posterior probability that H_0 is true is very high, and thus H_0 should be accepted.

Problems 10

10.1 A factory claims to produce ball bearings with an overall mean weight of $W = 250$ g and a standard deviation of 5 g. A quality control check of 100 bearings finds an average weight of $\overline{W} = 248$ g. Test at a 5% significance level the hypothesis $H_0 : W = 250$ g against the alternative $H_a : W \neq 250$ g. What is the critical region for $\overline{W}$ and what is the probability of accepting H_0 if the true value of the overall mean weight is 248 g?

10.2 A signal with constant strength S is sent from location A to location B. *En route* it is subject to random noise distributed as $N(0, 8)$. The signal is sent 10 times and the strength of the signals received at B are:

$$14 \quad 15 \quad 13 \quad 16 \quad 14 \quad 14 \quad 17 \quad 15 \quad 14 \quad 18$$

Test at the 5% level the hypothesis that the signal was transmitted with strength 13.

10.3 A supplier sells resistors of nominal value 5 ohms in packs of 10 and charges a premium by claiming that the average value of the resistors sold is not less than 5 ohms. A buyer tests this claim by measuring the values of the resistors in two packs with the results:

$$5.3 \quad 5.4 \quad 4.9 \quad 5.1 \quad 5.0 \quad 4.8 \quad 5.1 \quad 5.2 \quad 4.7 \quad 4.9$$
$$5.2 \quad 5.5 \quad 4.7 \quad 4.6 \quad 5.5 \quad 5.4 \quad 5.0 \quad 5.0 \quad 4.8 \quad 5.4$$

Test the supplier's claim at a 10% significance level.

10.4 Nails are sold in packets with an average weight of 100 g and the seller has priced the packets in the belief that 95% of them are within 4 g of the mean. A sample of 20 packets are weighed with the results:

$$100 \quad 97 \quad 89 \quad 93 \quad 103 \quad 105 \quad 93 \quad 110 \quad 101 \quad 102$$
$$98 \quad 99 \quad 105 \quad 106 \quad 89 \quad 103 \quad 90 \quad 93 \quad 92 \quad 106$$

Assuming an approximate normal distribution, test this hypothesis at a 10% significance level against the alternative that the variance is greater than expected.

10.5 A new production technique, M1, has been developed for the abrasive material of car brakes that is claimed reduces the spread in the lifetimes of the product. It is tested against the existing technique, M2, by measuring samples of the effective lifetimes of random samples of each production type. Use the lifetime data below to test the claim at a 5% significance level.

$$\text{M1} \quad 98 \quad 132 \quad 109 \quad 116 \quad 131 \quad 124 \quad 117 \quad 120 \quad 116 \quad 99 \quad 109 \quad 113$$
$$\text{M2} \quad 100 \quad 120 \quad 134 \quad 99 \quad 130 \quad 113 \quad 106 \quad 124 \quad 118$$

10.6 Two groups of students study for a physics examination. Group 1 of 15 students study full-time at college and achieve an average mark of 75% with a standard deviation of 3. Group 2 of 6 students study part-time at home and achieve an average mark of 70% with a standard deviation of 5. If the two populations are assumed to be normally distributed with equal variances, test at a 5% level of significance the claim that full-time study produces better results.

10.7 A manufacturer of an electrical device states that if it is stress tested with a high voltage, on average no more than 4% will fail. A buyer checks this by stress testing a random sample of 1000 units and finds that 50 fail. What can be said about the manufacturer's statement at a 10% significance level?

10.8 Six samples of the same radioisotope are obtained from four different sources and their activities in kilo Becquerel (kBq) measure are found to be:

S1	S2	S3	S4
91	129	119	100
100	127	141	93
99	100	123	89
89	98	137	110
110	97	132	132
96	113	124	116

Test at a 10% significance level, the hypothesis that the mean activity does not depend on the source of the supply.

10.9 The supplier in Example 10.7 claims that no more than 10 unacceptable devices are distributed on any day. This is checked by daily testing batches of devices. A 5-day run of such tests resulted in 13, 12, 15, 12, and 11 defective devices. Is the supplier's claim supported at the 5% significance level?

10.10 Using the data for Problem 10.4, and assuming an approximate normal distribution with unknown variance, test at a 10% significance level the hypothesis that the mean weight of the vendor's packets of nails is 100 g.

10.11 Suppose a larger number of packets was weighed in Problem 10.4 and the same sample mean and variance were found to be 98 and 30, respectively. How many packets must have been weighed for the null hypothesis to be rejected at the 10% significance level?

Hypothesis testing II: Other tests

Overview

Chapter 10 focused on testing hypotheses concerning parameters of distributions. In this chapter, we discuss hypotheses other than those about the values of parameters, starting with the important goodness-of-fit tests, including Pearson's χ^2 and the Kolmogorov-Smirnov, and tests for the independence of variables in a sample. Goodness-of-fit tests are discussed for discrete and continuous distributions. The rest of the chapter is about nonparametric tests. These do not involve any assumptions about the form of the distribution of the parent population. Specific tests briefly discussed are: The sign test, the signed-rank test, the rank-sum test, the runs test, and a test using the rank correlation coefficient.

In Chapter 10 we discussed how to test a statistical hypothesis H_0 about a single population parameter, such as whether its mean μ was equal to a specific value μ_0, against a definite alternative H_a, for example that μ was greater than μ_0. There was also given a brief explanation about how this may be extended to many parameters by the method known as analysis of variance. In this chapter we will discuss a range of other tests. Some of these address questions about the whole population without always referring to a specific alternative, which is left as implied. Examples are those that examine whether a set of observations is described by a specific probability density, or whether a sample of observations is random, or whether two sets of observations are compatible. We will also discuss some tests that are applicable to non-numeric data. We will start by looking at the first of these questions.

11.1 Goodness-of-fit tests

In Section 8.4.1 we introduced the method of estimation known as 'minimum chi-square', and at the beginning of that chapter we briefly discussed how the same technique could be used to test the compatibility of repeated measurements. In the latter applications we are testing the statistical hypothesis that estimates produced by the measurement process all come from the same population. Such procedures are called, for obvious reasons, *goodness-of-fit tests* and are widely used in physical science.

289

11.1.1 Discrete distributions

We will start by considering the case of a discrete random variable x that can take on a finite number of values $x_i (i = 1, 2, ..., k)$ with corresponding probabilities $p_i (i = 1, 2, ..., k)$. We will test the null hypothesis

$$H_0 : p_i = \pi_i, \quad i = 1, 2, ..., k, \tag{11.1a}$$

against the alternative

$$H_a : p_i \neq \pi_i, \tag{11.1b}$$

where the π_i are specified fixed values, and

$$\sum_{i=1}^{k} \pi_i = \sum_{i=1}^{k} p_i = 1.$$

To do this we will use the method of likelihood ratios that was developed in Chapter 10. The likelihood function for a sample of size k is

$$L(\mathbf{p}) = \prod_{i=1}^{k} p_i^{f_i},$$

where $f_i = f_i(\mathbf{x})$ is the observed frequency of the value x_i. The maximum value of $L(\mathbf{p})$ if H_0 is true is

$$\max L(\mathbf{p}) = L(\boldsymbol{\pi}) = \prod_{i=1}^{k} \pi_i^{f_i}. \tag{11.2}$$

To find the maximum value of $L(\mathbf{p})$ if H_a is true, we need to know the ML estimator of $\mathbf{p}$, that is, $\widehat{\mathbf{p}}$. So we must maximize

$$\ln L(\mathbf{p}) = \sum_{i=1}^{k} f_i \ln p_i, \tag{11.3a}$$

subject to the constraint

$$\sum_{i=1}^{k} p_i = 1. \tag{11.3b}$$

Introducing the Lagrange multiplier Λ, the variational function is

$$P = \ln L(\mathbf{p}) - \Lambda \left[\sum_{i=1}^{k} p_i - 1 \right],$$

and setting $\partial P / \partial p_i = 0$, gives $p_i = f_i / \Lambda$. Now since

$$\sum_{j=1}^{k} p_j = 1 = \frac{1}{\Lambda} \sum_{j=1}^{k} f_j = \frac{n}{\Lambda},$$

the required ML estimator is $p_i = f_i / n$. Thus, the maximum value of $L(\mathbf{p})$ if H_a is true, is

$$L(\widehat{\mathbf{p}}) = \prod_{i=1}^{k} \left(\frac{f_i}{n} \right)^{f_i}. \tag{11.4}$$

The likelihood ratio is therefore, from (11.2) and (11.4),

$$\lambda = \frac{L(\boldsymbol{\pi})}{L(\widehat{\mathbf{p}})} = n^n \prod_{i=1}^{k} \left(\frac{\pi_i}{f_i} \right)^{f_i}. \tag{11.5}$$

Finally, H_0 is accepted if $\lambda < \lambda_c$, where λ_c is a given fixed value of λ that depends on the confidence level of the test, and H_0 is rejected if $\lambda > \lambda_c$.

Example 11.1

A die is thrown 60 times and the resulting frequencies of the faces are shown in the table below.

Face	1	2	3	4	5	6
Frequency	9	8	12	11	6	14

Test whether the die is 'true' at a 10% significance level.
Solution: From Equation (11.5),

$$-2 \ln \lambda = -2 \left[n \ln n + \sum_{i=1}^{k} f_i \ln \left(\frac{\pi_i}{f_i} \right) \right], \tag{11.6}$$

where

$$\pi_i = 1/6; \quad n = 60; \quad \text{and} \quad k = 6.$$

Thus,

$$-2 \ln \lambda = -2[60 \ln 60 - 60 \ln 6 - 9 \ln 9 \ldots - 14 \ln 14] \approx 4.3.$$

We showed in Chapter 10 that $-2 \ln \lambda$ is approximately distributed as χ^2, with $(k-1)$ degrees of freedom, in this case five. (There are only 5 degrees of freedom, not 6, because of the constraint (11.3b).) From Table C.4 we find

$$\chi^2_{0.1}(5) = 9.24,$$

and since $\chi^2 < 9.24$, we can accept the hypothesis that the die is true, that is,

$$H_0 : p_i = 1/6, \quad i = 1, 2, ..., 6,$$

at a 10% significance level. $\alpha < 0.5$.

An alternative goodness-of-fit test is due Pearson. He considered the statistic

$$X^2 = \sum_{i=1}^{k} \frac{(f_i - e_i)^2}{e_i} = \sum_{i=1}^{k} \frac{(f_i - n\pi_i)^2}{n\pi_i}, \tag{11.7}$$

where f_i are the observed frequencies, and e_i are the expected frequencies under the null hypothesis. Pearson showed that X^2 is approximately distributed as χ^2 with $(k-1)$ degrees of freedom for large values of n. At first sight the two statistics X^2 and $-2 \ln \lambda$ appear to be unrelated, but in fact they can be shown to be equivalent asymptotically. To see this, we define

$$\Delta_i \equiv (f_i - n\pi_i)/n\pi_i,$$

and write (11.6) as

$$-2 \ln \lambda = 2 \sum_{i=1}^{k} \ln(1 + \Delta_i).$$

For small Δ_i, the logarithm can be expanded as

$$\ln(1 + \Delta_i) = \Delta_i - \Delta_i^2/2 + \cdots,$$

which gives

$$-2 \ln \lambda = 2 \sum_{i=1}^{k} \left[(f_i - n\pi_i) + n\pi_i \right] \left[\Delta_i - \frac{1}{2}\Delta_i^2 + O(n^{-3}) \right]$$

$$= 2 \sum_{i=1}^{k} \left[n\pi_i \Delta_i^2 + n\pi_i \Delta_i - \frac{1}{2} n\pi_i \Delta_i^2 + O(n^{-3}) \right].$$

Using the definition of Δ_i and the fact that

$$\sum_{i=1}^{k} \Delta_i \pi_i = 0,$$

gives

$$-2\ln\lambda = \sum_{i=1}^{k}\left[n\pi_i\Delta_i^2 + O(n^{-3})\right] = X^2\left[1 + O(n^{-3})\right].$$

So X^2 and $-2\ln\lambda$ are asymptotically equivalent statistics, although they will differ for small samples. Using the data of Example 11.1, gives $X^2 = 4.20$, compared to $-2\ln\lambda = 4.30$. The Pearson statistic is easy to calculate and in practice is widely used. It is usually written

$$\chi^2 = \sum_{i=1}^{k}\frac{(o_i - e_i)^2}{e_i}, \tag{11.8}$$

where o_i are the observed frequencies, and e_i are the expected frequencies under the null hypothesis.

Example 11.2

A manufacturing process for an electrical component results in four different quality levels which, in increasing order, are A, B, C, and D. The quality of each component is independent of the quality of any other component and the probabilities of a component having quality level A, B, C, or D are 0.20, 0.20, 0.25, and 0.35, respectively. A laboratory purchases several components and pays a lower price by agreeing to accept a sample chosen at random from examples with qualities A, B, C, and D. To test that the manufacturer is not in practice supplying more lower quality components than expected, the lab tests a random sample of 60 of them and finds 15, 7, 19, and 19 are of quality A, B, C, and D, respectively. At what significance level do the components not satisfy the manufacturer's claim?

Solution: The hypothesis being tested is

$$H_0 : p_i = \pi_i, \quad i = 1, 2, \ldots, k$$

where

$$\pi_i = 0.20, 0.20, 0.25, 0.35, \text{ for } i = 1, 2, 3, 4, \text{ respectively.}$$

We start by calculating the expected frequencies under H_0, that is, $e_i = 60\pi_i$. Thus

$$e_i = 12, 12, 15, 21, \text{ for } i = 1, 2, 3, 4, \text{ respectively.}$$

Then from (11.8) we find $\chi^2 = 4.09$ for 3 degrees of freedom. The p-value of the test is therefore $p = P[\chi^2 \geq 4.09 : H_0]$ and from Table C.4 this is approximately 0.25. Thus H_0 would be rejected at a significance level of 0.25. Since a probability of 0.25 for rejecting H_0 when it is true is not very useful, a lower significance level is likely to be chosen and the tested sample indicates the manufacturer is fulfilling the contract.

11.1.2 Continuous distributions

For the case of continuous distributions, the null hypothesis is usually that a population is described by a certain density function $f(x)$. This hypothesis may be tested by dividing the observations into k intervals and then comparing the observed interval frequencies $o_i(i = 1, 2, ..., k)$ with the expected values $e_i(i = 1, 2, ..., k)$ predicted by the postulated density function, using Equation (11.8). For practical work, a rule of thumb is that the number of observations in each bin should not be less than about 5. This may necessitate combining data from two or more bins until this criterion is satisfied.

If the expected frequencies π_i are unknown, but are estimated from the sample in terms of r parameters to be $\widehat{\pi}_i$ then the statistic

$$\chi^2 = \sum_{i=1}^{k} \frac{(o_i - n\widehat{\pi}_i)^2}{n\widehat{\pi}_i},\qquad(11.9)$$

is also distributed as χ^2 but now with $(k-1-r)$ degrees of freedom. In using the chi-square test of a continuous distribution with unknown parameters one has always to be careful that the method of estimating the parameters still leads to an asymptotic χ^2 distribution. In general, this will not be true if the parameters are estimated either from the original data *or* from the grouped data. The correct procedure is to estimate the parameters $\boldsymbol{\theta}$ by the ML method using the likelihood function

$$L(\boldsymbol{\theta}) = \prod_{i=1}^{k} [p_i(\mathbf{x}, \boldsymbol{\theta})]^{f_i},$$

where p_i is the appropriate density function. Such estimates are usually difficult to obtain, but if one uses the simple estimates then one may be working at a higher significance level than intended. This will happen, for example, in the case of the normal distribution.

Example 11.3

A quantity x is measured 100 times in the range 0–40 and the observed frequencies o in eight bins are given below.

x	0–5	5–10	10–15	15–20	20–25	25–30	30–35	35–40
o	5	9	15	26	18	14	8	5

Test the hypothesis that x has a normal distribution.

Solution: If the hypothesis is true, then the expected frequencies e have a normal distribution with the same mean and standard deviation as the observed data. The latter may be estimated from the sample and are $\bar{x} = 19.6$ and $s = 8.77$. Using these we can find the expected frequencies for each of the bins. This may be done either by finding the z value at each of the bin boundaries and using Table C.1 of the normal distribution function or by direct integration of the normal *pdf*. Either way, the expected frequencies are:

$$e \quad 3.5 \quad 8.9 \quad 16.3 \quad 21.8 \quad 21.3 \quad 15.1 \quad 7.8 \quad 2.9$$

Since the first and last bins have entries less than 5, we should combine these with the neighboring bins, so that we finally get

$$
\begin{array}{ccccccc}
o & 14 & 15 & 26 & 18 & 14 & 13 \\
e & 12.4 & 16.3 & 21.8 & 21.3 & 15.1 & 10.7
\end{array}
$$

Then from (11.9) we find $\chi^2 = 2.16$ and because we have estimated two parameters from the data, this is for 3 degrees of freedom. From Table C.4, there is a probability of greater than 50% of finding a value at least large as this for 3 degrees of freedom, so the hypothesis of normality is consistent with the observations.

The method described in Chapter 8 of using a χ^2 distribution to determine the quality of the fit of data $y_i(x)$ to an assumed function $f(x)$ (see Example 8.3) is often called the χ^2 goodness of fit test and is used like Pearson's χ^2 test. If the error is known to be normally distributed, with a known variance σ^2, then

$$\chi^2 = \frac{1}{\sigma^2} \sum_{i=1}^{n} \{y_i(x) - f(x)\}^2$$

has the χ^2 distribution with n degrees of freedom. The significance α determines the value of χ^2 defining the critical region for the hypothesis that $f(x)$ is the true value of the data, discounting a random error.

If $f(x)$ is calculated using parameters determined from the data, then the degrees of freedom of the distribution are reduced in the usual way. If the data elements do not have a common variance or are possibly not independent, the formula becomes

$$z = (\mathbf{y} - \mathbf{f})^{\mathrm{T}} \mathbf{V}^{-1} (\mathbf{y} - \mathbf{f}),$$

where $\mathbf{V}$ is the variance matrix, $\mathbf{y}$ is the vector of data, and $\mathbf{f}$ is the 'true' value corresponding to each data point.

A very common situation is two measurements of a parameter are measured or two samples are taken and the precision (variance) of the measurement is known. The χ^2 test can be used to test the hypothesis that the two measurements or two means of the sample are the same by comparing the difference of the measurements (or means) to a 'true' value of zero. In the case of two measurements, the variance is the sum of the variances of the two data points and in the case of two means, the variance is the sum of the sample variances of the means. That is, the variance of the measurements in the sample divided by the number of measurements. Problems 11.3 and 11.5 provide simple examples of the χ^2 goodness of fit test.

A disadvantage of Pearson's χ^2 test is that the data must be binned, which could be a problem in cases where the number of observations is small. One method that does not require binning is the *Kolmogorov-Smirnov test*. In this test we start with a sample y_i of size n from a

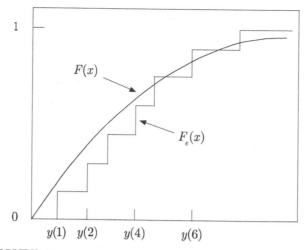

FIGURE 11.1   Construction of the Kolmogorov-Smirnov statistic.

continuous distribution and test the hypothesis H_0 that the distribution function is $F(x)$. This is done by defining a piecewise continuous function $F_e(x)$ as the proportion of the observed values that are less than or equal to x, that is, $(y_i \leq x)/n$. If H_0 is true, $F_e(x)$ will be close to $F(x)$ and a natural test statistic is the quantity $D \equiv \max|F_e(x) - F(x)|$, where the maximum is found by considering all values of x. This is called the Kolmogorov-Smirnov statistic.

To compute D, the sample is first ordered in increasing values and relabeled $x(1), x(2), ..., x(n)$. Then $F_e(x)$ consists of a series of steps with increases of $1/n$ at the points $x(1), x(2), ..., x(n)$ as shown in Figure 11.1. Now because D is defined in terms of a modulus, it can be written

$$D = \max\{|F(x) - F_e(x)|\} = \max\left\{\max\left(\left|F[x(j)] - \frac{j}{n}\right|, \left|F[x(j)] - \frac{j-1}{n}\right|\right)\right\}, \qquad (11.10)$$

where j takes all values from 1 to n. So, if the data lead to a value of $D = d$, the p-value of the test is

$$p = P[D \geq d : H_0].$$

Kolmogorov showed that in the limit as $n \to \infty$

$$P[\sqrt{n}\,D \leq z : H_0] \to L(z)$$

where $L(z)$, which does not depend on n, is given by

$$L(z) = 1 - 2\sum_{k=1}^{\infty} (-1)^{k-1} e^{-2k^2 z^2}.$$

There is no simple formula for the distribution of D for finite values of n but tables are available and most statistical packages have a method for calculating the distribution. It is important to note that the distribution F must be completely specified in advance. If the sample is used to estimate the parameters, then the distribution function of D depends on the distribution F and the Kolmogorov-Smirnov test does not apply.

It is often useful to compare two samples and test the hypothesis that both samples are from the same distribution, without knowing the form of the distribution. Smirnov showed that this hypothesis can be tested using the statistic

$$D = \max\{|F_{e1}(x) - F_{e2}(x)|\},$$

where $F_{e1}(x)$ and $F_{e2}(x)$ are the piecewise continuous distribution functions for the first and second samples, respectively and the maximum is taken over all values of x in both samples. The distribution function for D comparing two samples is the same as the distribution function D for comparing a sample to a known distribution.

For finite values of n, a test at significance level α can be found by considering the quantity

$$D^* \equiv (\sqrt{n} + 0.12 + 0.11/\sqrt{n})D, \tag{11.11}$$

which approximates to $\sqrt{n}\,D$ for large samples.[1] Then if $P[D^* \geq d^*] = \alpha$, the critical values of d_α^* are:

$$d_{0.1}^* = 1.224, \quad d_{0.05}^* = 1.358, \quad d_{0.025}^* = 1.480, \quad d_{0.01}^* = 1.626. \tag{11.12}$$

A test at a significance level α would reject H_0 if the observed value of D^* is at least as large as d_α^*.

Example 11.4

A sample of size six is drawn from a population and the values in ascending order are:

1	2	3	4	5	6
0.20	0.54	0.71	1.21	1.85	2.45

Test the hypothesis that the sample comes from a population with the distribution function of Example 3.2.

Solution: From the data and the distribution function of Example 3.2, we can construct the following table:

j	1	2	3	4	5	6
$F[x(j)]$	0.001	0.018	0.035	0.123	0.283	0.443
$j/n - F[x(j)]$	0.166	0.316	0.465	0.544	0.551	0.557
$F[x(j)] - (j-1)/n$	0.001	-0.149	-0.298	-0.377	-0.384	-0.390

From this table and Equation (11.10) we have $D = 0.557$ and hence from (11.11), $D^* = 1.455$. Finally, using the critical values given in (11.12), we reject the null hypothesis at the 0.05 level of significance, but not at the 0.025 level.

---

[1] This and similar approximations for several other useful statistics may be found in M.A. Stephens *Use of the Kolmogorov-Smirnov, Cramer-Von Mises and Related Statistics Without Extensive Tables*, Journal of the Royal Statistical Society. Series B (1970), vol. 32, no. 1, pp. 115–122.

11.1.3 Linear hypotheses

In Section 8.1.3 we briefly mentioned the use of the χ^2 and F distributions as goodness-of-fit tests in connection with the use of the linear least-squares method of estimation. These applications were designed to test hypotheses concerning the quality of the approximation of the observations by some assumed expression linear in the parameters. We shall generalize that discussion now to consider some other hypothesis tests that can be performed using the least-squares results.

We have seen in Section 8.1 that the weighted sum of residuals $S = \mathbf{R}^T \mathbf{V}^{-1} \mathbf{R}$, where

$$\mathbf{R} = \mathbf{Y} - \mathbf{\Phi}\mathbf{\Theta},$$

and $\mathbf{V}$ is the variance matrix of the observations, is distributed, for normal errors or large samples, as χ^2 with $(n-p)$ degrees of freedom, where n is the number of observations and p is the number of parameters $\theta_k (k = 1, 2, ..., p)$. We also saw (compare Equation (8.31)) that

$$\mathbf{R}^T \mathbf{V}^{-1} \mathbf{R} = \left(\mathbf{Y} - \mathbf{Y}^0\right)^T \mathbf{V}^{-1}\left(\mathbf{Y} - \mathbf{Y}^0\right) - \left(\widehat{\mathbf{\Theta}} - \mathbf{\Theta}\right)^T \mathbf{M}^{-1}\left(\widehat{\mathbf{\Theta}} - \mathbf{\Theta}\right),$$

where $\mathbf{M}$ is the variance matrix of the parameters. It follows from the additive property of χ^2 that since

$$\left(\mathbf{Y} - \mathbf{Y}^0\right)^T \mathbf{V}^{-1}\left(\mathbf{Y} - \mathbf{Y}^0\right)$$

is distributed as χ^2 with n degrees of freedom, the quantity

$$\left(\widehat{\mathbf{\Theta}} - \mathbf{\Theta}\right)^T \mathbf{M}^{-1}\left(\widehat{\mathbf{\Theta}} - \mathbf{\Theta}\right)$$

is distributed as χ^2 with p degrees of freedom. To test deviations from the least-squares estimates for the parameters we need to know the distribution of

$$\left(\widehat{\mathbf{\Theta}} - \mathbf{\Theta}\right)^T \mathbf{E}\left(\widehat{\mathbf{\Theta}} - \mathbf{\Theta}\right),$$

where $\mathbf{E}^{-1}$ is the error matrix of Equation (8.32). In the notation of Section 8.1.2,

$$\mathbf{E}^{-1} = \widehat{w}\left(\mathbf{\Phi}^T \mathbf{W} \mathbf{\Phi}\right)^{-1}, \tag{11.13}$$

and so

$$\left(\widehat{\mathbf{\Theta}} - \mathbf{\Theta}\right)^T \mathbf{E}\left(\widehat{\mathbf{\Theta}} - \mathbf{\Theta}\right) = \frac{1}{\widehat{w}}\left(\widehat{\mathbf{\Theta}} - \mathbf{\Theta}\right)^T \left(\mathbf{\Phi}^T \mathbf{W} \mathbf{\Phi}\right)\left(\widehat{\mathbf{\Theta}} - \mathbf{\Theta}\right).$$

That is,

$$\left(\widehat{\boldsymbol{\Theta}} - \boldsymbol{\Theta}\right)^{T} \mathbf{E}\left(\widehat{\boldsymbol{\Theta}} - \boldsymbol{\Theta}\right) = \left(\widehat{\boldsymbol{\Theta}} - \boldsymbol{\Theta}\right)^{T} \mathbf{M}^{-1}\left(\widehat{\boldsymbol{\Theta}} - \boldsymbol{\Theta}\right)\left(\frac{w}{\widehat{w}}\right).$$

But we know that

$$\left(\widehat{\boldsymbol{\Theta}} - \boldsymbol{\Theta}\right)^{T} \mathbf{M}^{-1}\left(\widehat{\boldsymbol{\Theta}} - \boldsymbol{\Theta}\right)$$

is distributed as χ^2 with p degrees of freedom and $\widehat{w}/w$ is distributed as $\chi^2(n-p)/(n-p)$, so

$$\left(\widehat{\boldsymbol{\Theta}} - \boldsymbol{\Theta}\right)^{T} \mathbf{E}\left(\widehat{\boldsymbol{\Theta}} - \boldsymbol{\Theta}\right)/p$$

is distributed as

$$\frac{\chi^2(p)/(p)}{\chi^2(n-p)/(n-p)} = F(p, n-p).$$

Thus, to test the hypothesis $H_0 : \boldsymbol{\Theta} = \boldsymbol{\Theta}_0$, we compute the test statistic

$$F_0 = \left(\widehat{\boldsymbol{\Theta}} - \boldsymbol{\Theta}\right)^{T} \mathbf{E}\left(\widehat{\boldsymbol{\Theta}} - \boldsymbol{\Theta}\right)/p \tag{11.14}$$

and reject the hypothesis at a significance level of α if $F_0 > F_\alpha(p, n-p)$.

The foregoing discussion is based on the work of Section 8.1.2, where we considered the least-squares method without constraints on the parameters. By analogy with the work of Section 8.2, we will now generalize the discussion to include the general linear hypothesis

$$H_0: \sum_{i=1}^{p} C_{li}\theta_i = Z_l, \quad l \leq p. \tag{11.15}$$

This may be a hypothesis about all the parameters or any subset of them. The null hypothesis H_0 may be tested by comparing the least-squares solution for the weighted sum of residuals when H_0 is true, that is, S_c, with the sum in the unconstrained situation, that is, S_u. Comparing with (8.57), where S_c is labeled S, S_u is the first term in the expected value on the right, and the second term is the additional sum of residuals $S_a = S_c - S_u$ which is present if the hypothesis H_0 is true. It is clear from the additive property of χ^2 that S_a is distributed as χ^2 with l degrees of freedom, independently of S_u, which itself is distributed as χ^2 with $(n-p)$ degrees of freedom. Thus, the ratio

$$F = \frac{S_a/l}{S_u/(n-p)}, \tag{11.16}$$

is distributed as $F(l, n-p)$. Noting that the numerator on the right-hand side of (8.59) is the right-hand side of (8.57) multiplied by w and that using (8.32) and the unbiased estimate $\widehat{w}$ of w, it is not difficult to show that $\widehat{w}\mathbf{N}^{-1} = \mathbf{E}^{-1}$, we can show that

$$F = \left(\mathbf{Z} - \mathbf{C}\widehat{\mathbf{\Theta}}\right)^T \left(\mathbf{C}\mathbf{E}^{-1}\mathbf{C}^T\right)^{-1} \left(\mathbf{Z} - \mathbf{C}\widehat{\mathbf{\Theta}}\right) / l. \tag{11.17}$$

Thus H_0 is rejected at the α significance level if $F > F_\alpha(l, n-p)$. (Compare the discussion at the end of Section 8.1.3.)

Example 11.5

An experiment based on 10 measurements results in the following estimates for 3 parameters,

$$\widehat{\theta}_1 = 2; \quad \widehat{\theta}_2 = 4; \quad \widehat{\theta}_3 = 1,$$

with an associated error matrix

$$\mathbf{E}^{-1} = \begin{pmatrix} 1 & 0 & 0 \\ 0 & 2 & 1 \\ 0 & 1 & 1 \end{pmatrix}.$$

Test the hypothesis

$$H_0: \theta_1 = 0; \quad \theta_2 = 0$$

at the 5% significance level.

Solution: For the above hypothesis,

$$\widehat{\mathbf{\Theta}} = \begin{pmatrix} 2 \\ 4 \\ 1 \end{pmatrix}; \quad \mathbf{C} = \begin{pmatrix} 1 & 0 & 0 \\ 0 & 1 & 0 \end{pmatrix}; \quad \mathbf{Z} = \begin{pmatrix} 0 \\ 0 \end{pmatrix},$$

and the calculated value of F from Equation (11.17) is 6. From Table C.5 of the F distribution, we find that

$$F_\alpha(l, n-p) = F_{0.05}(2, 7) = 4.74,$$

and so we can reject the hypothesis at a 5% significance level.

Finally, it is useful to consider the power of the test of the general linear hypothesis, that is, find the distribution of F if H_0 is not true. Now $S_u/(n-p)$ is distributed as $\chi^2/(n-p)$ regardless of whether H_0 is true or false, but S_a/l is only distributed as χ^2/l if H_0 is true. If H_0 is false, then S_a/l will, in general, be distributed as *noncentral* χ^2. It follows that F is distributed as a *noncentral F distribution*.[2] Tables of these distributions are available, and statistical libraries are

---

[2] Noncentral χ^2 and F distributions are discussed in detail in M. Kendall and A. Stuart *The Advanced Theory of Statistics, Volume 2, Griffin* (1979).

available to generate the distributions in several common computer languages including Python, R, and C++. Either may be used to construct the power curves. A feature of the noncentral F distribution is that the power of the test increases as λ increases.

11.2 Tests for independence

The χ^2 procedure of Section 11.1 can also be used to construct a test for the independence of variables. Suppose n observations have been made and the results are characterized by two random variables x and y that can take the discrete values $x_1, x_2, ..., x_r$ and $y_1, y_2, ..., y_c$, respectively. (Continuous variables can be accommodated by dividing the range into intervals, as described in Section 11.1.2.) If the number of times the value x_i is observed together with y_j is n_{ij}, then the data can be summarized in the matrix form shown in Table 11.1, called an $r \times c$ *contingency table*.

TABLE 11.1   An $r \times c$ contingency table.

	y_1	y_2	$\cdots$	y_c
x_1	n_{11}	n_{12}	$\cdots$	n_{1c}
x_2	n_{21}	n_{22}	$\cdots$	n_{2c}
$\vdots$	$\vdots$	$\vdots$		$\vdots$
x_r	n_{r1}	n_{r2}	$\cdots$	n_{rc}

If we denote by p_i the marginal probability for x_i, that is, the probability for x to have the value x_i, independent of the value of y, and likewise denote by q_j the marginal probability for y_j, then if the null hypothesis H_0 is that x and y are independent variables is true, the probability for observing x_i simultaneously with y_j is $p_i q_j$ ($i = 1, 2, ..., r; j = 1, 2, ..., c$). Since the marginal probabilities are not specified in the null hypothesis they must be estimated, and using the maximum likelihood method this gives

$$\widehat{p}_i = \frac{1}{n} \sum_{j=1}^{c} n_{ij}, \quad i = 1, 2, ..., r, \tag{11.18a}$$

and

$$\widehat{q}_j = \frac{1}{n} \sum_{i=1}^{c} n_{ij}, \quad j = 1, 2, ..., c, \tag{11.18b}$$

and if H_0 is true, the expected values for the elements of the contingency table are $n\widehat{p}_i\widehat{q}_j$. So H_0 can be tested by calculating the quantity

$$\chi^2 = \sum_{i=1}^{r} \sum_{j=1}^{c} \frac{\left(n_{ij} - n\widehat{p}_i\widehat{q}_j\right)^2}{n\widehat{p}_i\widehat{q}_j} \tag{11.19}$$

and comparing it with χ^2_α at a significance level α. It remains to find the number of degrees of freedom for the statistic χ^2. Firstly, we note that because

$$\sum_{i=1}^{r} \widehat{p}_i = \sum_{j=1}^{c} \widehat{q}_j = 1, \tag{11.20}$$

only $(r-1)$ parameters p_i and $(c-1)$ parameters q_i need to be estimated, that is, $(r+c-2)$ in total. Therefore, the number of degrees of freedom is

$$(\text{number of entries} - 1) - (r+c-2) = (r-1)(c-1). \tag{11.21}$$

Example 11.6

A laboratory has three pieces of test apparatus of the same type, and they are used by four technicians on a 1-month rota. A record is kept of the number of machine breakdowns for each month, and the average number is shown in the table according to which technician was using the machine.

Technician	Equipment			
	E1	E2	E3	Total
T1	3	6	1	10
T2	0	1	2	3
T3	3	2	4	9
T4	6	2	1	9
Total	12	11	8	31

Are the variables E, the equipment number, and T, the technician who used the equipment, independent random variables in determining the rate of breakdowns?

Solution: In the notation above, $r = 4, c = 3$ and $n = 31$. So, from (11.18), using the data given in the table,

$$\widehat{p}_1 = \frac{10}{31}, \ \widehat{p}_2 = \frac{3}{31}, \ \widehat{p}_3 = \frac{9}{31}, \ \widehat{p}_4 = \frac{9}{31},$$

and

$$\widehat{q}_1 = \frac{12}{31}, \ \widehat{q}_2 = \frac{11}{31}, \ \widehat{q}_3 = \frac{8}{31}.$$

Then from (11.19)

$$\chi^2 = \sum_{i=1}^{c} \sum_{j=1}^{r} \frac{\left(n_{ij} - n\widehat{p}_i\widehat{q}_j\right)^2}{n\widehat{p}_i\widehat{q}_j} = 10.7,$$

and this is for $(r-1)(c-1) = 6$ degrees of freedom. The p-value is

$$p = P\left[\chi_6^2 \geq 10.7 : H_0\right] = \alpha$$

and from Table C.4, $\alpha \approx 0.1$, so the data are consistent at the 10% significance level with the independence of the two variables in determining the number of breakdowns.

11.3 Nonparametric tests

All the tests that were discussed in Chapter 10 assumed that the population distribution was known. But there are cases where this is not true, and to deal with these situations we can order the observations by rank and apply tests that do not rely on information about the underlying distribution. These are variously called *nonparametric tests* or *distribution-free tests*. Not making assumptions about the population distribution is both a strength and a weakness of such tests; a strength because of their generality, a weakness because they do not use all the information contained in the data. An additional advantage is that such tests are usually quick and easy to implement, but because they are less powerful than more specific tests we have discussed, the latter should usually be used if there is a choice. In addition, although the data met in physical science are usually numeric, occasionally we must deal with nonnumeric data, for example, when testing a piece of equipment, the outcome could be 'pass' or 'fail'. Some nonparametric tests can also be applied to these situations. The discussion will be brief, and not all proofs of statements will be given.

11.3.1 Sign test

In this section we pose the question of whether it is possible to test hypotheses about the average of a population when its distribution is unknown. One simple test that can do this is the *sign test*, and as an example of its use we will test the null hypothesis $H_0 : \mu = \mu_0$ against some alternative, such as $H_a : \mu = \mu_a$ or $H_a : \mu > \mu_0$ using a random sample of size n in the case where n is small, so that the sampling distribution is not normal. In general, if we make no assumption about the form of the population distribution, then in the sign test and those that follow, μ refers to the median, but if we know that the population distribution is symmetric, then μ is the arithmetic mean. For simplicity the notation μ will be used for both cases. We start by assigning a plus sign to all data values that exceed μ_0 and a minus sign to all those that are less than μ_0. We would expect the plus and minus sign to be approximately equal and any deviation will lead to rejection of the null hypothesis at some significance level. In principle, because we are dealing with a continuous distribution, no observation can in principle be exactly equal to μ_0, but in practice approximate equality will occur depending on the precision with which the measurements are made. In these cases, the points of 'practical equality' are removed from the data set and the value of n reduced accordingly. The test statistic X is the number of plus signs in the sample (or equally we could use the number of minus signs). If H_0 is true, the probabilities of obtaining a plus or minus sign are equal to $1/2$ and so X has a binomial distribution with $p = p_0 = 1/2$. Significance levels can thus be obtained from the binomial distribution for one-sided and two-sided tests at any given level α.

For example, if the alternative hypothesis is $H_a : \mu > \mu_0$, then the largest critical region of size not exceeding α is obtained from the inequality $x \geq k_\alpha$, where

$$\sum_{x=k_\alpha}^{n} B(x : n, p_0) \leq \alpha, \qquad (11.22a)$$

and B is the binomial probability and $p_0 = p = \frac{1}{2}$ if H_0 is true. Similarly, if $H_a : \mu < \mu_0$, we form the inequality $x \leq k'_\alpha$, where k'_α is defined by

$$\sum_{x=0}^{k'_\alpha} B(x : n, p_0) \leq \alpha. \qquad (11.22b)$$

Finally, if $H_a : \mu \neq \mu_0$, that is, we have a two-tailed test, then the largest critical region is defined by

$$x \leq k'_{\alpha/2} \quad \text{and} \quad x \geq k_{\alpha/2}. \qquad (11.22c)$$

For sample sizes greater than about 10, the normal approximation to the binomial may be used with mean $\mu = np$ and $\sigma^2 = np(1-p)$.

Example 11.7

A mobile phone battery needs to be regularly recharged even if no calls are made. Over 12 periods when charging was required, it was found that the intervals in hours between charging were:

$$50 \quad 35 \quad 45 \quad 65 \quad 39 \quad 38 \quad 47 \quad 52 \quad 43 \quad 37 \quad 44 \quad 40$$

Use the sign test to test at a 10% significance level the hypothesis that the battery needs recharging on average every 45 hours.

Solution: We are testing the null hypothesis $H_0 : \mu_0 = 45$ against the alternative $H_a : \mu_0 \neq 45$. Firstly, we remove the data point with value 45, reducing n to 11, and then assign a plus sign to those measurements greater than 45 and a minus sign to those less than 45. This gives $x = 4$ as the number of plus signs. As this is a two-tailed test, we need to find the values of $k_{0.05}$ and $k'_{0.05}$ for $n = 11$. From Table C.2, these are: $k'_{0.05} = 3$ and $k_{0.05} = 9$. Since $x = 4$ lies in the acceptance region, we accept the null hypothesis at this significance level.

The sign test can be extended in a straightforward way to two-sample cases, for example, to test the hypothesis that $\mu_1 = \mu_2$ using samples of size n drawn from two nonnormal distributions. In this case the differences $(i = 1, 2, ..., n)$ of each pair of observations is replaced by a plus or minus sign depending on whether d_i is greater than or less than zero, respectively. If the null hypothesis instead of being $\mu_1 - \mu_2 = 0$ is instead $\mu_1 - \mu_2 = d$, then the procedure is the same, but the quantity d is subtracted from each d_i before the test is made.

11.3.2 Signed-rank test

The sign test uses only the positive and negative signs of the differences between the observations and μ_0 in a one-sample case (or the signs of the differences d_i between observations in a paired sample case). Because it ignores the magnitudes of the differences, it can, for example, lead to conclusions that differ from those obtained using the t-distribution, which assumes the population distribution is normal. There is however another test, called the *Wilcoxon signed-rank test*, or simply the *signed-rank test*, that does take account of the magnitude of these differences. The test proceeds as follows.

We will assume that the null hypothesis is $H_0 : \mu = \mu_0$ and that the distribution is symmetric. If the latter is not true, then the tests will refer to the median of the distribution. Firstly, the differences $d_i = x_i - \mu_0$ $(i = 1, 2, ..., n)$ or $d_i = x_i - y_i$, are found and any that are zero discarded. The absolute values of the remaining set of differences are then ranked in ascending order, that is, rank 1 assigned to the smallest absolute value of d_i, rank 2 assigned to the second smallest absolute value of d_i, etc. If $|d_i|$ is equal for two or more values, their rank is assigned to be the average of the ranks they would have had if they had been slightly different and so distinguishable. For example, if two differences are equal and notionally have rank 5, both would be assigned a rank of 5.5 and the next term would be assigned a rank 7.

We now define w_+ and w_-, respectively, to be the sum of the rank numbers corresponding to $d_i > 0$ and $d_i < 0$. For different samples of the same size n, the values of these statistics will vary, and for example to test $H_0 : \mu = \mu_0$ against the alternative $H_a : \mu < \mu_0$, that is, a one-tailed test, we would reject H_0 if w_+ is small and w_- large. (Similarly for testing $H_0 : \mu_1 = \mu_2$ against $H_a : \mu_1 < \mu_2$.) Likewise, if w_+ is large and w_- small, we would accept $H_a : \mu > \mu_0$ (and similarly $H_a : \mu_1 > \mu_2$). For two-tailed tests with $H_a : \mu \neq \mu_0$, or $H_a : \mu_1 \neq \mu_2$, H_0 would be rejected if either w_+, w_-, and hence w, were sufficiently small. If the two-tailed case is generalized to $H_0 : \mu_1 - \mu_2 = d_0$, then the same test can be used by subtracting d_0 from each of the differences d_i, as in the sign test, and in this case the distribution need not be symmetric.

The final question is to decide what is 'large' and 'small' in this context. For small $n < 5$, it can be shown that provided $\alpha < 0.05$ for a one-tailed test, or less than 0.10 for a two-tailed test, any value of w will lead to acceptance of the null hypothesis. For $5 \leq n \leq 30$, critical values of w are given in Appendix C.7. The test is summarized in Table 11.2.

For n greater than about 25, it can be shown that the sampling distribution of w_+, or w_-, tends to a normal distribution with

$$\text{mean } \mu = \frac{n(n-1)}{4} \quad \text{and variance } \sigma^2 = \frac{n(n+1)(2n+1)}{24} \tag{11.23}$$

TABLE 11.2   Signed-rank test.

H_0	H_a	Test statistic
$\mu = \mu_0$	$\mu \neq \mu_0$	w
	$\mu < \mu_0$	w_+
	$\mu > \mu_0$	w_-
$\mu_1 = \mu_2$	$\mu_1 \neq \mu_2$	w
	$\mu_1 < \mu_2$	w_+
	$\mu_1 > \mu_2$	w_-

so that the required probabilities may be found using the standardized variable $z_\pm = (w_\pm - \mu)/\sigma$. For the range $5 \leq n \leq 25$ it is necessary to calculate the explicit probabilities or obtain values from tabulations, such as given in Table C.7.

Example 11.8

Rework Example 11.7 *using the signed-rank test.*
Solution: Using the data of Example 11.7 with $\mu_0 = 45$, we have

x_i	50	35	45	65	39	38	47	52	43	37	44	40
d_i	5	−10	0	20	−6	−7	2	7	−2	−8	−1	−5
$|d_i|$	5	10		20	6	7	2	7	2	8	1	5
Rank	4.5	10		11	6	7.5	2.5	7.5	2.5	9	1	4.5

where the point with $d_i = 0$ has been discarded and the rank of ties have been averaged. From the rank numbers we calculate $w_+ = 25.5$, $w_- = 40.5$, and so $w = \min\{w_+, w_-\} = 25.5$. From Table C.7, the critical region for $n = 11$ in a two-tailed test with $\alpha = 0.10$ is $w \leq 13$, so again we accept the null hypothesis at this significance level.

11.3.3 Rank-sum test

This is also due to Wilcoxon, and so is sometimes called the *Wilcoxon rank-sum test*, which is used to compare the means of two continuous distributions. When applied to the case of non-normal distributions, it is more powerful than the two-sample t-test discussed in Chapter 10.

We will use it to test the null hypothesis $H_0 : \mu_1 = \mu_2$ against some suitable alternative. Firstly, samples of sizes n_1 and n_2 are selected from the two populations, and if $n_1 \neq n_2$, we assume that $n_2 > n_1$. Then the $n_1 + n_2$ observations are arranged in ascending order and a rank number $1, 2, ..., n$ is assigned to each. In the case of ties, the rank number is taken to be the

TABLE 11.3   Rank-sum test.

H_0	H_a	Test statistic
$\mu_1 = \mu_2$	$\mu_1 \neq \mu_2$	u
	$\mu_1 < \mu_2$	u_1
	$\mu_1 > \mu_2$	u_2

mean of the rank numbers that the observations would have had if they had been distinguish-able, just as in the signed-rank test. We now proceed in a similar way to that used in the signed-rank test. Let $w_{1,2}$ be the sums of the rank numbers corresponding to the $n_{1,2}$ sets of observations.

With repeated samples, the values of $w_{1,2}$ will vary and may be viewed as random variables. So, just as for the signed-rank test, $H_0 : \mu_1 = \mu_2$ will be rejected in favor of $H_a : \mu_1 < \mu_2$ if w_1 is small and w_2 is large, and similarly it will be rejected in favor of $H_a : \mu_1 > \mu_2$ if w_1 is large and w_2 is small. For a two-tailed test, H_0 is rejected in favor of $H_a : \mu_1 \neq \mu_2$ if $w = \min\{w_1, w_2\}$ is sufficiently small. It is common practice to work with the statistics

$$u_{1,2} = w_{1,2} - \frac{n_{1,2}(n_{1,2} + 1)}{2}. \tag{11.24}$$

The test is summarized in Table 11.3.

For large sample sizes the variables $u_{1,2}$ are approximately normally distributed with

$$\text{mean } \mu = \frac{n_1 n_2}{2} \quad \text{and} \quad \text{variance } \sigma^2 = \frac{n_1 n_2 (n_1 + n_2 + 1)}{12}. \tag{11.25}$$

Conventionally this approximation is used for $n_1 \geq 10$ and $n_2 \geq 20$. Then, as usual, the statistics $z_{1,2} = (u_{1,2} - \mu)/\sigma$ are standardized normal variates and may be used to find suitable critical regions for the test. For smaller samples, critical values of u may be found from tables, such as Appendix C.8. The rank-sum test may be extended to accommodate many populations. This extension is called the Kruskal-Wallis test.

Example 11.9

A laboratory buys power packs from two sources S1 and S2. Both types have a nominal rating of 10 volts. The results of testing a sample of each type yield the following data for the actual voltage:

| S1 | 9 | 8 | 7 | 8 | | |
| S2 | 10 | 9 | 8 | 11 | 7 | 9 |

Use the rank-sum test to test the hypothesis that the two types of pack supply the same average voltage at the 10% significance level.

Solution: The data are first ranked as follows:

S1	9		8		7		8			
S2		10		9		8		11	7	9
Rank	7	9	4	7	1.5	4	4	10	1.5	7

Where we have averaged the ranking numbers for ties. From this table we calculate $w_1 = 16.5$, $w_2 = 38.5$, and hence, in the notations above, $u_1 = 10.5$ and $u_2 = 23.5$ so that $u = 10.5$. From Table C.8, the critical region u, for a two-tailed test at significance level 0.10 and sample sizes $n_1 = 4$ and $n_2 = 6$ is $u \leq 3$. Since the calculated value of u is greater than 3, we accept the hypothesis at the 10% significance level.

11.3.4 Run test

It is a basic assumption in much of statistics that a data set constitutes a simple random sample from some underlying population distribution. But we have noted in Section 5.1.1 that it is often difficult to ensure that randomness has been achieved and so a test for this is desirable. The *run test* does this. In this test, the null hypothesis H_0 is that the observed data set is a random sample.

To derive the test, we shall consider a simple case where the elements of the sample $x_i (i = 1, 2, ..., N)$ can only take the values A or B. (In a nonnumeric situation these might be 'pass' and 'fail'.) A *run* is then defined as a consecutive sequence of either A's or B's, for example

$$AAABAABBBBBAABBBABBAA.$$

In this example, there are $n = 10$ A's, $m = 11$ B's, with $N = n + m = 21$, and $r = 9$ runs. The total number of permutations of the N points may be found by asking how many ways the m B's can be located in the list of N items. Clearly, this is

$$_N C_m \equiv \binom{N}{m} = \frac{N!}{(N-m)!m!},$$

and if H_0 were true, each would be equally likely. The probability mass function for R, the number of runs (noting that if $n = N - m$ then $_N C_m = _N C_n$) is

$$P[R = k : H_0] \equiv \frac{\text{number of ways of obtaining } k \text{ runs}}{_N C_n}. \tag{11.26}$$

The numerator can be found by firstly considering only the n A's and finding how many ways the B's can be placed between them to result in k runs. If k is even, and the first run consists of A's, B's must be inserted in $(k/2 - 1)$ places between A's, and a run of B's must follow the last A. There are $(n - 1)$ locations where runs of B's can be inserted, giving $_{n-1} C_{k/2-1}$ distinct arrangements of runs of A's. There is a corresponding factor for the possible arrangements of the B's and the product of these two terms must be multiplied by a factor of two because the entire sequence could start with either an A or a B. Thus

$$P[R = 2k\colon H_0] = 2\frac{{}_{(n-1)}C_{(k-1)} \times {}_{(m-1)}C_{(k-1)}}{{}_N C_n}. \tag{11.27a}$$

An analogous argument for k odd gives

$$P[R = 2k+1\colon H_0] = \frac{\left({}_{(n-1)}C_{(k-1)} \times {}_{(m-1)}C_k\right) + \left({}_{(n-1)}C_k \times {}_{(m-1)}C_{(k-1)}\right)}{{}_N C_n}. \tag{11.27b}$$

For small values of n and m these expressions can be used to calculate the required probabilities, but it is rather tedious. Usually one consults tabulations, an example of which is Table C.9. This table gives critical values of the number of runs r for given values of n and m. For example, for a two-tailed test, the critical region is defined by the inequalities $r \le a$ and $r \ge b$, where a is the largest value of r in the table for which

$$P[r \le a\colon H_0] \le \alpha/2, \tag{11.28a}$$

and b is the smallest value of r for which

$$P[r \ge b\colon H_0] \le \alpha/2 \tag{11.28b}$$

For large n and m, it can be shown that R is approximately normally distributed with mean and variance given by

$$\mu = \frac{2nm}{n+m} + 1 \text{ and } \sigma^2 = \frac{2nm\,(2nm - n - m)}{(n+m)^2(n+m-1)}, \tag{11.29}$$

so the variable $z \equiv (r-\mu)/\sigma$ will be approximately distributed as a standardized normal distribution. In this case the p-value is approximately

$$p\text{-value} = 2\min\{N(z), 1 - N(z)\}. \tag{11.30}$$

One use of the run test is to supplement the χ^2 goodness-of-fit test described in Section 11.1.1. An example of this is given below.

Example 11.10

Figure 11.2 *shows a linear least-squares fit to a set of 24 data points. The χ^2 value is 19.8 for 22 degrees of freedom, which is acceptable, but there is some suggestion that the data are not randomly distributed about the fitted line. Use the run test to test this hypothesis at a 10% significance level.*

Solution: The data have $n = 7, m = 16$ (one point is on the fitted line) and for a two-tailed test at a 10% significance level, the critical regions may be found using (11.28) and Table C.9. They are

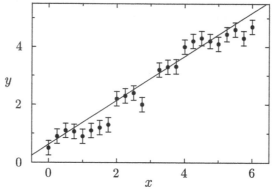

FIGURE 11.2   Linear fit to data.

$r \leq 6$ or $r \geq 15$. Since the observed value of r is 9, the hypothesis that the data are randomly distributed about the best-fit line must be accepted.

11.3.5 Rank correlation coefficient

In previous chapters we have used the sample (Pearson's) correlation coefficient ρ to measure the correlation between two sets of continuous random variables x and y. If instead we replace their numerical values by their rankings, then we obtain the *rank correlation coefficient* (this is due to Spearman and so is also called the *Spearman rank correlation coefficient* or *Spearman's correlation coefficient*) denoted ρ_R. It is given by

$$\rho_R = 1 - \frac{6}{n(n^2-1)} \sum_{i=1}^{n} d_i^2, \tag{11.31}$$

where n is the number of data pairs and d_i are the differences in the ranks of x_i and y_i. In practice, ties are treated as in the previous tests, that is, the differences are averaged as if the ties could be distinguished. The rank correlation coefficient ρ_R is similar to ρ. It is a number between $+1$ and -1 with the extreme values indicating complete positive or negative correlation. For example, $\rho_R = 1$ implies that the ranking numbers of x_i and y_i are identical and a value close to zero indicates that the ranking numbers are uncorrelated. The advantages of using the Spearman correlation coefficient rather than Pearson's are the usual ones: No assumptions need to be made about the distribution of the x and y variables, and the test can be applied to nonnumeric data.

The significance of the rank correlation coefficient is found by considering the distribution of ρ_R under the assumption that x and y are independent. In this case $\rho_R = 0$ and critical values can be calculated. An example is Table C.10. Note that the distribution of values of ρ_R is symmetric about $\rho_R = 0$, so left-tailed areas are equal to right-tailed areas for a given significance level α and for a two-tailed test the critical regions are equal in the two tails of the distribution. For larger values $n \geq 30$, the normal distribution may be used with mean zero and variance $(n-1)^{-1/2}$, so that $z = \rho_R\sqrt{n-1}$ is a standard normal variable.

Example 11.11

A laboratory manager is doubtful whether regular preventive maintenance is leading to fewer breakdowns of his portfolio of equipment, so he records details of the annual rate of breakdowns (B) and the interval in months between services (I) for 10 similar machines (M), with the results shown below. Test the hypothesis that there is no correlation between B and I at a 5% significance level.

M	1	2	3	4	5	6	7	8	9	10
B	2	5	4	3	6	8	9	10	1	12
I	3	4	3	5	6	4	6	5	3	6

Solution: We start by rank ordering B and I to give B_R, I_R and $d^2 \equiv (B_R - I_R)^2$:

M	B	I	B_R	I_R	$d^2 \equiv (B_R - I_R)^2$
1	2	3	2	2	0
2	5	4	5	4.5	0.25
3	4	3	4	2	4
4	3	5	3	6.5	12.25
5	6	6	6	9	9
6	8	4	7	4.5	6.25
7	9	6	8	9	1
8	10	5	9	6.5	6.25
9	1	3	1	2	1
10	12	6	10	9	1

Then from (11.31) we find $\rho_R = 0.75$. We will test the null hypothesis $H_0 : \rho_R = 0$ against the alternative $H_0 : \rho_R > 0$. From Table C.10, we find that we would reject H_0 at the 0.05 significance level because $\rho_R > 0.564$. Recall from the remarks in Section 1.3.3 that this does *not* mean that preventive maintenance *causes* fewer breakdowns, it simply means that there is a significant correlation between the two variables so that this assumption cannot be ruled out.

Problems 11

11.1 A radioactive source was observed for successive intervals of 1 minute and the number of particles emitted of a specific type during 500 intervals recorded. The resulting observations o_i are shown below.

Counts	0	1	2	3	4	5	6	7	8	9	10	11	12
o_i	1	8	38	67	75	85	89	66	39	15	10	5	2

Use the method of likelihood ratios to test the hypothesis that the number of particles emitted has a Poisson distribution with parameter $\lambda = 5$.

11.2 Repeat problem 11.1 using the Pearson goodness of fit test and estimating the parameter λ as the sample mean.

11.3 Two experiments determine the value of a parameter. In the first experiment, 5 measurements are made with a mean of 2.05 and in the second experiment, 10 measurements are made with a mean of 2.09. It is known that the errors are normally distributed, with standard deviation 0.01 in the first experiment and 0.02 in the second experiment. Can the hypothesis that the value of the parameter is the same in the two experiments be accepted at 0.05% significance?

11.4 A sample of size 10 is drawn from a population and the values are:

1	2	3	4	5	6	7	8	9	10
40	55	65	77	94	118	135	150	167	190

Use the Kolmogorov-Smirnov technique to test at a 10% significance level the hypothesis that the sample comes from an exponential population with mean 100.

11.5 Four measurements of a quantity give values 1.12, 1.13, 1.10, and 1.09. If they all come from normal populations with $\sigma^2 = 4 \times 10^{-4}$, test at a 5% significance level the hypothesis that the populations are identical and have a common mean $\mu_0 = 1.09$ against the alternative that $\mu_0 \neq 1.09$.

11.6 Three operatives (O) are given the task of testing identical electrical components for a fixed period of time in the morning (M), afternoon (A), and evening (E) and the numbers successfully tested are given below.

	M	A	E
O1	65	70	75
O2	95	89	70
O3	85	70	58

Test at the 10% significance level the hypothesis that the variables O and the time of day are independent variables in determining the number of components tested.

11.7 Use the sign test to calculate the p-value for the hypothesis that the median of the following numbers is 35.

9	21	34	47	54	55	53
47	38	28	21	15	11	8

Use the normal approximation to the binomial to calculate the p-value.

11.8 The sample of size 10 shown below was drawn from a symmetric population.

x	12	11	18	17	15	19	19	20	17	16

Use the signed-rank test to examine at a 5% significance level the hypothesis that the population mean is 15.5.

11.9 Two samples of weights are measured and give the following data:

S1	9	6	7	8	7		
S2	8	7	9	10	9	11	8

Use the rank-sum test to test at a 5% significance level the hypothesis that the mean value obtained from S1 is smaller than that from S2.

11.10 Test at a 10% significance level whether the following numbers, in the order first row then second row, are randomly distributed about their mean.

2	4	7	3	7	9	5	9	3	6
7	5	2	8	9	7	3	4	2	1

APPENDIX

A

Miscellaneous mathematics

A.1   Matrix algebra

A *matrix* **A** is a two-dimensional array of numbers (taken to be real in these notes), which is written as

$$\mathbf{A} = \begin{pmatrix} a_{11} & a_{12} & \cdots & a_{1n} \\ a_{21} & a_{22} & \cdots & a_{2n} \\ \vdots & \vdots & & \vdots \\ a_{m1} & a_{m2} & \cdots & a_{mn} \end{pmatrix}$$

where the general element in the ith row and jth column is denoted by a_{ij}. A matrix with m rows and n columns is said to be of *order* $(m \times n)$. For the cases $m = 1$ and $n = 1$, we have the *row* and *column* matrices

$$(a_1 \quad a_2 \quad \cdots \quad a_n) \text{ and } \begin{pmatrix} a_1 \\ a_2 \\ \vdots \\ a_m \end{pmatrix},$$

respectively.

Matrices are used in many places in the book, but particularly in Chapter 8, to write a set of n linear equations in p unknowns,

$$\sum_{j=1}^{p} a_{ij}x_j = b_i, \qquad i = 1, 2, \ldots, n,$$

in the compact form

$$\mathbf{AX} = \mathbf{B},$$

where $\mathbf{A}$ is a matrix of order $(n \times p)$, $\mathbf{X}$ is a $(p \times 1)$ column vector and $\mathbf{B}$ is an $(n \times 1)$ column vector. The set of n column vectors $\mathbf{X}_i(i = 1, 2, ..., n)$, all of the same order, are said to be *linearly dependent* if there exist n scalars $\alpha_i(i = 1, 2, ..., n)$, not all zero, such that

$$\sum_{i=1}^{n} \alpha_i \mathbf{X}_i = \mathbf{0},$$

where $\mathbf{0}$ is the *null matrix*, that is, a matrix with all its elements zero. If no such set of scalars exists, then the set of vectors is said to be *linearly independent*.

The *transpose* of the matrix $\mathbf{A}$, denoted by $\mathbf{A}^T$, is obtained by interchanging the rows and columns of $\mathbf{A}$, so

$$\mathbf{A}^T = \begin{pmatrix} a_{11} & a_{21} & \cdots & a_{m1} \\ a_{12} & a_{22} & \cdots & a_{m2} \\ \vdots & \vdots & & \vdots \\ a_{1n} & a_{2n} & \cdots & a_{mn} \end{pmatrix}$$

is an $(n \times m)$ matrix.

A matrix with an equal number of rows and columns is called a *square* matrix, and if a square matrix $\mathbf{A}$ has elements such that $a_{ji} = a_{ij}$, it is said to be *symmetric*. A particular example of a symmetric matrix is the *unit-matrix* $\mathbf{1}$, with elements equal to unity for $i = j$, and zero otherwise. A symmetric matrix $\mathbf{A}$ is said to be *positive definite* if for any vector $\mathbf{V}$, (i) $\mathbf{V}^T \mathbf{A} \mathbf{V} \geq 0$ and (ii) $\mathbf{V}^T \mathbf{A} \mathbf{V} = 0$ implies $\mathbf{V} = \mathbf{0}$. A square matrix with elements $a_{ij} \neq 0$ only if $i = j$ is called *diagonal*; the unit matrix is an example of such a matrix. The line containing the elements $a_{11}, a_{22}, ..., a_{nn}$ is called the *principal*, or *main*, *diagonal* and the sum of its terms is the *trace* of the matrix, written as

$$\mathrm{Tr}\mathbf{A} = \sum_{i=1}^{n} a_{ii}.$$

The *determinant* of a square $(n \times n)$ matrix $\mathbf{A}$ is defined by

$$\det \mathbf{A} = |\mathbf{A}| \equiv \sum \left(\pm a_{1i} a_{2j} \ldots a_{nk} \right), \tag{A.1}$$

where the summation is taken over all permutations of $i, j, ..., k$ where these indices are the integers $1, 2, ..., n$. The positive sign is used for even permutations and the negative sign for odd permutations. The *minor* m_{ij} of the element a_{ij} is defined as the determinant obtained from $\mathbf{A}$ by deleting the ith row and the jth column, and the *cofactor* of a_{ij} is defined as $(-1)^{i+j}$ times the minor m_{ij}. The determinant of $\mathbf{A}$ may also be written in terms of its cofactors. For example, if

$$\mathbf{A} = \begin{pmatrix} 2 & 1 & 3 \\ 1 & 2 & 4 \\ 2 & 1 & 1 \end{pmatrix}, \tag{A.2}$$

then

$$\det \mathbf{A} = 2 \times \begin{vmatrix} 2 & 4 \\ 1 & 1 \end{vmatrix} - \begin{vmatrix} 1 & 4 \\ 2 & 1 \end{vmatrix} + 3 \times \begin{vmatrix} 1 & 2 \\ 2 & 1 \end{vmatrix}$$

$$= 2 \times (2 - 4) - (1 - 8) + 3 \times (1 - 4) = -6.$$

The *adjoint* matrix is defined as the transposed matrix of cofactors and is denoted by $\mathbf{A}^\dagger$. Thus the adjoint of the matrix in (A.2) is

$$\mathbf{A}^\dagger = \begin{pmatrix} -2 & 2 & -2 \\ 7 & -4 & -5 \\ -3 & 0 & 3 \end{pmatrix}. \tag{A.3}$$

A matrix $\mathbf{A}$ with real elements that satisfies the condition $\mathbf{A}^T = \mathbf{A}^{-1}$ is said to be *orthogonal* and its complex analogue is a *unitary* matrix for which $\mathbf{A}^\dagger = \mathbf{A}^{-1}$. For any symmetric matrix $\mathbf{A}$, a unitary matrix $\mathbf{U}$ may be found that when multiplying $\mathbf{A}$ transforms it to diagonal form, that is, the matrix $\mathbf{UA}$ has zeros everywhere except on the principal diagonal.

One particular determinant we have met is the *Jacobian J*. Consider n random variables $x_i (i = 1, 2, ..., n)$ which are themselves a function of n other linearly independent random variables $y_i (i = 1, 2, ..., n)$, and assume that the relations can be inverted to give $x_i(y_1, y_2, \cdots, y_n)$. If the partial derivatives $\partial y_i / \partial x_j$ are continuous for all i and j, then J is defined by (some authors use the term Jacobian to mean the determinant of J, i.e., $|J|$)

$$J \equiv \frac{\partial(x_1, x_2, \cdots x_n)}{\partial(y_1, y_2, \cdots y_n)} \equiv \begin{vmatrix} \dfrac{\partial x_1}{\partial y_1} & \dfrac{\partial x_2}{\partial y_1} & \cdots & \dfrac{\partial x_n}{\partial y_1} \\ \dfrac{\partial x_1}{\partial y_2} & \dfrac{\partial x_2}{\partial y_2} & \cdots & \dfrac{\partial x_n}{\partial y_2} \\ \vdots & \vdots & & \vdots \\ \dfrac{\partial x_1}{\partial y_n} & \dfrac{\partial x_2}{\partial y_n} & \cdots & \dfrac{\partial x_n}{\partial y_n} \end{vmatrix}.$$

The Jacobian has been used, for example, in Section 3.4, when discussing functions of a random variable. In this case if n random variables $x_i (i = 1, 2, ..., n)$ have a joint probability density $f(x_1, x_2, ..., x_n)$, then the joint probability density $g(y_1, y_2, ..., y_n)$ of a new set of variable y_i, which are themselves a function of the n variables $x_i (i = 1, 2, ..., n)$ defined by $y_i = y_i(x_1, x_2, ..., x_n)$, is given by

$$g(y_1, y_2, ..., y_n) = f(x_1, x_2, ..., x_n)|J|.$$

Two matrices may be added and subtracted if they contain the same number of rows and columns, and such addition is both commutative and associative. The (inner) product $\mathbf{A} = \mathbf{BC}$ of two matrices $\mathbf{B}$ and $\mathbf{C}$ has elements given by

$$a_{ij} = \sum_k b_{ik} c_{kj}$$

and so is defined only if the number of columns in the first matrix $\mathbf{B}$ is equal to the number of rows in the second matrix $\mathbf{C}$. Matrix multiplication is not, in general, commutative but is associative.

Division of matrices is more complicated and needs some preliminary definitions. If we form all possible square submatrices of the matrix $\mathbf{A}$ (not necessarily square) and find that at least one determinant of order r is nonzero, but all determinants of order $(r+1)$ are zero, then the matrix is said to be of *rank r*. A square matrix of order n with rank $r < n$ has det $\mathbf{A} = 0$ and is said to be *singular*. The rank of a matrix may thus be expressed as the greatest number of linearly independent rows or columns existing in the matrix, and so, for example, a nonsingular square matrix of order $(n \times n)$ must have rank n. Conversely, if a square matrix $\mathbf{A}$, of order $(n \times n)$, has rank $r = n$, then it is nonsingular and there exists a matrix $\mathbf{A}^{-1}$, known as the *inverse* matrix, such that

$$\mathbf{A}\mathbf{A}^{-1} = \mathbf{A}^{-1}\mathbf{A} = 1, \tag{A.4}$$

where 1 is a diagonal matrix with elements of unity along the principal diagonal and zeros elsewhere. This is the analogous process to division in scalar algebra. The inverse is given by

$$\mathbf{A}^{-1} = \mathbf{A}^\dagger |\mathbf{A}|^{-1}. \tag{A.5}$$

Thus if

$$\mathbf{A} = \begin{pmatrix} 1 & 2 \\ 3 & 4 \end{pmatrix},$$

then

$$\mathbf{A}^\dagger = \begin{pmatrix} 4 & -2 \\ -3 & 1 \end{pmatrix} \quad \text{and} \quad |\mathbf{A}| = (1 \times 4) - (3 \times 2) = -2,$$

so that

$$\mathbf{A}^{-1} = -\frac{1}{2} \begin{pmatrix} 4 & -2 \\ -3 & 1 \end{pmatrix},$$

and to check

$$\mathbf{A}\mathbf{A}^{-1} = -\frac{1}{2}\begin{pmatrix} 1 & 2 \\ 3 & 4 \end{pmatrix}\begin{pmatrix} 4 & -2 \\ -3 & 1 \end{pmatrix} = \begin{pmatrix} 1 & 0 \\ 0 & 1 \end{pmatrix} = \mathbf{1}.$$

Finally, for products of matrices,

$$(\mathbf{ABC}\dots\mathbf{D})^T = \mathbf{D}^T\dots\mathbf{C}^T\mathbf{B}^T\mathbf{A}^T \tag{A.6}$$

and, if $\mathbf{A}, \mathbf{B}, \mathbf{C}, \dots, \mathbf{D}$ are all square nonsingular matrices,

$$(\mathbf{ABC}\dots\mathbf{D})^{-1} = \mathbf{D}^{-1}\dots\mathbf{C}^{-1}\mathbf{B}^{-1}\mathbf{A}^{-1}. \tag{A.7}$$

A point worth remarking on is that in practice equations (A.1) and (A.5) are rarely useful for the practical evaluation of the determinant and inverse of a matrix. For example, in the least-squares method where the matrix of the normal equations (which is positive definite) has to be inverted, the most efficient methods in common use are those based either on the so-called Cholesky's decomposition of a positive definite matrix, or on Golub's factorization by orthogonal matrices, the details of which may be found in any modern textbook on numerical methods.

A.2 Classical theory of minima

If $f(x)$ is a function of the single variable x, which in a certain interval possesses continuous derivatives

$$f^{(j)}(x) \equiv \frac{d^j f(x)}{dx^j}, \quad (j = 1, 2, \dots, n+1),$$

then *Taylor's Theorem* states that if x and $(x+h)$ belong to this interval then

$$f(x+h) = \sum_{j=0}^{n}\frac{h^j}{j!}f^{(j)}(x) + R_n,$$

where $f^{(0)}(x) = f(x)$, and the remainder term is given by

$$R_n = \frac{h^{n+1}}{(n+1)!}f^{(n+1)}(x+\theta h), \quad 0 < \theta < 1.$$

For a function of p-variables Taylor's expansion becomes

$$f(\mathbf{x}+t\mathbf{h}) = \sum_{i=0}^{n}\frac{t^j}{j!}(\mathbf{h}\nabla)^j f(\mathbf{x}) + R_n,$$

where $\mathbf{h}$ is the row vector $(h_1, h_2, \dots, h_p)$, ∇^T is the row vector

$$\left(\frac{\partial}{\partial x_1} \quad \frac{\partial}{\partial x_2} \quad \cdots \quad \frac{\partial}{\partial x_p} \right)$$

and

$$R_n = \frac{t^{n+1}}{(n+1)!}(\mathbf{h}\nabla)^{n+1}f(\mathbf{x}+\theta t\mathbf{h}), \ 0 < \theta < 1.$$

A *necessary* condition for a *turning point* (maximum, minimum, or saddle point) of $f(\mathbf{x})$ to exist is that

$$\frac{\partial f(\mathbf{x})}{\partial x_i} = 0$$

for all $i = 1, 2, ..., p$. A *sufficient* condition for this point to be a minimum is that the second partial derivatives exist and that $D_i > 0$ for all $i = 1, 2, ..., p$, where

$$D_i = \begin{vmatrix} \dfrac{\partial^2 f}{\partial x_1^2} & \dfrac{\partial^2 f}{\partial x_1 \partial x_2} & \cdots & \dfrac{\partial^2 f}{\partial x_1 \partial x_i} \\[2mm] \dfrac{\partial^2 f}{\partial x_2 \partial x_1} & \dfrac{\partial^2 f}{\partial x_2^2} & \cdots & \dfrac{\partial^2 f}{\partial x_2 \partial x_i} \\[2mm] \vdots & \vdots & & \vdots \\[2mm] \dfrac{\partial^2 f}{\partial x_i \partial x_1} & \dfrac{\partial^2 f}{\partial x_i \partial x_2} & \cdots & \dfrac{\partial^2 f}{\partial x_i^2} \end{vmatrix}.$$

If we seek a minimum of $f(\mathbf{x})$ subject to the s *equality constraints*

$$e_j(\mathbf{x}) = 0, \ j = 1, 2, ..., s,$$

then the quantity to consider is the *Lagrangian form*

$$L(\mathbf{x}, \boldsymbol{\lambda}) = f(\mathbf{x}) + \sum_{j=1}^{s} \lambda_j e_j(\mathbf{x}),$$

where the constants λ_i are the so-called *Lagrange multipliers*. If the first partial derivatives of $e_j(\mathbf{x})$ exist, then the required minimum is the unconstrained solution of the equations

$$e_j(\mathbf{x}) = 0, \ j = 1, 2, ..., s$$

and

$$\frac{\partial f(\mathbf{x})}{\partial x_i} + \sum_{j=1}^{s} \lambda_j \frac{\partial e_j(\mathbf{x})}{\partial x_i} = 0, \ i = 1, 2, ..., p.$$

This technique has been used in several places and extensively in the discussion of the least squares method of estimation in Chapter 8.

A.3 Delta function

The *delta function* $\delta(x)$ was first introduced by the theoretical physicist Paul Dirac in applications of Fourier transforms and so is often referred to as the Dirac delta function. It is defined by the relation

$$\int f(x)\delta(x-a)\mathrm{d}x \equiv f(a), \tag{A.8}$$

where a is a real constant that lies within the range of integration so that

$$\delta(x-a) = 0, \quad x \neq a,$$

and

$$\int_b^c \delta(x-a)\mathrm{d}x = 1, \quad b < a < c.$$

If a is outside the range of integration, the integral is defined to be zero. In particular,

$$\int_{-\infty}^{\infty} \delta(x)\,\mathrm{d}x = 1 \tag{A.9}$$

and $\delta(x) = 0$ except at the single point $x = 0$.

This property of the δ-function is unlike that of any 'normal' function, and for this reason it was viewed with some skepticism by mathematicians when Dirac first introduced it in his classic formulation of quantum mechanics. It is now recognized as the first of a new class of functions called *generalized functions*[a] to which a set of well-behaved functions $f_n(x)$ approximates ever more closely as the integer $n \to \infty$. The choice of these functions is not unique, and one example for the δ-function is

$$f_n = \frac{n}{\pi}\frac{1}{1 + n^2 x^2}. \tag{A.10}$$

Other limiting processes are also possible. For example, the *pdf* of the normal distribution of zero mean approximates the δ-function ever more closely as the variance approaches zero.

The δ-function can be conveniently used to convert a discrete probability mass function to an entirely equivalent continuous probability density function. This allows a discrete

---

[a] see, for example, M.J. Lighthill, *An Introduction to Fourier Analysis and Generalised Functions*, Cambridge University Press [1958]

probability mass function (*pmf*) to be used unambiguously in integral formulas, for example, in determining characteristic functions. As a simple example, consider an arbitrary *pmf*

$$P[X = x_i] = p_i$$

for n values $x_1, ..., x_n$ of the random variable X, and n associate probability values $p_1, ..., p_n$. The equivalent *pdf* $f_X(x)$ is

$$f_X(x) = \sum_{i=1}^{n} \delta(x - x_i)p_i.$$

To demonstrate the equivalence, we note that the expected value $E[X]$ has the same value when calculated from either the *pmf* or the *pdf*. From the *pdf* we have

$$E[X] = \int_{-\infty}^{\infty} x f_X(x) dx = \sum_{i=1}^{n} \int_{-\infty}^{\infty} x \delta(x - x_i)p_i dx.$$

Using the definition of the δ-function to evaluate the integral we find

$$E[X] = \sum_{i=1}^{n} x_i p_i,$$

just as we would if we had evaluated the expected value of the *pmf* directly.

A more useful example is the determination of the characteristic function of the Poisson probability mass function:

$$P[X = k] = \frac{1}{k!} e^{-\lambda} \lambda^k,$$

where k is any nonnegative integer and λ is a constant parameter. The *cf* of a *pdf* is the Fourier transform. To apply the Fourier transform formula, we first cast the Poisson *pmf* as a *pdf*:

$$f_X(x) = \sum_{k=0}^{\infty} \delta(x - k) \frac{1}{k!} e^{-\lambda} \lambda^k.$$

The *cf* is then

$$\phi_X(t) = \int_{-\infty}^{\infty} e^{itx} \sum_{k=0}^{\infty} \delta(x - k) \frac{1}{k!} e^{-\lambda} \lambda^k dx,$$

which, using the definition of the δ-function and summing the series for the exponential function, gives the result

$$\phi_X(t) = e^{-\lambda} \exp(\lambda e^{it}).$$

A.4 Distribution of the mean of a Poisson sample

The *cf* of a Poisson distribution is given in Appendix A.3. Let the random variable Y be the sum of a sample of n realizations of a Poisson random variable with parameter λ. Since the *cf* of the *pdf* of a sum of identical independent random variables is the product of the *cf*s of the individual *pdf*s, we have immediately that the *cf* of the *pdf* $f_Y(y)$ of the random variable Y is

$$\phi_Y(t) = \prod_{j=1}^{n} e^{-\lambda} \exp\left(\lambda e^{it}\right) = e^{-n\lambda} \exp\left(n\lambda e^{it}\right).$$

But this is the *cf* of a Poisson random variable with parameter $n\lambda$, so the *pmf* of Y is

$$P[Y = k] = \frac{1}{k!}e^{-n\lambda}(n\lambda)^k$$

in which k is any nonnegative integer.

We can determine the *pmf* of the sample mean directly from this *pdf* by noting that the value $\bar{y}$ of the sample mean random variable $\bar{Y}$ for a sample with sum k is k/n. Thus

$$P\left[\bar{Y} = \frac{k}{n}\right] = P[Y = k] = \frac{1}{k!}e^{-n\lambda}(n\lambda)^k.$$

In this case the integer k is $n\bar{y}$, so

$$P\left[\bar{Y} = \bar{y}\right] = \frac{1}{(n\bar{y})!}e^{-n\lambda}(n\lambda)^{n\bar{y}}.$$

We conclude that although the sum of Poisson variables is a Poisson variable, the mean of such a sum is not a Poisson variable.

A.5 The Chebyshev inequality

It is intuitive that the smaller the variance of a random variable, the less probable it is that a value will be far from the mean. Chebyshev's inequality is an important and easily understood upper bound on this probability (see equation 5.19 in Chapter 5).

Let Z be a random variable with mean μ. We are interested in the probability that $|Z - \mu|$ is larger than some positive number c. Let F be the event (collection of outcomes) in our sample space of all outcomes ω for which this occurs. That is

$$F = \{\omega : |Z(\omega) - \mu| \geq c\}.$$

Let $I_F(\omega)$ be a discrete random variable that has the value 1 for any $\omega \in F$ and 0 for any $w \notin F$. The expected value of I_F is

$$E[I_F] = 1 \times P[\omega \in F] + 0 \times P[\omega \notin F] = P[\omega \in F].$$

Chebyshev's inequality is a direct consequence of the inequality

$$(Z(\omega) - \mu)^2 \geq c^2 I_F(\omega).$$

This inequality is obviously true if $\omega \notin F$ because the left-hand side is a nonnegative real number and the right-hand side is zero. If $\omega \in F$, the right-hand side is c^2 and the left-hand side is no smaller than c^2 by the definition of F, so it is true in this case also.

We arrive at Chebyshev's inequality by taking the expected value of each side of the above inequality. The expected value of the left-hand side is the variance of Z, so we have

$$var(Z) \geq c^2 E[I_F] = c^2 P[\omega \in F] = c^2 P[|Z(\omega) - \mu| \geq c].$$

Dividing by c^2, we finally have

$$P[|Z(\omega) - \mu| \geq c] \leq \frac{var(Z)}{c^2}.$$

This is Chebyshev's inequality, showing that the smaller the variance, the more likely the values of the random variable will be close to the mean. A second, equivalent, interpretation is that the further from the mean the value of a random variable is, the smaller the probability that the random variable will have that value.

B

Optimization of nonlinear functions[1]

In Chapters 7 and 8 we encountered the problem of finding the maxima, or minima, of nonlinear functions, sometimes of several variables. These are examples of a more general class of optimization problems, which, although occurring frequently in statistical estimation procedures, are not of a statistical nature. In practice, computer codes exist to tackle these problems, and it is not suggested that the reader write their own optimization code except in the simplest of circumstances, but nevertheless, it is useful to know a little of the theory on which the methods are based to better appreciate their limitations. The discussion of this appendix is therefore confined to the main ideas involved, illustrated by one or two examples. Fuller accounts are given in the books cited in the bibliography, from which these brief notes draw extensively.

B.1   General principles

We consider only *minimization* problems since

$$\min f(\mathbf{x}) = \max[-f(\mathbf{x})].$$

The general problem to be solved is then to minimize the function $f(x_1, x_2, ..., x_p) \equiv f(\mathbf{x})$, subject to the *m inequality constraints*

$$c_i(\mathbf{x}) \geq 0, \qquad i = 1, 2, ..., m$$

and the s equality constraints,

$$e_j(\mathbf{x}) = 0, \qquad j = 1, 2, ..., s.$$

All other constraints can be reduced to either of the above forms by suitable transformations. We will discuss firstly the features of methods of optimization in general and then describe in more detail a few of the most successful methods in current use.

---

[1] The matrix notations used in this appendix are reviewed in Appendix A.

Any point that satisfies all the constraints is called *feasible,* and the entire set of such points is called the *feasible region.* Points lying outside the feasible region are said to be *nonfeasible.* Nearly all practical methods of optimization are *iterative* in the sense that an initial feasible vector $\mathbf{x}^{(0)}$ must be specified from which the method will generate a series of vectors $\mathbf{x}^{(1)}, \mathbf{x}^{(2)}, \ldots, \mathbf{x}^{(n)}$ etc., which represent improved approximations to the solution.

The iterative procedure may be expressed by the equation

$$\mathbf{x}^{(n+1)} = \mathbf{x}^{(n)} + h_n \mathbf{d}_n, \tag{B.1}$$

where $\mathbf{d}_n$ is a *p*-dimensional *directional vector,* and h_n is the distance moved along it. The basic problem is to determine the most suitable vector $\mathbf{d}_n$, since once it is chosen the function $f(\mathbf{x})$ can be calculated and a suitable value of h_n found. Iterative techniques fall naturally into two classes: (a) *direct search* methods and (b) *gradient* methods.

Direct search methods are based on a sequential examination of a series of trial solutions produced from an initial feasible point. Based on the examinations, the strategy for further searching is determined. These methods are characterized by the fact that they only explicitly require values of the function, and knowledge of the derivatives of $f(\mathbf{x})$ is not required. The latter fact is both a strength and weakness of the methods, for although in problems involving many variables, the calculation of derivatives can be difficult, and/or time-consuming, it is clear that more efficient methods should be possible if the information contained in the derivatives is used. In practice, direct search methods are most useful for situations involving a few parameters, or where the calculation of derivatives is very difficult, or for finding promising regions in the parameter space where optima might reasonably be located.

Gradient methods make explicit use of the partial derivatives of the function, in addition to values of the function itself. The *gradient direction* at any point is that direction whose components are proportional to the first-order partial derivatives of the function at the point. The importance of this quantity will be seen as follows. If we make small perturbations $\delta \mathbf{x}$ from the current point $\mathbf{x}$, then to first order in $\delta \mathbf{x}$,

$$\delta f = \sum_{j=1}^{p} \frac{\partial f}{\partial x_j} \delta x_j. \tag{B.2}$$

To obtain the perturbation giving the greatest change in the function, we need to consider the Lagrangian form

$$F(\mathbf{x}, \lambda) = \delta f + \lambda \left(\sum_{j=1}^{p} \delta x_j^2 - \Delta^2 \right), \tag{B.3}$$

where λ is a Lagrange multiplier and Δ is the magnitude of the perturbations, that is,

$$\Delta = \left[\sum_{j=1}^{p} \delta x_j^2 \right]^{1/2}.$$

Using (B.2) in (B.3), and forming the differential with respect to δx_j, gives

$$\frac{\partial f}{\partial x_j} + 2\lambda \delta x_j = 0, \quad j = 1, 2\cdots, p$$

and hence

$$\frac{\delta x_1}{\partial f/\partial x_1} = \frac{\delta x_2}{\partial f/\partial x_2} = \cdots = \frac{\delta x_p}{\partial f/\partial x_p}.$$

That is, for any Δ the greatest value of δf is obtained if the perturbations δx_j are chosen to be proportional to $\partial f/\partial x_j$, and that, further, if $\delta f < 0$, that is, the search is to converge to a minimum, the constant of proportionality must be negative. This direction is called the *direction of steepest descent*. It follows that the function can always be reduced by following the direction of steepest descent, although this may only be true for a short distance.

One remark that is worth making about gradient methods concerns the actual calculation of the derivatives. Although gradient methods are in general more efficient than direct search methods, their efficiency can drop considerably if the derivatives are not obtained analytically, and so if numerical methods are used to calculate these quantities, great care must be taken to ensure that inaccuracies do not result.

So far, we have not specified the form of the function to be minimized, except that it is nonlinear in its variables. However, in many practical problems involving unconstrained functions it is found that the function can be well approximated by a quadratic form in the neighborhood of the minimum. There is therefore considerable interest in methods that *guarantee* to find the minimum of a quadratic in a specified number of steps. Such methods are said to be *quadratically convergent*, and the hope is that problems that are not strictly quadratic may still be tractable by such methods, a hope that is borne out rather well in practice.

The most useful of the methods having the property of quadratic convergence are those making use of the so-called *conjugate directions*, defined as follows. Two direction vectors $\mathbf{d}_1$ and $\mathbf{d}_2$ are said to be conjugate with respect to the positive definite matrix $\mathbf{G}$ if $\mathbf{d}_1^T \mathbf{G} \, \mathbf{d}_2 = 0$. The importance of conjugate directions in optimization problems stems from the following theorem.

If $\mathbf{d}_i (i = 1, 2, ..., p)$ is a set of vectors mutually conjugated with respect to the positive definite matrix $\mathbf{G}$, the minimum of the quadratic form

$$f(\mathbf{x}) = \frac{1}{2}\mathbf{x}^T \mathbf{G} \mathbf{x} + \mathbf{b}^T \mathbf{x} + a, \tag{B.4}$$

where a is a constant and $\mathbf{b}$ a constant vector, can be found from an arbitrary point $\mathbf{x}^{(0)}$ by a finite descent calculation in which each of the vectors $\mathbf{d}_i$ is used as a descent direction only once, their order of use being arbitrary.

The proof of this important result is as follows. Since the vectors $\mathbf{d}_i$ are linearly independent, any arbitrary vector $\mathbf{v}$ may be written in the form

$$\mathbf{v} = \sum_{i=1}^{p} \alpha_i \mathbf{d}_i, \tag{B.5}$$

where, because of the conjugacy of the vectors $\mathbf{d}_i$,

$$\alpha_i = \frac{\mathbf{d}_i^T \mathbf{G} \mathbf{v}}{\mathbf{d}_i^T \mathbf{G} \mathbf{d}_i}. \tag{B.6}$$

If the general iterative equation (B.1) is applied repeatedly, then at the nth stage we have

$$\mathbf{x}^{(n)} = \mathbf{x}^{(0)} + \sum_{i=1}^{n} h_i \mathbf{d}_i, \tag{B.7}$$

and in the $(n+1)$th stage, from $\mathbf{x}^{(n)} \to \mathbf{x}^{(n+1)}$, the distance h_{n+1} along the direction $\mathbf{d}_{n+1}$ is found from the equation

$$\mathbf{d}_{n+1}^T \nabla f \left[\mathbf{x}^{(n+1)} \right] = 0.$$

Using (B.4) this gives

$$\mathbf{d}_{n+1}^T \left[\mathbf{G} \left(\mathbf{x}^{(0)} + \sum_{i=1}^{n} h_i \mathbf{d}_i + h_{n+1} \mathbf{d}_{n+1} \right) + \mathbf{b} \right] = 0,$$

and so

$$h_{n+1} = - \frac{\mathbf{d}_{n+1}^T [\mathbf{G}\mathbf{x}^{(0)} + \mathbf{b}]}{\mathbf{d}_{n+1}^T \mathbf{G} \mathbf{d}_{n+1}}, \tag{B.8}$$

which depends only on $\mathbf{x}^{(0)}$ and not on the path by which $\mathbf{x}^{(n+1)}$ is reached from $\mathbf{x}^{(0)}$. Using (B.8) in (B.7) for p-steps gives

$$\mathbf{x}^{(p)} = \mathbf{x}^{(0)} - \sum_{i=1}^{p} \frac{\mathbf{d}_i^T [\mathbf{G}\mathbf{x}^{(0)} + \mathbf{b}] \mathbf{d}_i}{\mathbf{d}_i^T \mathbf{G} \mathbf{d}_i},$$

and, using (B.5) and (B.6), this becomes

$$\mathbf{x}^{(p)} = \mathbf{x}^{(0)} - \mathbf{x}^{(0)} - \mathbf{G}^{-1}\mathbf{b} = -\mathbf{G}^{-1}\mathbf{b}, \tag{B.9}$$

which from (B.4) shows that the minimum has been reached in p iterations.

Although methods having the property of quadratic convergence will guarantee to converge to the exact minimum of a quadratic in p steps, where p is the dimensionality of the problem, when applied to functions that are not strictly quadratic, the problem arises of determining when convergence has taken place. A suitable practical criterion is to consider that convergence has been achieved if, for given small values of ε and ε'

$$f[\mathbf{x}^{(n)}] - f[\mathbf{x}^{(n+1)}] < \varepsilon,$$

and/or

$$\left| \mathbf{x}^{(n)} - \mathbf{x}^{(n+1)} \right| < \varepsilon',$$

for a sequence of q successive iterations, where q is a number which will vary with the type of function being minimized. A generous overestimate is $q \sim p$, the number of variables, but a considerably smaller value of q is usually sufficient.

Finally, it should be mentioned that all present techniques for optimizing nonlinear functions locate only *local optima*, that is, points x_m at which $f(x_m) < f(x)$ for all x in a region in the neighborhood of x_m. For multivariate problems there may well be better local optima located at some distance from x_m. At present there are no general methods for locating the *global optimum* (i.e., the absolute optimum) of a function, and so it is essential to restart the search procedure from different initial points $x^{(0)}$ to ensure that the full p-dimensional space has been explored.

B.2 Unconstrained minimization of functions of one variable

The problem of minimizing a function of one variable is very important in practice, because many methods for optimizing multivariate functions proceed by a series of searches along a line in the parameter space, and each of these searches is equivalent to a univariate search. The latter fall into two groups: (a) those which specify an interval within which the minimum lies, and (b) those which specify the minimum by a point approximating it. The latter methods are the most useful in practice and we shall only consider them here. The basic procedure is as follows. Proceeding from an initial point $x^{(0)}$, a systematic search technique is applied to find a region containing the minimum. This bracket is then refined by fitting a quadratic interpolation polynomial to the three points making up the bracket, and the minimum of this polynomial found. As a result of this evaluation a new bracket is formed, and the procedure is repeated. The method is both simple and very safe in practice.

A practical implementation of this procedure is as follows. The function is first evaluated at the points $x^{(0)}$ and $(x^{(0)} + h)$. If $f(x^{(0)} + h) \leq f(x^{(0)})$, then $f(x^{(0)} + 2h)$ is evaluated. This doubling of the step-length h is repeated until a value of $f(x)$ is found such that $f(x^{(0)} + 2^n h) > f(x^{(0)} + 2^{n-1} h)$. At this point the step-length is halved and a step again taken from the last successive point, that is, the $(n-1)$th. This procedure produces four points equally spaced along the axis of search, at each of which the function has been evaluated. The end point furthest from the point corresponding to the smallest function value is rejected, and the remaining three points used for quadratic interpolation. Had the first step failed, then the search is continued by reversing the sign of the step length. If the first step in this direction also fails, then the minimum has been bracketed and the interpolation may be made. If the three points used for the interpolation are x_1, x_2, x_3 with $x_1 < x_2 < x_3$ and $x_3 - x_2 = x_2 - x_1 = l$, then the minimum of the fitted quadratic is at

$$x_m = x_2 + \frac{l[f(x_1) - f(x_3)]}{2[f(x_1) - 2f(x_2) + f(x_3)]}.$$

An iteration is completed by evaluating $f(x_m)$. Convergence tests are now applied and, if required, a further iteration is performed, with a reduced step length, using as the initial point whichever of x_2 or x_m corresponds to the smaller function value.

B.3   Unconstrained minimization of multivariable functions

Many methods for locating optima of multivariate functions are based on a series of linear searches along a line in the parameter space. By a *linear method* we will therefore mean any technique which uses a set of direction vectors in the search, and which proceeds by explorations along these directions, deciding future strategy by the results obtained in previous searches.

B.3.1   Direct search methods

The simplest of all possible *direct search* methods would be to keep $(p-1)$ of the parameters fixed and find a minimum with respect to the pth parameter, doing this in turn for each variable. In general, the contours of equal function value will be aligned along the so-called *principal axes*, which are not parallel to the coordinate axes, so only very small steps will be taken at each stage and the technique is very inefficient. Moreover, the inefficiency increases as the number of variables increases. It would clearly be very much more efficient to reorientate the direction vectors along more advantageous directions and this is done in several techniques, the most successful of which is due to Powell.

The method uses conjugate directions and utilizes the fact that, for a positive definite quadratic form, if searches for minima are made along p conjugate directions, then the join of these minima is conjugate to all of those directions, a result that follows from the definition of conjugate directions. The procedure is to start from $x^{(0)}$ and locate the minimum in the direction $d_1^{(n)}$. Then from the new minimum point $x^{(1)}$ locate the minimum in the direction $d_2^{(n)}$, etc., until the minimum in the direction $d_p^{(n)}$ is found. The direction of total progress made during this cycle is then

$$d = x^{(p)} - x^{(0)}.$$

New search directions are now constructed, and care must be taken to ensure that the new direction vectors are always linearly independent. Powell showed that for the quadratic form of (B.4), if $d_i^{(n)}$ is scaled so that

$$d_i^{(n)T} G d_i^{(n)} = 1, \quad i = 1, 2, ..., p$$

then the determinant D of the matrix whose columns are $d_i^{(n)}$ has a maximum if, and only if, the vectors are mutually conjugate with respect to G. Thus the direction d only replaces an existing search direction if by so doing D is increased. In this case the minimum in the direction d is found and used as a starting point in the next iteration, the list of direction vectors being updated as follows

$$\left(d_1^{(n+1)}, d_2^{(n+1)}, ..., d_p^{(n+1)} \right) = \left(d_1^{(n)}, d_2^{(n)}, ..., d_{j-1}^{(n)}, d_{j+1}^{(n)}, ..., d_p^{(n)}, d \right),$$

where $d_j^{(n)}$ is that direction vector along which the greatest reduction in the function value occurred during the nth stage.

B.3.2 Gradient methods

The simplest technique using gradients is that of steepest descent mentioned above. In this method the normalized gradient vector at the current point is first found, and then using a step-length h_i, a new point is generated via the general iterative equation. This procedure is continued until a function value is found which has not decreased. The step-length is then reduced, and the search restarted from the best previous point. If the actual minimum along each search direction is located then the performance of this method is similar in appearance to an alternating variable search, and is rather erratic, the search directions oscillating about the principal axes. A method that in principle is far better based on an examination of the second derivatives of the function.

(a) Newton's Method

A second-order Taylor expansion of the function $f(x)$ about the minimum point x_{min} is

$$f(\mathbf{x}) = f(\mathbf{x}_{min}) + \sum_{j=1}^{p} h_j \left(\frac{\partial f}{\partial x_j} \right)_{\mathbf{x}=\mathbf{x}_{min}} + \frac{1}{2} \sum_{j=1}^{p} \sum_{k=1}^{p} h_j h_k \left(\frac{\partial^2 f}{\partial x_j \partial x_k} \right)_{\mathbf{x}=\mathbf{x}_{min}}.$$

Differentiating this equation gives

$$g_l \equiv \frac{\partial f}{\partial x_l} = \sum_{j=1}^{p} h_j \left(\frac{\partial^2 f}{\partial x_j \partial x_l} \right)_{\mathbf{x}=\mathbf{x}_{min}}, \qquad i = 1, 2, ..., p. \tag{B.10}$$

The minimum is therefore obtained in one step by the move $\mathbf{x}_{min} = \mathbf{x} - \mathbf{h}$, where the components of $\mathbf{h}$ are found by solving the p linear equations (B.10). If we define

$$G_{jk} \equiv \left(\frac{\partial^2 f}{\partial x_j \partial x_k} \right),$$

then

$$\mathbf{x}_{min} = \mathbf{x} - \mathbf{G}_{min}^{-1} \mathbf{g},$$

where, again, $\mathbf{G}_{min}$ means that $\mathbf{G}$ is evaluated at $\mathbf{x}_{min}$. Since $\mathbf{G}_{min}^{-1}$ will not of course be known, it is usual to replace it by $\mathbf{G}^{-1}$ evaluated at the current point $\mathbf{x}^{(i)}$ and use the iterative equation

$$\mathbf{x}^{(n+1)} = \mathbf{x}^{(n)} - \mathbf{G}_n^{-1} \mathbf{g}_n. \tag{B.11}$$

The method is clearly quadratically convergent but suffers from severe difficulties.

Firstly, there is the numerical problem of calculating the inverse matrix of second derivatives, and secondly, and more seriously, for a general function, $\mathbf{G}^{-1}$ is not guaranteed to be positive definite, and in this case the method will diverge. Thus, while Newton's method is efficient in the immediate neighborhood of a minimum, away from this point it has little to recommend it, the method of steepest descent being far preferable.

In view of the above remarks an efficient method would be one that starts by using the method of steepest descent and, at a later stage, uses Newton's method. A method that does this automatically is due to Davidon and represents the most powerful method currently available for optimizing unconstrained functions.

(b) Davidon's method

This method is an iterative scheme based on successive approximations to the matrix $\mathbf{G}_{\min}^{-1}$. The best approximation to this matrix, say $\mathbf{H}_n$, is used to define a new search direction by a modification of equation (B.12), that is,

$$\mathbf{x}^{(n+1)} = \mathbf{x}^{(n)} - h_n \mathbf{H}_n \mathbf{g}_n,$$

where $\mathbf{g}_n$ is the vector of first derivatives of $f(\mathbf{x}^{(n)})$ with respect to $\mathbf{x}^{(n)}$. The step length h_n is that necessary to find the minimum in the search direction $\mathbf{d}_n = -\mathbf{H}_n \mathbf{g}_n$ and may be found by any univariate search procedure. If the sequence $\{\mathbf{H}_n\}$ is positive definite, it can be shown that the convergence of this method is guaranteed. Furthermore, if the search directions $\mathbf{d}_n$ are mutually conjugate, then the method is quadratically convergent.

Davidon has shown that both conditions can be met if, at each stage of the iteration, the matrix $\mathbf{H}_n$ is updated according to the relation

$$\mathbf{H}_{n+1} = \mathbf{H}_n + \mathbf{A}_n + \mathbf{B}_n,$$

where the matrices $\mathbf{A}_n$ and $\mathbf{B}_n$ are given by

$$\mathbf{A}_n = \frac{-h_n \left[\mathbf{H}_n \mathbf{g}_n \mathbf{g}_n^T \mathbf{H}_n^T\right]}{\left(\mathbf{H}_n \mathbf{g}_n\right)^T \mathbf{V}} \quad \text{and} \quad \mathbf{B}_n = \frac{-\mathbf{H}_n \mathbf{V} \mathbf{V}^T \mathbf{H}_n^T}{\mathbf{V}^T \mathbf{H}_n \mathbf{V}},$$

with $\mathbf{V} = \mathbf{g}_{n+1} - \mathbf{g}_n$, where $\mathbf{g}_{n+1}$ is the gradient at $\mathbf{x}_{n+1}$. It is usual to start the iteration from the unit matrix $\mathbf{H}_0 = \mathbf{1}$. The matrix $\mathbf{A}_n$ ensures that the sequence $\{\mathbf{H}_n\}$ converges to $\mathbf{G}_{\min}^{-1}$, and $\mathbf{B}_n$ ensures that each $\mathbf{H}_n$ is positive definite. The derivation of these expressions may be found in the book by Kowalik and Osborne, cited in the bibliography.[2]

B.4   Constrained optimization

Constrained optimization is a more difficult problem than unconstrained optimization, and only a very brief discussion will be given here.

Firstly, an obvious remark: if the constraints can be removed by suitable transformations, then this should be done. For example, many problems involve simple constraints on the parameters that can be expressed in the form

$$l \leq x \leq u,$$

[2] A computer program (called Minuit) that utilized the Davidon approach was first written by a CERN staff member, Frederick James, in the 1970s. The latest version, called Minuit2, now exists in both Fortran and C++ languages. It is used by particle physicists worldwide and has been used in thousands of published research papers.

which can be removed completely by the transformation

$$x = l + (u - l)\sin^2 y,$$

thereby enabling an unconstrained minimization to be performed with respect to y. Such transformations *cannot* produce additional local optima. If the constraints cannot be removed, then one of the simplest ways of incorporating them is to arrange that the production of nonfeasible points is unattractive. This is the basis of a practical technique involving *penalty functions*, where the function to be minimized is modified by additional terms designed to achieve this.

The general problem was stated in Section B.1. Using the notation for constraints given there, we consider the function

$$F(\mathbf{x}) = f(\mathbf{x}) + \sum_{i=1}^{n} \lambda_i \, c_i^2(\mathbf{x}) S[c_i(\mathbf{x})] + \sum_{j=1}^{g} \lambda_j' \, e_j^2(\mathbf{x}), \qquad (B.12)$$

where $S(q)$ is the function

$$S(q) = \begin{cases} 0, & q \geq 0 \\ 1, & q < 0 \end{cases}$$

and λ_i, λ_i' are positive scale factors, chosen so that the contributions of the various terms to (B.13) are approximately equal. The 'penalty', that is, the sum of the second and third terms on the right-hand side of (B.12), is thus the weighted sum of squares of the amounts by which the constraints are violated.

This method works reasonably well in practice but has the disadvantage of requiring that values of $f(\mathbf{x})$ be calculated at nonfeasible points, and this may not always be possible, leading to program failure. A method that restricts the search to feasible points is due to Carroll and is known as *Carroll's created response surface technique*. In this method, if the constraints are inequalities, the surface

$$F(\mathbf{x}, k) = f(\mathbf{x}) + k \sum_{i=1}^{m} \frac{w_i}{c_i(\mathbf{x})}$$

is considered, where $k > 0$, and the w_i are positive constants. A minimum is found as a function of $\mathbf{x}$ and this is then used as the starting value for a new minimization for a reduced value of k, and the procedure repeated until $k = 0$ is reached. In all minimizations nonfeasible points are excluded. The theoretical development of this method and its extension to incorporate equality constraints may be found in the book of Kowalik and Osborne.

C

Statistical tables

The tables in this appendix are included as a convenience for the student and practitioner. However, it should be noted that all these tables can be generated using readily available software and programming languages. In fact, all the tables in this appendix were generated using either an Excel spreadsheet or simple programs in the Python or R programming languages. The specific functions and libraries from which they are obtained are mentioned within each table.

C.1 Normal distribution

These tables give values of N, the standardized cumulative distribution function:

$$N(x) = \frac{1}{\sqrt{2\pi}} \int_{-\infty}^{x} \exp\left(\frac{t^2}{2}\right) dt.$$

Note that $N(-x) = 1 - N(x)$.
The data in these tables were produced using the Microsoft Excel function NORM.DIST().

x	0	0.01	0.02	0.03	0.04	0.05	0.06	0.07	0.08	0.09
0	0.5000	0.5040	0.5080	0.5120	0.5160	0.5199	0.5239	0.5279	0.5319	0.5359
0.1	0.5398	0.5438	0.5478	0.5517	0.5557	0.5596	0.5636	0.5675	0.5714	0.5753
0.2	0.5793	0.5832	0.5871	0.5910	0.5948	0.5987	0.6026	0.6064	0.6103	0.6141
0.3	0.6179	0.6217	0.6255	0.6293	0.6331	0.6368	0.6406	0.6443	0.6480	0.6517
0.4	0.6554	0.6591	0.6628	0.6664	0.6700	0.6736	0.6772	0.6808	0.6844	0.6879
0.5	0.6915	0.6950	0.6985	0.7019	0.7054	0.7088	0.7123	0.7157	0.7190	0.7224
0.6	0.7257	0.7291	0.7324	0.7357	0.7389	0.7422	0.7454	0.7486	0.7517	0.7549
0.7	0.7580	0.7611	0.7642	0.7673	0.7704	0.7734	0.7764	0.7794	0.7823	0.7852

(Continued)

x	0	0.01	0.02	0.03	0.04	0.05	0.06	0.07	0.08	0.09
0.8	0.7881	0.7910	0.7939	0.7967	0.7995	0.8023	0.8051	0.8078	0.8106	0.8133
0.9	0.8159	0.8186	0.8212	0.8238	0.8264	0.8289	0.8315	0.8340	0.8365	0.8389
1	0.8413	0.8438	0.8461	0.8485	0.8508	0.8531	0.8554	0.8577	0.8599	0.8621
1.1	0.8643	0.8665	0.8686	0.8708	0.8729	0.8749	0.8770	0.8790	0.8810	0.8830
1.2	0.8849	0.8869	0.8888	0.8907	0.8925	0.8944	0.8962	0.8980	0.8997	0.9015
1.3	0.9032	0.9049	0.9066	0.9082	0.9099	0.9115	0.9131	0.9147	0.9162	0.9177
1.4	0.9192	0.9207	0.9222	0.9236	0.9251	0.9265	0.9279	0.9292	0.9306	0.9319
1.5	0.9332	0.9345	0.9357	0.9370	0.9382	0.9394	0.9406	0.9418	0.9429	0.9441
1.6	0.9452	0.9463	0.9474	0.9484	0.9495	0.9505	0.9515	0.9525	0.9535	0.9545
1.7	0.9554	0.9564	0.9573	0.9582	0.9591	0.9599	0.9608	0.9616	0.9625	0.9633
1.8	0.9641	0.9649	0.9656	0.9664	0.9671	0.9678	0.9686	0.9693	0.9699	0.9706
1.9	0.9713	0.9719	0.9726	0.9732	0.9738	0.9744	0.9750	0.9756	0.9761	0.9767
2	0.9772	0.9778	0.9783	0.9788	0.9793	0.9798	0.9803	0.9808	0.9812	0.9817
2.1	0.9821	0.9826	0.9830	0.9834	0.9838	0.9842	0.9846	0.9850	0.9854	0.9857
2.2	0.9861	0.9864	0.9868	0.9871	0.9875	0.9878	0.9881	0.9884	0.9887	0.9890
2.3	0.9893	0.9896	0.9898	0.9901	0.9904	0.9906	0.9909	0.9911	0.9913	0.9916
2.4	0.9918	0.9920	0.9922	0.9925	0.9927	0.9929	0.9931	0.9932	0.9934	0.9936
2.5	0.9938	0.9940	0.9941	0.9943	0.9945	0.9946	0.9948	0.9949	0.9951	0.9952
2.6	0.9953	0.9955	0.9956	0.9957	0.9959	0.9960	0.9961	0.9962	0.9963	0.9964
2.7	0.9965	0.9966	0.9967	0.9968	0.9969	0.9970	0.9971	0.9972	0.9973	0.9974
2.8	0.9974	0.9975	0.9976	0.9977	0.9977	0.9978	0.9979	0.9979	0.9980	0.9981
2.9	0.9981	0.9982	0.9982	0.9983	0.9984	0.9984	0.9985	0.9985	0.9986	0.9986
3	0.9987	0.9987	0.9987	0.9988	0.9988	0.9989	0.9989	0.9989	0.9990	0.9990
3.1	0.9990	0.9991	0.9991	0.9991	0.9992	0.9992	0.9992	0.9992	0.9993	0.9993
3.2	0.9993	0.9993	0.9994	0.9994	0.9994	0.9994	0.9994	0.9995	0.9995	0.9995
3.3	0.9995	0.9995	0.9995	0.9996	0.9996	0.9996	0.9996	0.9996	0.9996	0.9997
3.4	0.9997	0.9997	0.9997	0.9997	0.9997	0.9997	0.9997	0.9997	0.9997	0.9998

	Commonly used values of the standard normal distribution.								
x	1.282	1.645	1.960	2.326	2.576	3.090	3.291	3.891	4.417
$N(x)$	0.900	0.950	0.975	0.990	0.995	0.999	1.000	1.000	1.000
$2(1 - N(x))$	0.200	0.100	0.050	0.020	0.010	0.002	0.001	0.0001	0.00001

C.2 Binomial distribution

This table gives the right tail of the cumulative binomial distribution, that is, the probability $B_p(s)$ of obtaining s or more successes in n independent Bernoulli trials, where the probability of success in each trial is p:

$$B_p(s) = \sum_{i=s}^{n} \binom{n}{i} p^i (1-p)^{n-i}.$$

If $p > 0.5$, the values of B may be obtained from

$$B_p(s) = 1 - B_q(n - s + 1) = \sum_{i=n-s+1}^{n} \binom{n}{i} q^i (1-q)^{n-i},$$

where $q = 1 - p$.
The data in this table were produced using the Microsoft Excel function BINOM.DIST().

						p					
n	s	0.05	0.1	0.15	0.2	0.25	0.3	0.35	0.4	0.45	0.5
2	1	0.0975	0.1900	0.2775	0.3600	0.4375	0.5100	0.5775	0.6400	0.6975	0.7500
	2	0.0025	0.0100	0.0225	0.0400	0.0625	0.0900	0.1225	0.1600	0.2025	0.2500
3	1	0.1426	0.2710	0.3859	0.4880	0.5781	0.6570	0.7254	0.7840	0.8336	0.8750
	2	0.0072	0.0280	0.0608	0.1040	0.1563	0.2160	0.2818	0.3520	0.4253	0.5000
	3	0.0001	0.0010	0.0034	0.0080	0.0156	0.0270	0.0429	0.0640	0.0911	0.1250
4	1	0.1855	0.3439	0.4780	0.5904	0.6836	0.7599	0.8215	0.8704	0.9085	0.9375
	2	0.0140	0.0523	0.1095	0.1808	0.2617	0.3483	0.4370	0.5248	0.6090	0.6875
	3	0.0005	0.0037	0.0120	0.0272	0.0508	0.0837	0.1265	0.1792	0.2415	0.3125
	4	0.0000	0.0001	0.0005	0.0016	0.0039	0.0081	0.0150	0.0256	0.0410	0.0625

(Continued)

n	s	0.05	0.1	0.15	0.2	0.25	0.3	0.35	0.4	0.45	0.5
5	1	0.2262	0.4095	0.5563	0.6723	0.7627	0.8319	0.8840	0.9222	0.9497	0.9688
	2	0.0226	0.0815	0.1648	0.2627	0.3672	0.4718	0.5716	0.6630	0.7438	0.8125
	3	0.0012	0.0086	0.0266	0.0579	0.1035	0.1631	0.2352	0.3174	0.4069	0.5000
	4	0.0000	0.0005	0.0022	0.0067	0.0156	0.0308	0.0540	0.0870	0.1312	0.1875
	5	0.0000	0.0000	0.0001	0.0003	0.0010	0.0024	0.0053	0.0102	0.0185	0.0313
6	1	0.2649	0.4686	0.6229	0.7379	0.8220	0.8824	0.9246	0.9533	0.9723	0.9844
	2	0.0328	0.1143	0.2235	0.3446	0.4661	0.5798	0.6809	0.7667	0.8364	0.8906
	3	0.0022	0.0159	0.0473	0.0989	0.1694	0.2557	0.3529	0.4557	0.5585	0.6563
	4	0.0001	0.0013	0.0059	0.0170	0.0376	0.0705	0.1174	0.1792	0.2553	0.3438
	5	0.0000	0.0001	0.0004	0.0016	0.0046	0.0109	0.0223	0.0410	0.0692	0.1094
	6	0.0000	0.0000	0.0000	0.0001	0.0002	0.0007	0.0018	0.0041	0.0083	0.0156
7	1	0.3017	0.5217	0.6794	0.7903	0.8665	0.9176	0.9510	0.9720	0.9848	0.9922
	2	0.0444	0.1497	0.2834	0.4233	0.5551	0.6706	0.7662	0.8414	0.8976	0.9375
	3	0.0038	0.0257	0.0738	0.1480	0.2436	0.3529	0.4677	0.5801	0.6836	0.7734
	4	0.0002	0.0027	0.0121	0.0333	0.0706	0.1260	0.1998	0.2898	0.3917	0.5000
	5	0.0000	0.0002	0.0012	0.0047	0.0129	0.0288	0.0556	0.0963	0.1529	0.2266
	6	0.0000	0.0000	0.0001	0.0004	0.0013	0.0038	0.0090	0.0188	0.0357	0.0625
	7	0.0000	0.0000	0.0000	0.0000	0.0001	0.0002	0.0006	0.0016	0.0037	0.0078
8	1	0.3366	0.5695	0.7275	0.8322	0.8999	0.9424	0.9681	0.9832	0.9916	0.9961
	2	0.0572	0.1869	0.3428	0.4967	0.6329	0.7447	0.8309	0.8936	0.9368	0.9648
	3	0.0058	0.0381	0.1052	0.2031	0.3215	0.4482	0.5722	0.6846	0.7799	0.8555
	4	0.0004	0.0050	0.0214	0.0563	0.1138	0.1941	0.2936	0.4059	0.5230	0.6367
	5	0.0000	0.0004	0.0029	0.0104	0.0273	0.0580	0.1061	0.1737	0.2604	0.3633
	6	0.0000	0.0000	0.0002	0.0012	0.0042	0.0113	0.0253	0.0498	0.0885	0.1445
	7	0.0000	0.0000	0.0000	0.0001	0.0004	0.0013	0.0036	0.0085	0.0181	0.0352
	8	0.0000	0.0000	0.0000	0.0000	0.0000	0.0001	0.0002	0.0007	0.0017	0.0039

(Continued)

n	s	p									
		0.05	0.1	0.15	0.2	0.25	0.3	0.35	0.4	0.45	0.5
9	1	0.3698	0.6126	0.7684	0.8658	0.9249	0.9596	0.9793	0.9899	0.9954	0.9980
	2	0.0712	0.2252	0.4005	0.5638	0.6997	0.8040	0.8789	0.9295	0.9615	0.9805
	3	0.0084	0.0530	0.1409	0.2618	0.3993	0.5372	0.6627	0.7682	0.8505	0.9102
	4	0.0006	0.0083	0.0339	0.0856	0.1657	0.2703	0.3911	0.5174	0.6386	0.7461
	5	0.0000	0.0009	0.0056	0.0196	0.0489	0.0988	0.1717	0.2666	0.3786	0.5000
	6	0.0000	0.0001	0.0006	0.0031	0.0100	0.0253	0.0536	0.0994	0.1658	0.2539
	7	0.0000	0.0000	0.0000	0.0003	0.0013	0.0043	0.0112	0.0250	0.0498	0.0898
	8	0.0000	0.0000	0.0000	0.0000	0.0001	0.0004	0.0014	0.0038	0.0091	0.0195
	9	0.0000	0.0000	0.0000	0.0000	0.0000	0.0000	0.0001	0.0003	0.0008	0.0020
10	1	0.4013	0.6513	0.8031	0.8926	0.9437	0.9718	0.9865	0.9940	0.9975	0.9990
	2	0.0861	0.2639	0.4557	0.6242	0.7560	0.8507	0.9140	0.9536	0.9767	0.9893
	3	0.0115	0.0702	0.1798	0.3222	0.4744	0.6172	0.7384	0.8327	0.9004	0.9453
	4	0.0010	0.0128	0.0500	0.1209	0.2241	0.3504	0.4862	0.6177	0.7340	0.8281
	5	0.0001	0.0016	0.0099	0.0328	0.0781	0.1503	0.2485	0.3669	0.4956	0.6230
	6	0.0000	0.0001	0.0014	0.0064	0.0197	0.0473	0.0949	0.1662	0.2616	0.3770
	7	0.0000	0.0000	0.0001	0.0009	0.0035	0.0106	0.0260	0.0548	0.1020	0.1719
	8	0.0000	0.0000	0.0000	0.0001	0.0004	0.0016	0.0048	0.0123	0.0274	0.0547
	9	0.0000	0.0000	0.0000	0.0000	0.0000	0.0001	0.0005	0.0017	0.0045	0.0107
	10	0.0000	0.0000	0.0000	0.0000	0.0000	0.0000	0.0000	0.0001	0.0003	0.0010
11	1	0.4312	0.6862	0.8327	0.9141	0.9578	0.9802	0.9912	0.9964	0.9986	0.9995
	2	0.1019	0.3026	0.5078	0.6779	0.8029	0.8870	0.9394	0.9698	0.9861	0.9941
	3	0.0152	0.0896	0.2212	0.3826	0.5448	0.6873	0.7999	0.8811	0.9348	0.9673
	4	0.0016	0.0185	0.0694	0.1611	0.2867	0.4304	0.5744	0.7037	0.8089	0.8867
	5	0.0001	0.0028	0.0159	0.0504	0.1146	0.2103	0.3317	0.4672	0.6029	0.7256
	6	0.0000	0.0003	0.0027	0.0117	0.0343	0.0782	0.1487	0.2465	0.3669	0.5000
	7	0.0000	0.0000	0.0003	0.0020	0.0076	0.0216	0.0501	0.0994	0.1738	0.2744
	8	0.0000	0.0000	0.0000	0.0002	0.0012	0.0043	0.0122	0.0293	0.0610	0.1133
	9	0.0000	0.0000	0.0000	0.0000	0.0001	0.0006	0.0020	0.0059	0.0148	0.0327

(Continued)

n	s	\multicolumn{10}{c}{p}
		0.05	0.1	0.15	0.2	0.25	0.3	0.35	0.4	0.45	0.5
	10	0.0000	0.0000	0.0000	0.0000	0.0000	0.0000	0.0002	0.0007	0.0022	0.0059
	11	0.0000	0.0000	0.0000	0.0000	0.0000	0.0000	0.0000	0.0000	0.0002	0.0005
12	1	0.4596	0.7176	0.8578	0.9313	0.9683	0.9862	0.9943	0.9978	0.9992	0.9998
	2	0.1184	0.3410	0.5565	0.7251	0.8416	0.9150	0.9576	0.9804	0.9917	0.9968
	3	0.0196	0.1109	0.2642	0.4417	0.6093	0.7472	0.8487	0.9166	0.9579	0.9807
	4	0.0022	0.0256	0.0922	0.2054	0.3512	0.5075	0.6533	0.7747	0.8655	0.9270
	5	0.0002	0.0043	0.0239	0.0726	0.1576	0.2763	0.4167	0.5618	0.6956	0.8062
	6	0.0000	0.0005	0.0046	0.0194	0.0544	0.1178	0.2127	0.3348	0.4731	0.6128
	7	0.0000	0.0001	0.0007	0.0039	0.0143	0.0386	0.0846	0.1582	0.2607	0.3872
	8	0.0000	0.0000	0.0001	0.0006	0.0028	0.0095	0.0255	0.0573	0.1117	0.1938
	9	0.0000	0.0000	0.0000	0.0001	0.0004	0.0017	0.0056	0.0153	0.0356	0.0730
	10	0.0000	0.0000	0.0000	0.0000	0.0000	0.0002	0.0008	0.0028	0.0079	0.0193
	11	0.0000	0.0000	0.0000	0.0000	0.0000	0.0000	0.0001	0.0003	0.0011	0.0032
	12	0.0000	0.0000	0.0000	0.0000	0.0000	0.0000	0.0000	0.0000	0.0001	0.0002
13	1	0.4867	0.7458	0.8791	0.9450	0.9762	0.9903	0.9963	0.9987	0.9996	0.9999
	2	0.1354	0.3787	0.6017	0.7664	0.8733	0.9363	0.9704	0.9874	0.9951	0.9983
	3	0.0245	0.1339	0.3080	0.4983	0.6674	0.7975	0.8868	0.9421	0.9731	0.9888
	4	0.0031	0.0342	0.1180	0.2527	0.4157	0.5794	0.7217	0.8314	0.9071	0.9539
	5	0.0003	0.0065	0.0342	0.0991	0.2060	0.3457	0.4995	0.6470	0.7721	0.8666
	6	0.0000	0.0009	0.0075	0.0300	0.0802	0.1654	0.2841	0.4256	0.5732	0.7095
	7	0.0000	0.0001	0.0013	0.0070	0.0243	0.0624	0.1295	0.2288	0.3563	0.5000
	8	0.0000	0.0000	0.0002	0.0012	0.0056	0.0182	0.0462	0.0977	0.1788	0.2905
	9	0.0000	0.0000	0.0000	0.0002	0.0010	0.0040	0.0126	0.0321	0.0698	0.1334
	10	0.0000	0.0000	0.0000	0.0000	0.0001	0.0007	0.0025	0.0078	0.0203	0.0461
	11	0.0000	0.0000	0.0000	0.0000	0.0000	0.0001	0.0003	0.0013	0.0041	0.0112
	12	0.0000	0.0000	0.0000	0.0000	0.0000	0.0000	0.0000	0.0001	0.0005	0.0017
	13	0.0000	0.0000	0.0000	0.0000	0.0000	0.0000	0.0000	0.0000	0.0000	0.0001

(Continued)

n	s	\multicolumn{10}{c}{p}
		0.05	0.1	0.15	0.2	0.25	0.3	0.35	0.4	0.45	0.5
14	1	0.5123	0.7712	0.8972	0.9560	0.9822	0.9932	0.9976	0.9992	0.9998	0.9999
	2	0.1530	0.4154	0.6433	0.8021	0.8990	0.9525	0.9795	0.9919	0.9971	0.9991
	3	0.0301	0.1584	0.3521	0.5519	0.7189	0.8392	0.9161	0.9602	0.9830	0.9935
	4	0.0042	0.0441	0.1465	0.3018	0.4787	0.6448	0.7795	0.8757	0.9368	0.9713
	5	0.0004	0.0092	0.0467	0.1298	0.2585	0.4158	0.5773	0.7207	0.8328	0.9102
	6	0.0000	0.0015	0.0115	0.0439	0.1117	0.2195	0.3595	0.5141	0.6627	0.7880
	7	0.0000	0.0002	0.0022	0.0116	0.0383	0.0933	0.1836	0.3075	0.4539	0.6047
	8	0.0000	0.0000	0.0003	0.0024	0.0103	0.0315	0.0753	0.1501	0.2586	0.3953
	9	0.0000	0.0000	0.0000	0.0004	0.0022	0.0083	0.0243	0.0583	0.1189	0.2120
	10	0.0000	0.0000	0.0000	0.0000	0.0003	0.0017	0.0060	0.0175	0.0426	0.0898
	11	0.0000	0.0000	0.0000	0.0000	0.0000	0.0002	0.0011	0.0039	0.0114	0.0287
	12	0.0000	0.0000	0.0000	0.0000	0.0000	0.0000	0.0001	0.0006	0.0022	0.0065
	13	0.0000	0.0000	0.0000	0.0000	0.0000	0.0000	0.0000	0.0001	0.0003	0.0009
	14	0.0000	0.0000	0.0000	0.0000	0.0000	0.0000	0.0000	0.0000	0.0000	0.0001
15	1	0.5367	0.7941	0.9126	0.9648	0.9866	0.9953	0.9984	0.9995	0.9999	1.0000
	2	0.1710	0.4510	0.6814	0.8329	0.9198	0.9647	0.9858	0.9948	0.9983	0.9995
	3	0.0362	0.1841	0.3958	0.6020	0.7639	0.8732	0.9383	0.9729	0.9893	0.9963
	4	0.0055	0.0556	0.1773	0.3518	0.5387	0.7031	0.8273	0.9095	0.9576	0.9824
	5	0.0006	0.0127	0.0617	0.1642	0.3135	0.4845	0.6481	0.7827	0.8796	0.9408
	6	0.0001	0.0022	0.0168	0.0611	0.1484	0.2784	0.4357	0.5968	0.7392	0.8491
	7	0.0000	0.0003	0.0036	0.0181	0.0566	0.1311	0.2452	0.3902	0.5478	0.6964
	8	0.0000	0.0000	0.0006	0.0042	0.0173	0.0500	0.1132	0.2131	0.3465	0.5000
	9	0.0000	0.0000	0.0001	0.0008	0.0042	0.0152	0.0422	0.0950	0.1818	0.3036
	10	0.0000	0.0000	0.0000	0.0001	0.0008	0.0037	0.0124	0.0338	0.0769	0.1509
	11	0.0000	0.0000	0.0000	0.0000	0.0001	0.0007	0.0028	0.0093	0.0255	0.0592
	12	0.0000	0.0000	0.0000	0.0000	0.0000	0.0001	0.0005	0.0019	0.0063	0.0176
	13	0.0000	0.0000	0.0000	0.0000	0.0000	0.0000	0.0001	0.0003	0.0011	0.0037
	14	0.0000	0.0000	0.0000	0.0000	0.0000	0.0000	0.0000	0.0000	0.0001	0.0005
	15	0.0000	0.0000	0.0000	0.0000	0.0000	0.0000	0.0000	0.0000	0.0000	0.0000

(Continued)

n	s	0.05	0.1	0.15	0.2	0.25	0.3	0.35	0.4	0.45	0.5
							p				
16	1	0.5599	0.8147	0.9257	0.9719	0.9900	0.9967	0.9990	0.9997	0.9999	1.0000
	2	0.1892	0.4853	0.7161	0.8593	0.9365	0.9739	0.9902	0.9967	0.9990	0.9997
	3	0.0429	0.2108	0.4386	0.6482	0.8029	0.9006	0.9549	0.9817	0.9934	0.9979
	4	0.0070	0.0684	0.2101	0.4019	0.5950	0.7541	0.8661	0.9349	0.9719	0.9894
	5	0.0009	0.0170	0.0791	0.2018	0.3698	0.5501	0.7108	0.8334	0.9147	0.9616
	6	0.0001	0.0033	0.0235	0.0817	0.1897	0.3402	0.5100	0.6712	0.8024	0.8949
	7	0.0000	0.0005	0.0056	0.0267	0.0796	0.1753	0.3119	0.4728	0.6340	0.7728
	8	0.0000	0.0001	0.0011	0.0070	0.0271	0.0744	0.1594	0.2839	0.4371	0.5982
	9	0.0000	0.0000	0.0002	0.0015	0.0075	0.0257	0.0671	0.1423	0.2559	0.4018
	10	0.0000	0.0000	0.0000	0.0002	0.0016	0.0071	0.0229	0.0583	0.1241	0.2272
	11	0.0000	0.0000	0.0000	0.0000	0.0003	0.0016	0.0062	0.0191	0.0486	0.1051
	12	0.0000	0.0000	0.0000	0.0000	0.0000	0.0003	0.0013	0.0049	0.0149	0.0384
	13	0.0000	0.0000	0.0000	0.0000	0.0000	0.0000	0.0002	0.0009	0.0035	0.0106
	14	0.0000	0.0000	0.0000	0.0000	0.0000	0.0000	0.0000	0.0001	0.0006	0.0021
	15	0.0000	0.0000	0.0000	0.0000	0.0000	0.0000	0.0000	0.0000	0.0001	0.0003
	16	0.0000	0.0000	0.0000	0.0000	0.0000	0.0000	0.0000	0.0000	0.0000	0.0000
17	1	0.5819	0.8332	0.9369	0.9775	0.9925	0.9977	0.9993	0.9998	1.0000	1.0000
	2	0.2078	0.5182	0.7475	0.8818	0.9499	0.9807	0.9933	0.9979	0.9994	0.9999
	3	0.0503	0.2382	0.4802	0.6904	0.8363	0.9226	0.9673	0.9877	0.9959	0.9988
	4	0.0088	0.0826	0.2444	0.4511	0.6470	0.7981	0.8972	0.9536	0.9816	0.9936
	5	0.0012	0.0221	0.0987	0.2418	0.4261	0.6113	0.7652	0.8740	0.9404	0.9755
	6	0.0001	0.0047	0.0319	0.1057	0.2347	0.4032	0.5803	0.7361	0.8529	0.9283
	7	0.0000	0.0008	0.0083	0.0377	0.1071	0.2248	0.3812	0.5522	0.7098	0.8338
	8	0.0000	0.0001	0.0017	0.0109	0.0402	0.1046	0.2128	0.3595	0.5257	0.6855
	9	0.0000	0.0000	0.0003	0.0026	0.0124	0.0403	0.0994	0.1989	0.3374	0.5000
	10	0.0000	0.0000	0.0000	0.0005	0.0031	0.0127	0.0383	0.0919	0.1834	0.3145
	11	0.0000	0.0000	0.0000	0.0001	0.0006	0.0032	0.0120	0.0348	0.0826	0.1662
	12	0.0000	0.0000	0.0000	0.0000	0.0001	0.0007	0.0030	0.0106	0.0301	0.0717

(Continued)

		p									
n	s	0.05	0.1	0.15	0.2	0.25	0.3	0.35	0.4	0.45	0.5
	13	0.0000	0.0000	0.0000	0.0000	0.0000	0.0001	0.0006	0.0025	0.0086	0.0245
	14	0.0000	0.0000	0.0000	0.0000	0.0000	0.0000	0.0001	0.0005	0.0019	0.0064
	15	0.0000	0.0000	0.0000	0.0000	0.0000	0.0000	0.0000	0.0001	0.0003	0.0012
	16	0.0000	0.0000	0.0000	0.0000	0.0000	0.0000	0.0000	0.0000	0.0000	0.0001
	17	0.0000	0.0000	0.0000	0.0000	0.0000	0.0000	0.0000	0.0000	0.0000	0.0000
18	1	0.6028	0.8499	0.9464	0.9820	0.9944	0.9984	0.9996	0.9999	1.0000	1.0000
	2	0.2265	0.5497	0.7759	0.9009	0.9605	0.9858	0.9954	0.9987	0.9997	0.9999
	3	0.0581	0.2662	0.5203	0.7287	0.8647	0.9400	0.9764	0.9918	0.9975	0.9993
	4	0.0109	0.0982	0.2798	0.4990	0.6943	0.8354	0.9217	0.9672	0.9880	0.9962
	5	0.0015	0.0282	0.1206	0.2836	0.4813	0.6673	0.8114	0.9058	0.9589	0.9846
	6	0.0002	0.0064	0.0419	0.1329	0.2825	0.4656	0.6450	0.7912	0.8923	0.9519
	7	0.0000	0.0012	0.0118	0.0513	0.1390	0.2783	0.4509	0.6257	0.7742	0.8811
	8	0.0000	0.0002	0.0027	0.0163	0.0569	0.1407	0.2717	0.4366	0.6085	0.7597
	9	0.0000	0.0000	0.0005	0.0043	0.0193	0.0596	0.1391	0.2632	0.4222	0.5927
	10	0.0000	0.0000	0.0001	0.0009	0.0054	0.0210	0.0597	0.1347	0.2527	0.4073
	11	0.0000	0.0000	0.0000	0.0002	0.0012	0.0061	0.0212	0.0576	0.1280	0.2403
	12	0.0000	0.0000	0.0000	0.0000	0.0002	0.0014	0.0062	0.0203	0.0537	0.1189
	13	0.0000	0.0000	0.0000	0.0000	0.0000	0.0003	0.0014	0.0058	0.0183	0.0481
	14	0.0000	0.0000	0.0000	0.0000	0.0000	0.0000	0.0003	0.0013	0.0049	0.0154
	15	0.0000	0.0000	0.0000	0.0000	0.0000	0.0000	0.0000	0.0002	0.0010	0.0038
	16	0.0000	0.0000	0.0000	0.0000	0.0000	0.0000	0.0000	0.0000	0.0001	0.0007
	17	0.0000	0.0000	0.0000	0.0000	0.0000	0.0000	0.0000	0.0000	0.0000	0.0001
	18	0.0000	0.0000	0.0000	0.0000	0.0000	0.0000	0.0000	0.0000	0.0000	0.0000
19	1	0.6226	0.8649	0.9544	0.9856	0.9958	0.9989	0.9997	0.9999	1.0000	1.0000
	2	0.2453	0.5797	0.8015	0.9171	0.9690	0.9896	0.9969	0.9992	0.9998	1.0000
	3	0.0665	0.2946	0.5587	0.7631	0.8887	0.9538	0.9830	0.9945	0.9985	0.9996
	4	0.0132	0.1150	0.3159	0.5449	0.7369	0.8668	0.9409	0.9770	0.9923	0.9978
	5	0.0020	0.0352	0.1444	0.3267	0.5346	0.7178	0.8500	0.9304	0.9720	0.9904

(Continued)

n	s					p					
		0.05	0.1	0.15	0.2	0.25	0.3	0.35	0.4	0.45	0.5
	6	0.0002	0.0086	0.0537	0.1631	0.3322	0.5261	0.7032	0.8371	0.9223	0.9682
	7	0.0000	0.0017	0.0163	0.0676	0.1749	0.3345	0.5188	0.6919	0.8273	0.9165
	8	0.0000	0.0003	0.0041	0.0233	0.0775	0.1820	0.3344	0.5122	0.6831	0.8204
	9	0.0000	0.0000	0.0008	0.0067	0.0287	0.0839	0.1855	0.3325	0.5060	0.6762
	10	0.0000	0.0000	0.0001	0.0016	0.0089	0.0326	0.0875	0.1861	0.3290	0.5000
	11	0.0000	0.0000	0.0000	0.0003	0.0023	0.0105	0.0347	0.0885	0.1841	0.3238
	12	0.0000	0.0000	0.0000	0.0000	0.0005	0.0028	0.0114	0.0352	0.0871	0.1796
	13	0.0000	0.0000	0.0000	0.0000	0.0001	0.0006	0.0031	0.0116	0.0342	0.0835
	14	0.0000	0.0000	0.0000	0.0000	0.0000	0.0001	0.0007	0.0031	0.0109	0.0318
	15	0.0000	0.0000	0.0000	0.0000	0.0000	0.0000	0.0001	0.0006	0.0028	0.0096
	16	0.0000	0.0000	0.0000	0.0000	0.0000	0.0000	0.0000	0.0001	0.0005	0.0022
	17	0.0000	0.0000	0.0000	0.0000	0.0000	0.0000	0.0000	0.0000	0.0001	0.0004
	18	0.0000	0.0000	0.0000	0.0000	0.0000	0.0000	0.0000	0.0000	0.0000	0.0000
	19	0.0000	0.0000	0.0000	0.0000	0.0000	0.0000	0.0000	0.0000	0.0000	0.0000
20	1	0.6415	0.8784	0.9612	0.9885	0.9968	0.9992	0.9998	1.0000	1.0000	1.0000
	2	0.2642	0.6083	0.8244	0.9308	0.9757	0.9924	0.9979	0.9995	0.9999	1.0000
	3	0.0755	0.3231	0.5951	0.7939	0.9087	0.9645	0.9879	0.9964	0.9991	0.9998
	4	0.0159	0.1330	0.3523	0.5886	0.7748	0.8929	0.9556	0.9840	0.9951	0.9987
	5	0.0026	0.0432	0.1702	0.3704	0.5852	0.7625	0.8818	0.9490	0.9811	0.9941
	6	0.0003	0.0113	0.0673	0.1958	0.3828	0.5836	0.7546	0.8744	0.9447	0.9793
	7	0.0000	0.0024	0.0219	0.0867	0.2142	0.3920	0.5834	0.7500	0.8701	0.9423
	8	0.0000	0.0004	0.0059	0.0321	0.1018	0.2277	0.3990	0.5841	0.7480	0.8684
	9	0.0000	0.0001	0.0013	0.0100	0.0409	0.1133	0.2376	0.4044	0.5857	0.7483
	10	0.0000	0.0000	0.0002	0.0026	0.0139	0.0480	0.1218	0.2447	0.4086	0.5881
	11	0.0000	0.0000	0.0000	0.0006	0.0039	0.0171	0.0532	0.1275	0.2493	0.4119
	12	0.0000	0.0000	0.0000	0.0001	0.0009	0.0051	0.0196	0.0565	0.1308	0.2517
	13	0.0000	0.0000	0.0000	0.0000	0.0002	0.0013	0.0060	0.0210	0.0580	0.1316
	14	0.0000	0.0000	0.0000	0.0000	0.0000	0.0003	0.0015	0.0065	0.0214	0.0577
	15	0.0000	0.0000	0.0000	0.0000	0.0000	0.0000	0.0003	0.0016	0.0064	0.0207

(Continued)

| | | | | | | p | | | | | |
n	s	0.05	0.1	0.15	0.2	0.25	0.3	0.35	0.4	0.45	0.5
	16	0.0000	0.0000	0.0000	0.0000	0.0000	0.0000	0.0000	0.0003	0.0015	0.0059
	17	0.0000	0.0000	0.0000	0.0000	0.0000	0.0000	0.0000	0.0000	0.0003	0.0013
	18	0.0000	0.0000	0.0000	0.0000	0.0000	0.0000	0.0000	0.0000	0.0000	0.0002
	19	0.0000	0.0000	0.0000	0.0000	0.0000	0.0000	0.0000	0.0000	0.0000	0.0000
	20	0.0000	0.0000	0.0000	0.0000	0.0000	0.0000	0.0000	0.0000	0.0000	0.0000

C.3   Poisson distribution

This table gives the right tail P of the cumulative Poisson distribution, that is, the probability:

$$P = \sum_{n=k}^{\infty} \frac{\lambda^n}{n!} \exp(-\lambda)$$

that k or more events will occur in a Poisson distribution with average λ, for specified values of k and λ.

The data in these tables were produced using the Microsoft Excel function POISSON.DIST().

| | | | | | λ | | | | | |
k	0.1	0.2	0.3	0.4	0.5	0.6	0.7	0.8	0.9	1.0
0	1.0000	1.0000	1.0000	1.0000	1.0000	1.0000	1.0000	1.0000	1.0000	1.0000
1	0.0952	0.1813	0.2592	0.3297	0.3935	0.4512	0.5034	0.5507	0.5934	0.6321
2	0.0047	0.0175	0.0369	0.0616	0.0902	0.1219	0.1558	0.1912	0.2275	0.2642
3	0.0002	0.0011	0.0036	0.0079	0.0144	0.0231	0.0341	0.0474	0.0629	0.0803
4	0.0000	0.0001	0.0003	0.0008	0.0018	0.0034	0.0058	0.0091	0.0135	0.0190
5	0.0000	0.0000	0.0000	0.0001	0.0002	0.0004	0.0008	0.0014	0.0023	0.0037
6	0.0000	0.0000	0.0000	0.0000	0.0000	0.0000	0.0001	0.0002	0.0003	0.0006
7	0.0000	0.0000	0.0000	0.0000	0.0000	0.0000	0.0000	0.0000	0.0000	0.0001

(Continued)

						λ				
k	1.1	1.2	1.3	1.4	1.5	1.6	1.7	1.8	1.9	2.0
0	1.0000	1.0000	1.0000	1.0000	1.0000	1.0000	1.0000	1.0000	1.0000	1.0000
1	0.6671	0.6988	0.7275	0.7534	0.7769	0.7981	0.8173	0.8347	0.8504	0.8647
2	0.3010	0.3374	0.3732	0.4082	0.4422	0.4751	0.5068	0.5372	0.5663	0.5940
3	0.0996	0.1205	0.1429	0.1665	0.1912	0.2166	0.2428	0.2694	0.2963	0.3233
4	0.0257	0.0338	0.0431	0.0537	0.0656	0.0788	0.0932	0.1087	0.1253	0.1429
5	0.0054	0.0077	0.0107	0.0143	0.0186	0.0237	0.0296	0.0364	0.0441	0.0527
6	0.0010	0.0015	0.0022	0.0032	0.0045	0.0060	0.0080	0.0104	0.0132	0.0166
7	0.0001	0.0003	0.0004	0.0006	0.0009	0.0013	0.0019	0.0026	0.0034	0.0045
8	0.0000	0.0000	0.0001	0.0001	0.0002	0.0003	0.0004	0.0006	0.0008	0.0011
9	0.0000	0.0000	0.0000	0.0000	0.0000	0.0000	0.0001	0.0001	0.0002	0.0002

						λ				
k	2.1	2.2	2.3	2.4	2.5	2.6	2.7	2.8	2.9	3.0
0	1.0000	1.0000	1.0000	1.0000	1.0000	1.0000	1.0000	1.0000	1.0000	1.0000
1	0.8775	0.8892	0.8997	0.9093	0.9179	0.9257	0.9328	0.9392	0.9450	0.9502
2	0.6204	0.6454	0.6691	0.6916	0.7127	0.7326	0.7513	0.7689	0.7854	0.8009
3	0.3504	0.3773	0.4040	0.4303	0.4562	0.4816	0.5064	0.5305	0.5540	0.5768
4	0.1614	0.1806	0.2007	0.2213	0.2424	0.2640	0.2859	0.3081	0.3304	0.3528
5	0.0621	0.0725	0.0838	0.0959	0.1088	0.1226	0.1371	0.1523	0.1682	0.1847
6	0.0204	0.0249	0.0300	0.0357	0.0420	0.0490	0.0567	0.0651	0.0742	0.0839
7	0.0059	0.0075	0.0094	0.0116	0.0142	0.0172	0.0206	0.0244	0.0287	0.0335
8	0.0015	0.0020	0.0026	0.0033	0.0042	0.0053	0.0066	0.0081	0.0099	0.0119
9	0.0003	0.0005	0.0006	0.0009	0.0011	0.0015	0.0019	0.0024	0.0031	0.0038
10	0.0001	0.0001	0.0001	0.0002	0.0003	0.0004	0.0005	0.0007	0.0009	0.0011
11	0.0000	0.0000	0.0000	0.0000	0.0001	0.0001	0.0001	0.0002	0.0002	0.0003
12	0.0000	0.0000	0.0000	0.0000	0.0000	0.0000	0.0000	0.0000	0.0001	0.0001

(Continued)

λ

k	3.1	3.2	3.3	3.4	3.5	3.6	3.7	3.8	3.9	4.0
0	1.0000	1.0000	1.0000	1.0000	1.0000	1.0000	1.0000	1.0000	1.0000	1.0000
1	0.9550	0.9592	0.9631	0.9666	0.9698	0.9727	0.9753	0.9776	0.9798	0.9817
2	0.8153	0.8288	0.8414	0.8532	0.8641	0.8743	0.8838	0.8926	0.9008	0.9084
3	0.5988	0.6201	0.6406	0.6603	0.6792	0.6973	0.7146	0.7311	0.7469	0.7619
4	0.3752	0.3975	0.4197	0.4416	0.4634	0.4848	0.5058	0.5265	0.5468	0.5665
5	0.2018	0.2194	0.2374	0.2558	0.2746	0.2936	0.3128	0.3322	0.3516	0.3712
6	0.0943	0.1054	0.1171	0.1295	0.1424	0.1559	0.1699	0.1844	0.1994	0.2149
7	0.0388	0.0446	0.0510	0.0579	0.0653	0.0733	0.0818	0.0909	0.1005	0.1107
8	0.0142	0.0168	0.0198	0.0231	0.0267	0.0308	0.0352	0.0401	0.0454	0.0511
9	0.0047	0.0057	0.0069	0.0083	0.0099	0.0117	0.0137	0.0160	0.0185	0.0214
10	0.0014	0.0018	0.0022	0.0027	0.0033	0.0040	0.0048	0.0058	0.0069	0.0081
11	0.0004	0.0005	0.0006	0.0008	0.0010	0.0013	0.0016	0.0019	0.0023	0.0028
12	0.0001	0.0001	0.0002	0.0002	0.0003	0.0004	0.0005	0.0006	0.0007	0.0009
13	0.0000	0.0000	0.0000	0.0001	0.0001	0.0001	0.0001	0.0002	0.0002	0.0003
14	0.0000	0.0000	0.0000	0.0000	0.0000	0.0000	0.0000	0.0000	0.0001	0.0001

λ

k	4.1	4.2	4.3	4.4	4.5	4.6	4.7	4.8	4.9	5.0
0	1.0000	1.0000	1.0000	1.0000	1.0000	1.0000	1.0000	1.0000	1.0000	1.0000
1	0.9834	0.9850	0.9864	0.9877	0.9889	0.9899	0.9909	0.9918	0.9926	0.9933
2	0.9155	0.9220	0.9281	0.9337	0.9389	0.9437	0.9482	0.9523	0.9561	0.9596
3	0.7762	0.7898	0.8026	0.8149	0.8264	0.8374	0.8477	0.8575	0.8667	0.8753
4	0.5858	0.6046	0.6228	0.6406	0.6577	0.6743	0.6903	0.7058	0.7207	0.7350
5	0.3907	0.4102	0.4296	0.4488	0.4679	0.4868	0.5054	0.5237	0.5418	0.5595
6	0.2307	0.2469	0.2633	0.2801	0.2971	0.3142	0.3316	0.3490	0.3665	0.3840
7	0.1214	0.1325	0.1442	0.1564	0.1689	0.1820	0.1954	0.2092	0.2233	0.2378
8	0.0573	0.0639	0.0710	0.0786	0.0866	0.0951	0.1040	0.1133	0.1231	0.1334
9	0.0245	0.0279	0.0317	0.0358	0.0403	0.0451	0.0503	0.0558	0.0618	0.0681

(*Continued*)

	λ									
k	4.1	4.2	4.3	4.4	4.5	4.6	4.7	4.8	4.9	5.0
10	0.0095	0.0111	0.0129	0.0149	0.0171	0.0195	0.0222	0.0251	0.0283	0.0318
11	0.0034	0.0041	0.0048	0.0057	0.0067	0.0078	0.0090	0.0104	0.0120	0.0137
12	0.0011	0.0014	0.0017	0.0020	0.0024	0.0029	0.0034	0.0040	0.0047	0.0055
13	0.0003	0.0004	0.0005	0.0007	0.0008	0.0010	0.0012	0.0014	0.0017	0.0020
14	0.0001	0.0001	0.0002	0.0002	0.0003	0.0003	0.0004	0.0005	0.0006	0.0007
15	0.0000	0.0000	0.0000	0.0001	0.0001	0.0001	0.0001	0.0001	0.0002	0.0002
16	0.0000	0.0000	0.0000	0.0000	0.0000	0.0000	0.0000	0.0000	0.0001	0.0001

	λ									
k	5.1	5.2	5.3	5.4	5.5	5.6	5.7	5.8	5.9	6.0
0	1.0000	1.0000	1.0000	1.0000	1.0000	1.0000	1.0000	1.0000	1.0000	1.0000
1	0.9939	0.9945	0.9950	0.9955	0.9959	0.9963	0.9967	0.9970	0.9973	0.9975
2	0.9628	0.9658	0.9686	0.9711	0.9734	0.9756	0.9776	0.9794	0.9811	0.9826
3	0.8835	0.8912	0.8984	0.9052	0.9116	0.9176	0.9232	0.9285	0.9334	0.9380
4	0.7487	0.7619	0.7746	0.7867	0.7983	0.8094	0.8200	0.8300	0.8396	0.8488
5	0.5769	0.5939	0.6105	0.6267	0.6425	0.6578	0.6728	0.6873	0.7013	0.7149
6	0.4016	0.4191	0.4365	0.4539	0.4711	0.4881	0.5050	0.5217	0.5381	0.5543
7	0.2526	0.2676	0.2829	0.2983	0.3140	0.3297	0.3456	0.3616	0.3776	0.3937
8	0.1440	0.1551	0.1665	0.1783	0.1905	0.2030	0.2159	0.2290	0.2424	0.2560
9	0.0748	0.0819	0.0894	0.0973	0.1056	0.1143	0.1234	0.1328	0.1426	0.1528
10	0.0356	0.0397	0.0441	0.0488	0.0538	0.0591	0.0648	0.0708	0.0772	0.0839
11	0.0156	0.0177	0.0200	0.0225	0.0253	0.0282	0.0314	0.0349	0.0386	0.0426
12	0.0063	0.0073	0.0084	0.0096	0.0110	0.0125	0.0141	0.0159	0.0179	0.0201
13	0.0024	0.0028	0.0033	0.0038	0.0045	0.0051	0.0059	0.0068	0.0078	0.0088
14	0.0008	0.0010	0.0012	0.0014	0.0017	0.0020	0.0023	0.0027	0.0031	0.0036
15	0.0003	0.0003	0.0004	0.0005	0.0006	0.0007	0.0009	0.0010	0.0012	0.0014
16	0.0001	0.0001	0.0001	0.0002	0.0002	0.0002	0.0003	0.0004	0.0004	0.0005
17	0.0000	0.0000	0.0000	0.0001	0.0001	0.0001	0.0001	0.0001	0.0001	0.0002
18	0.0000	0.0000	0.0000	0.0000	0.0000	0.0000	0.0000	0.0000	0.0000	0.0001

(Continued)

	λ									
k	6.1	6.2	6.3	6.4	6.5	6.6	6.7	6.8	6.9	7.0
0	1.0000	1.0000	1.0000	1.0000	1.0000	1.0000	1.0000	1.0000	1.0000	1.0000
1	0.9978	0.9980	0.9982	0.9983	0.9985	0.9986	0.9988	0.9989	0.9990	0.9991
2	0.9841	0.9854	0.9866	0.9877	0.9887	0.9897	0.9905	0.9913	0.9920	0.9927
3	0.9423	0.9464	0.9502	0.9537	0.9570	0.9600	0.9629	0.9656	0.9680	0.9704
4	0.8575	0.8658	0.8736	0.8811	0.8882	0.8948	0.9012	0.9072	0.9129	0.9182
5	0.7281	0.7408	0.7531	0.7649	0.7763	0.7873	0.7978	0.8080	0.8177	0.8270
6	0.5702	0.5859	0.6012	0.6163	0.6310	0.6453	0.6594	0.6730	0.6863	0.6993
7	0.4098	0.4258	0.4418	0.4577	0.4735	0.4892	0.5047	0.5201	0.5353	0.5503
8	0.2699	0.2840	0.2983	0.3127	0.3272	0.3419	0.3567	0.3715	0.3864	0.4013
9	0.1633	0.1741	0.1852	0.1967	0.2084	0.2204	0.2327	0.2452	0.2580	0.2709
10	0.0910	0.0984	0.1061	0.1142	0.1226	0.1314	0.1404	0.1498	0.1595	0.1695
11	0.0469	0.0514	0.0563	0.0614	0.0668	0.0726	0.0786	0.0849	0.0916	0.0985
12	0.0224	0.0250	0.0277	0.0307	0.0339	0.0373	0.0409	0.0448	0.0490	0.0533
13	0.0100	0.0113	0.0127	0.0143	0.0160	0.0179	0.0199	0.0221	0.0245	0.0270
14	0.0042	0.0048	0.0055	0.0063	0.0071	0.0080	0.0091	0.0102	0.0115	0.0128
15	0.0016	0.0019	0.0022	0.0026	0.0030	0.0034	0.0039	0.0044	0.0050	0.0057
16	0.0006	0.0007	0.0008	0.0010	0.0012	0.0014	0.0016	0.0018	0.0021	0.0024
17	0.0002	0.0003	0.0003	0.0004	0.0004	0.0005	0.0006	0.0007	0.0008	0.0010
18	0.0001	0.0001	0.0001	0.0001	0.0002	0.0002	0.0002	0.0003	0.0003	0.0004
19	0.0000	0.0000	0.0000	0.0000	0.0001	0.0001	0.0001	0.0001	0.0001	0.0001

	λ									
k	7.1	7.2	7.3	7.4	7.5	7.6	7.7	7.8	7.9	8.0
0	1.0000	1.0000	1.0000	1.0000	1.0000	1.0000	1.0000	1.0000	1.0000	1.0000
1	0.9992	0.9993	0.9993	0.9994	0.9994	0.9995	0.9995	0.9996	0.9996	0.9997
2	0.9933	0.9939	0.9944	0.9949	0.9953	0.9957	0.9961	0.9964	0.9967	0.9970
3	0.9725	0.9745	0.9764	0.9781	0.9797	0.9812	0.9826	0.9839	0.9851	0.9862
4	0.9233	0.9281	0.9326	0.9368	0.9409	0.9446	0.9482	0.9515	0.9547	0.9576

(Continued)

	λ									
k	7.1	7.2	7.3	7.4	7.5	7.6	7.7	7.8	7.9	8.0
5	0.8359	0.8445	0.8527	0.8605	0.8679	0.8751	0.8819	0.8883	0.8945	0.9004
6	0.7119	0.7241	0.7360	0.7474	0.7586	0.7693	0.7797	0.7897	0.7994	0.8088
7	0.5651	0.5796	0.5940	0.6080	0.6218	0.6354	0.6486	0.6616	0.6743	0.6866
8	0.4162	0.4311	0.4459	0.4607	0.4754	0.4900	0.5044	0.5188	0.5330	0.5470
9	0.2840	0.2973	0.3108	0.3243	0.3380	0.3518	0.3657	0.3796	0.3935	0.4075
10	0.1798	0.1904	0.2012	0.2123	0.2236	0.2351	0.2469	0.2589	0.2710	0.2834
11	0.1058	0.1133	0.1212	0.1293	0.1378	0.1465	0.1555	0.1648	0.1743	0.1841
12	0.0580	0.0629	0.0681	0.0735	0.0792	0.0852	0.0915	0.0980	0.1048	0.1119
13	0.0297	0.0327	0.0358	0.0391	0.0427	0.0464	0.0504	0.0546	0.0591	0.0638
14	0.0143	0.0159	0.0176	0.0195	0.0216	0.0238	0.0261	0.0286	0.0313	0.0342
15	0.0065	0.0073	0.0082	0.0092	0.0103	0.0114	0.0127	0.0141	0.0156	0.0173
16	0.0028	0.0031	0.0036	0.0041	0.0046	0.0052	0.0059	0.0066	0.0074	0.0082
17	0.0011	0.0013	0.0015	0.0017	0.0020	0.0022	0.0026	0.0029	0.0033	0.0037
18	0.0004	0.0005	0.0006	0.0007	0.0008	0.0009	0.0011	0.0012	0.0014	0.0016
19	0.0002	0.0002	0.0002	0.0003	0.0003	0.0004	0.0004	0.0005	0.0006	0.0007

	λ									
k	8.1	8.2	8.3	8.4	8.5	8.6	8.7	8.8	8.9	9.0
0	1.0000	1.0000	1.0000	1.0000	1.0000	1.0000	1.0000	1.0000	1.0000	1.0000
1	0.9997	0.9997	0.9998	0.9998	0.9998	0.9998	0.9998	0.9998	0.9999	0.9999
2	0.9972	0.9975	0.9977	0.9979	0.9981	0.9982	0.9984	0.9985	0.9986	0.9988
3	0.9873	0.9882	0.9891	0.9900	0.9907	0.9914	0.9921	0.9927	0.9932	0.9938
4	0.9604	0.9630	0.9654	0.9677	0.9699	0.9719	0.9738	0.9756	0.9772	0.9788
5	0.9060	0.9113	0.9163	0.9211	0.9256	0.9299	0.9340	0.9379	0.9416	0.9450
6	0.8178	0.8264	0.8347	0.8427	0.8504	0.8578	0.8648	0.8716	0.8781	0.8843
7	0.6987	0.7104	0.7219	0.7330	0.7438	0.7543	0.7645	0.7744	0.7840	0.7932
8	0.5609	0.5746	0.5881	0.6013	0.6144	0.6272	0.6398	0.6522	0.6643	0.6761
9	0.4214	0.4353	0.4493	0.4631	0.4769	0.4906	0.5042	0.5177	0.5311	0.5443

(*Continued*)

	λ									
k	8.1	8.2	8.3	8.4	8.5	8.6	8.7	8.8	8.9	9.0
10	0.2959	0.3085	0.3212	0.3341	0.3470	0.3600	0.3731	0.3863	0.3994	0.4126
11	0.1942	0.2045	0.2150	0.2257	0.2366	0.2478	0.2591	0.2706	0.2822	0.2940
12	0.1193	0.1269	0.1348	0.1429	0.1513	0.1600	0.1689	0.1780	0.1874	0.1970
13	0.0687	0.0739	0.0793	0.0850	0.0909	0.0971	0.1035	0.1102	0.1171	0.1242
14	0.0372	0.0405	0.0439	0.0476	0.0514	0.0555	0.0597	0.0642	0.0689	0.0739
15	0.0190	0.0209	0.0229	0.0251	0.0274	0.0299	0.0325	0.0353	0.0383	0.0415
16	0.0092	0.0102	0.0113	0.0125	0.0138	0.0152	0.0168	0.0184	0.0202	0.0220
17	0.0042	0.0047	0.0053	0.0059	0.0066	0.0074	0.0082	0.0091	0.0101	0.0111
18	0.0018	0.0021	0.0023	0.0027	0.0030	0.0034	0.0038	0.0043	0.0048	0.0053
19	0.0008	0.0009	0.0010	0.0011	0.0013	0.0015	0.0017	0.0019	0.0022	0.0024

	λ									
k	9.1	9.2	9.3	9.4	9.5	9.6	9.7	9.8	9.9	10.0
0	1.0000	1.0000	1.0000	1.0000	1.0000	1.0000	1.0000	1.0000	1.0000	1.0000
1	0.9999	0.9999	0.9999	0.9999	0.9999	0.9999	0.9999	0.9999	0.9999	1.0000
2	0.9989	0.9990	0.9991	0.9991	0.9992	0.9993	0.9993	0.9994	0.9995	0.9995
3	0.9942	0.9947	0.9951	0.9955	0.9958	0.9962	0.9965	0.9967	0.9970	0.9972
4	0.9802	0.9816	0.9828	0.9840	0.9851	0.9862	0.9871	0.9880	0.9889	0.9897
5	0.9483	0.9514	0.9544	0.9571	0.9597	0.9622	0.9645	0.9667	0.9688	0.9707
6	0.8902	0.8959	0.9014	0.9065	0.9115	0.9162	0.9207	0.9250	0.9290	0.9329
7	0.8022	0.8108	0.8192	0.8273	0.8351	0.8426	0.8498	0.8567	0.8634	0.8699
8	0.6877	0.6990	0.7100	0.7208	0.7313	0.7416	0.7515	0.7612	0.7706	0.7798
9	0.5574	0.5704	0.5832	0.5958	0.6082	0.6204	0.6324	0.6442	0.6558	0.6672
10	0.4258	0.4389	0.4521	0.4651	0.4782	0.4911	0.5040	0.5168	0.5295	0.5421
11	0.3059	0.3180	0.3301	0.3424	0.3547	0.3671	0.3795	0.3920	0.4045	0.4170
12	0.2068	0.2168	0.2270	0.2374	0.2480	0.2588	0.2697	0.2807	0.2919	0.3032
13	0.1316	0.1393	0.1471	0.1552	0.1636	0.1721	0.1809	0.1899	0.1991	0.2084
14	0.0790	0.0844	0.0900	0.0958	0.1019	0.1081	0.1147	0.1214	0.1284	0.1355

(*Continued*)

					λ					
k	9.1	9.2	9.3	9.4	9.5	9.6	9.7	9.8	9.9	10.0
15	0.0448	0.0483	0.0520	0.0559	0.0600	0.0643	0.0688	0.0735	0.0784	0.0835
16	0.0240	0.0262	0.0285	0.0309	0.0335	0.0362	0.0391	0.0421	0.0454	0.0487
17	0.0122	0.0135	0.0148	0.0162	0.0177	0.0194	0.0211	0.0230	0.0249	0.0270
18	0.0059	0.0066	0.0073	0.0081	0.0089	0.0098	0.0108	0.0119	0.0130	0.0143
19	0.0027	0.0031	0.0034	0.0038	0.0043	0.0048	0.0053	0.0059	0.0065	0.0072
20	0.0012	0.0014	0.0015	0.0017	0.0020	0.0022	0.0025	0.0028	0.0031	0.0035
21	0.0005	0.0006	0.0007	0.0008	0.0009	0.0010	0.0011	0.0013	0.0014	0.0016
22	0.0002	0.0002	0.0003	0.0003	0.0004	0.0004	0.0005	0.0005	0.0006	0.0007
23	0.0001	0.0001	0.0001	0.0001	0.0001	0.0002	0.0002	0.0002	0.0003	0.0003
24	0.0000	0.0000	0.0000	0.0000	0.0001	0.0001	0.0001	0.0001	0.0001	0.0001

C.4 χ^2 distribution

This table gives the values of χ^2 squared for n degrees of freedom and quantile Q. That is:

$$Q(\chi^2, n) = \frac{1}{2^{n/2}\Gamma(n/2)} \int_0^{\chi^2} t^{n/2-1} \exp(-t/2)dt.$$

The data for this table were generated using the function scipy.stats.chi2.ppf() in a Python program.

χ^2

n	0.995	0.99	0.975	0.95	0.9	0.75	0.5	0.25	0.1	0.05	0.025	0.01	0.005
1	7.879	6.635	5.024	3.841	2.706	1.323	0.455	0.102	0.0158	0.00393	0.000982	0.000157	0.0000393
2	10.5966	9.2103	7.3778	5.9915	4.6052	2.7726	1.3863	0.5754	0.2107	0.1026	0.0506	0.0201	0.0100
3	12.8382	11.3449	9.3484	7.8147	6.2514	4.1083	2.3660	1.2125	0.5844	0.3518	0.2158	0.1148	0.0717
4	14.8603	13.2767	11.1433	9.4877	7.7794	5.3853	3.3567	1.9226	1.0636	0.7107	0.4844	0.2971	0.2070
5	16.7496	15.0863	12.8325	11.0705	9.2364	6.6257	4.3515	2.6746	1.6103	1.1455	0.8312	0.5543	0.4117
6	18.5476	16.8119	14.4494	12.5916	10.6446	7.8408	5.3481	3.4546	2.2041	1.6354	1.2373	0.8721	0.6757
7	20.2777	18.4753	16.0128	14.0671	12.0170	9.0371	6.3458	4.2549	2.8331	2.1673	1.6899	1.2390	0.9893
8	21.9550	20.0902	17.5345	15.5073	13.3616	10.2189	7.3441	5.0706	3.4895	2.7326	2.1797	1.6465	1.3444
9	23.5894	21.6660	19.0228	16.9190	14.6837	11.3888	8.3428	5.8988	4.1682	3.3251	2.7004	2.0879	1.7349
10	25.1882	23.2093	20.4832	18.3070	15.9872	12.5489	9.3418	6.7372	4.8652	3.9403	3.2470	2.5582	2.1559
11	26.7568	24.7250	21.9200	19.6751	17.2750	13.7007	10.3410	7.5841	5.5778	4.5748	3.8157	3.0535	2.6032
12	28.2995	26.2170	23.3367	21.0261	18.5493	14.8454	11.3403	8.4384	6.3038	5.2260	4.4038	3.5706	3.0738
13	29.8195	27.6882	24.7356	22.3620	19.8119	15.9839	12.3398	9.2991	7.0415	5.8919	5.0088	4.1069	3.5650
14	31.3193	29.1412	26.1189	23.6848	21.0641	17.1169	13.3393	10.1653	7.7895	6.5706	5.6287	4.6604	4.0747
15	32.8013	30.5779	27.4884	24.9958	22.3071	18.2451	14.3389	11.0365	8.5468	7.2609	6.2621	5.2293	4.6009
16	34.2672	31.9999	28.8454	26.2962	23.5418	19.3689	15.3385	11.9122	9.3122	7.9616	6.9077	5.8122	5.1422
17	35.7185	33.4087	30.1910	27.5871	24.7690	20.4887	16.3382	12.7919	10.0852	8.6718	7.5642	6.4078	5.6972
18	37.1565	34.8053	31.5264	28.8693	25.9894	21.6049	17.3379	13.6753	10.8649	9.3905	8.2307	7.0149	6.2648

(Continued)

χ^2

n	0.005	0.01	0.025	0.05	0.1	0.25	0.5	0.75	0.9	0.95	0.975	0.99	0.995
19	6.8440	7.6327	8.9065	10.1170	11.6509	14.5620	18.3377	22.7178	27.2036	30.1435	32.8523	36.1909	38.5823
20	7.4338	8.2604	9.5908	10.8508	12.4426	15.4518	19.3374	23.8277	28.4120	31.4104	34.1696	37.5662	39.9968
21	8.0337	8.8972	10.2829	11.5913	13.2396	16.3444	20.3372	24.9348	29.6151	32.6706	35.4789	38.9322	41.4011
22	8.6427	9.5425	10.9823	12.3380	14.0415	17.2396	21.3370	26.0393	30.8133	33.9244	36.7807	40.2894	42.7957
23	9.2604	10.1957	11.6886	13.0905	14.8480	18.1373	22.3369	27.1413	32.0069	35.1725	38.0756	41.6384	44.1813
24	9.8862	10.8564	12.4012	13.8484	15.6587	19.0373	23.3367	28.2412	33.1962	36.4150	39.3641	42.9798	45.5585
25	10.5197	11.5240	13.1197	14.6114	16.4734	19.9393	24.3366	29.3389	34.3816	37.6525	40.6465	44.3141	46.9279
26	11.1602	12.1981	13.8439	15.3792	17.2919	20.8434	25.3365	30.4346	35.5632	38.8851	41.9232	45.6417	48.2899
27	11.8076	12.8785	14.5734	16.1514	18.1139	21.7494	26.3363	31.5284	36.7412	40.1133	43.1945	46.9629	49.6449
28	12.4613	13.5647	15.3079	16.9279	18.9392	22.6572	27.3362	32.6205	37.9159	41.3371	44.4608	48.2782	50.9934
29	13.1211	14.2565	16.0471	17.7084	19.7677	23.5666	28.3361	33.7109	39.0875	42.5570	45.7223	49.5879	52.3356
30	13.78672	14.95346	16.79077	18.49266	20.59923	24.47761	29.33603	34.79974	40.25602	43.77297	46.97924	50.89218	53.67196

C.5   Student's t distribution

This table gives the values of Student's t for specified n degrees of freedom and specified quantile T. That is:

$$T = \frac{1}{\sqrt{n\pi}} \frac{\Gamma((n+1)/2)}{\Gamma(n/2)} \int_{-\infty}^{t} \left(1 + \frac{x^2}{n}\right)^{-\frac{n+1}{2}} dx.$$

Note that $T(-t) = 1 - T(t)$.

The data in this table were produced using the function scipy.stats.t.ppf() in a Python program and, for $n = \infty$, scipy.stats.norm.ppf().ppf was used.

					T			
n	0.6	0.75	0.9	0.95	0.975	0.99	0.995	0.9995
1	0.325	1.000	3.078	6.314	12.706	31.821	63.657	636.619
2	0.289	0.816	1.886	2.920	4.303	6.965	9.925	31.599
3	0.277	0.765	1.638	2.353	3.182	4.541	5.841	12.924
4	0.271	0.741	1.533	2.132	2.776	3.747	4.604	8.610
5	0.267	0.727	1.476	2.015	2.571	3.365	4.032	6.869
6	0.265	0.718	1.440	1.943	2.447	3.143	3.707	5.959
7	0.263	0.711	1.415	1.895	2.365	2.998	3.499	5.408
8	0.262	0.706	1.397	1.860	2.306	2.896	3.355	5.041
9	0.261	0.703	1.383	1.833	2.262	2.821	3.250	4.781
10	0.260	0.700	1.372	1.812	2.228	2.764	3.169	4.587
11	0.260	0.697	1.363	1.796	2.201	2.718	3.106	4.437
12	0.259	0.695	1.356	1.782	2.179	2.681	3.055	4.318
13	0.259	0.694	1.350	1.771	2.160	2.650	3.012	4.221
14	0.258	0.692	1.345	1.761	2.145	2.624	2.977	4.140
15	0.258	0.691	1.341	1.753	2.131	2.602	2.947	4.073
16	0.258	0.690	1.337	1.746	2.120	2.583	2.921	4.015
17	0.257	0.689	1.333	1.740	2.110	2.567	2.898	3.965
18	0.257	0.688	1.330	1.734	2.101	2.552	2.878	3.922
19	0.257	0.688	1.328	1.729	2.093	2.539	2.861	3.883
20	0.257	0.687	1.325	1.725	2.086	2.528	2.845	3.850
21	0.257	0.686	1.323	1.721	2.080	2.518	2.831	3.819

(Continued)

	T							
n	**0.6**	**0.75**	**0.9**	**0.95**	**0.975**	**0.99**	**0.995**	**0.9995**
22	0.256	0.686	1.321	1.717	2.074	2.508	2.819	3.792
23	0.256	0.685	1.319	1.714	2.069	2.500	2.807	3.768
24	0.256	0.685	1.318	1.711	2.064	2.492	2.797	3.745
25	0.256	0.684	1.316	1.708	2.060	2.485	2.787	3.725
26	0.256	0.684	1.315	1.706	2.056	2.479	2.779	3.707
27	0.256	0.684	1.314	1.703	2.052	2.473	2.771	3.690
28	0.256	0.683	1.313	1.701	2.048	2.467	2.763	3.674
29	0.256	0.683	1.311	1.699	2.045	2.462	2.756	3.659
30	0.256	0.683	1.310	1.697	2.042	2.457	2.750	3.646
40	0.255	0.681	1.303	1.684	2.021	2.423	2.704	3.551
60	0.254	0.679	1.296	1.671	2.000	2.390	2.660	3.460
120	0.254	0.677	1.289	1.658	1.980	2.358	2.617	3.373
∞	0.253	0.674	1.282	1.645	1.960	2.326	2.576	3.291

C.6 *F* distribution

These tables give the quantile $F_\alpha(n, m)$ of the F distribution for specified values of α, the numerator degrees of freedom, n, and the denominator degrees of freedom, m. That is, for specified values of α, n, and m, F_α is determined by the formula:

$$\alpha = \frac{\Gamma[(n+m)/2]}{\Gamma(n/2)\Gamma(m/2)}\left(\frac{n}{m}\right)^{n/2}\int_0^{F_\alpha}\frac{x^{(n-2)/2}}{(1+nx/m)^{(n+m)/2}}\mathrm{d}x.$$

Limiting values for infinite n or m are calculated using (6.30) and (6.32).

Note that values for $\alpha = 0.1, 0.05$, and 0.025 may be found using the reciprocal relation $F_{1-\alpha}(m, n) = 1/F_\alpha(n, m)$.

The data in these tables were produced using the function scipy.stats.f.ppf() in a Python program and, for $n = \infty$ and $m = \infty$, scipy.stats.chi2.ppf(). For $n = m = \infty$ $F_\alpha(n, m) \rightarrow 1.0$ for any α.

$\alpha = 0.90$

m	\multicolumn{19}{c}{n}
	1	2	3	4	5	6	7	8	9	10	12	15	20	24	30	40	60	120	∞
1	39.86	49.50	53.59	55.83	57.24	58.20	58.91	59.44	59.86	60.19	60.71	61.22	61.74	62.00	62.26	62.53	62.79	63.06	63.33
2	8.53	9.00	9.16	9.24	9.29	9.33	9.35	9.37	9.38	9.39	9.41	9.42	9.44	9.45	9.46	9.47	9.47	9.48	9.49
3	5.54	5.46	5.39	5.34	5.31	5.28	5.27	5.25	5.24	5.23	5.22	5.20	5.18	5.18	5.17	5.16	5.15	5.14	5.13
4	4.54	4.32	4.19	4.11	4.05	4.01	3.98	3.95	3.94	3.92	3.90	3.87	3.84	3.83	3.82	3.80	3.79	3.78	3.76
5	4.06	3.78	3.62	3.52	3.45	3.40	3.37	3.34	3.32	3.30	3.27	3.24	3.21	3.19	3.17	3.16	3.14	3.12	3.10
6	3.78	3.46	3.29	3.18	3.11	3.05	3.01	2.98	2.96	2.94	2.90	2.87	2.84	2.82	2.80	2.78	2.76	2.74	2.72
7	3.59	3.26	3.07	2.96	2.88	2.83	2.78	2.75	2.72	2.70	2.67	2.63	2.59	2.58	2.56	2.54	2.51	2.49	2.47
8	3.46	3.11	2.92	2.81	2.73	2.67	2.62	2.59	2.56	2.54	2.50	2.46	2.42	2.40	2.38	2.36	2.34	2.32	2.29
9	3.36	3.01	2.81	2.69	2.61	2.55	2.51	2.47	2.44	2.42	2.38	2.34	2.30	2.28	2.25	2.23	2.21	2.18	2.16
10	3.29	2.92	2.73	2.61	2.52	2.46	2.41	2.38	2.35	2.32	2.28	2.24	2.20	2.18	2.16	2.13	2.11	2.08	2.06
11	3.23	2.86	2.66	2.54	2.45	2.39	2.34	2.30	2.27	2.25	2.21	2.17	2.12	2.10	2.08	2.05	2.03	2.00	1.97
12	3.18	2.81	2.61	2.48	2.39	2.33	2.28	2.24	2.21	2.19	2.15	2.10	2.06	2.04	2.01	1.99	1.96	1.93	1.90
13	3.14	2.76	2.56	2.43	2.35	2.28	2.23	2.20	2.16	2.14	2.10	2.05	2.01	1.98	1.96	1.93	1.90	1.88	1.85
14	3.10	2.73	2.52	2.39	2.31	2.24	2.19	2.15	2.12	2.10	2.05	2.01	1.96	1.94	1.91	1.89	1.86	1.83	1.80
15	3.07	2.70	2.49	2.36	2.27	2.21	2.16	2.12	2.09	2.06	2.02	1.97	1.92	1.90	1.87	1.85	1.82	1.79	1.76
16	3.05	2.67	2.46	2.33	2.24	2.18	2.13	2.09	2.06	2.03	1.99	1.94	1.89	1.87	1.84	1.81	1.78	1.75	1.72
17	3.03	2.64	2.44	2.31	2.22	2.15	2.10	2.06	2.03	2.00	1.96	1.91	1.86	1.84	1.81	1.78	1.75	1.72	1.69
18	3.01	2.62	2.42	2.29	2.20	2.13	2.08	2.04	2.00	1.98	1.93	1.89	1.84	1.81	1.78	1.75	1.72	1.69	1.66

(Continued)

m	1	2	3	4	5	6	7	8	9	10	12	15	20	24	30	40	60	120	∞
19	2.99	2.61	2.40	2.27	2.18	2.11	2.06	2.02	1.98	1.96	1.91	1.86	1.81	1.79	1.76	1.73	1.70	1.67	1.63
20	2.97	2.59	2.38	2.25	2.16	2.09	2.04	2.00	1.96	1.94	1.89	1.84	1.79	1.77	1.74	1.71	1.68	1.64	1.61
21	2.96	2.57	2.36	2.23	2.14	2.08	2.02	1.98	1.95	1.92	1.87	1.83	1.78	1.75	1.72	1.69	1.66	1.62	1.59
22	2.95	2.56	2.35	2.22	2.13	2.06	2.01	1.97	1.93	1.90	1.86	1.81	1.76	1.73	1.70	1.67	1.64	1.60	1.57
23	2.94	2.55	2.34	2.21	2.11	2.05	1.99	1.95	1.92	1.89	1.84	1.80	1.74	1.72	1.69	1.66	1.62	1.59	1.55
24	2.93	2.54	2.33	2.19	2.10	2.04	1.98	1.94	1.91	1.88	1.83	1.78	1.73	1.70	1.67	1.64	1.61	1.57	1.53
25	2.92	2.53	2.32	2.18	2.09	2.02	1.97	1.93	1.89	1.87	1.82	1.77	1.72	1.69	1.66	1.63	1.59	1.56	1.52
26	2.91	2.52	2.31	2.17	2.08	2.01	1.96	1.92	1.88	1.86	1.81	1.76	1.71	1.68	1.65	1.61	1.58	1.54	1.50
27	2.90	2.51	2.30	2.17	2.07	2.00	1.95	1.91	1.87	1.85	1.80	1.75	1.70	1.67	1.64	1.60	1.57	1.53	1.49
28	2.89	2.50	2.29	2.16	2.06	2.00	1.94	1.90	1.87	1.84	1.79	1.74	1.69	1.66	1.63	1.59	1.56	1.52	1.48
29	2.89	2.50	2.28	2.15	2.06	1.99	1.93	1.89	1.86	1.83	1.78	1.73	1.68	1.65	1.62	1.58	1.55	1.51	1.47
30	2.88	2.49	2.28	2.14	2.05	1.98	1.93	1.88	1.85	1.82	1.77	1.72	1.67	1.64	1.61	1.57	1.54	1.50	1.46
40	2.84	2.44	2.23	2.09	2.00	1.93	1.87	1.83	1.79	1.76	1.71	1.66	1.61	1.57	1.54	1.51	1.47	1.42	1.38
60	2.79	2.39	2.18	2.04	1.95	1.87	1.82	1.77	1.74	1.71	1.66	1.60	1.54	1.51	1.48	1.44	1.40	1.35	1.29
120	2.75	2.35	2.13	1.99	1.90	1.82	1.77	1.72	1.68	1.65	1.60	1.55	1.48	1.45	1.41	1.37	1.32	1.26	1.19
∞	2.71	2.30	2.08	1.94	1.85	1.77	1.72	1.67	1.63	1.60	1.55	1.49	1.42	1.38	1.34	1.30	1.24	1.17	1.00

$\alpha = 0.95$

m \ n	1	2	3	4	5	6	7	8	9	10	12	15	20	24	30	40	60	120	∞
1	161.45	199.50	215.71	224.58	230.16	233.99	236.77	238.88	240.54	241.88	243.91	245.95	248.01	249.05	250.10	251.14	252.20	253.25	254.31
2	18.51	19.00	19.16	19.25	19.30	19.33	19.35	19.37	19.38	19.40	19.41	19.43	19.45	19.45	19.46	19.47	19.48	19.49	19.50
3	10.1	9.55	9.28	9.12	9.01	8.94	8.89	8.85	8.81	8.79	8.74	8.70	8.66	8.64	8.62	8.59	8.57	8.55	8.53
4	7.71	6.94	6.59	6.39	6.26	6.16	6.09	6.04	6.00	5.96	5.91	5.86	5.80	5.77	5.75	5.72	5.69	5.66	5.63
5	6.61	5.79	5.41	5.19	5.05	4.95	4.88	4.82	4.77	4.74	4.68	4.62	4.56	4.53	4.50	4.46	4.43	4.40	4.36
6	5.99	5.14	4.76	4.53	4.39	4.28	4.21	4.15	4.10	4.06	4.00	3.94	3.87	3.84	3.81	3.77	3.74	3.70	3.67
7	5.59	4.74	4.35	4.12	3.97	3.87	3.79	3.73	3.68	3.64	3.57	3.51	3.44	3.41	3.38	3.34	3.30	3.27	3.23
8	5.32	4.46	4.07	3.84	3.69	3.58	3.50	3.44	3.39	3.35	3.28	3.22	3.15	3.12	3.08	3.04	3.01	2.97	2.93
9	5.12	4.26	3.86	3.63	3.48	3.37	3.29	3.23	3.18	3.14	3.07	3.01	2.94	2.90	2.86	2.83	2.79	2.75	2.71
10	4.96	4.10	3.71	3.48	3.33	3.22	3.14	3.07	3.02	2.98	2.91	2.85	2.77	2.74	2.70	2.66	2.62	2.58	2.54
11	4.84	3.98	3.59	3.36	3.20	3.09	3.01	2.95	2.90	2.85	2.79	2.72	2.65	2.61	2.57	2.53	2.49	2.45	2.40
12	4.75	3.89	3.49	3.26	3.11	3.00	2.91	2.85	2.80	2.75	2.69	2.62	2.54	2.51	2.47	2.43	2.38	2.34	2.30
13	4.67	3.81	3.41	3.18	3.03	2.92	2.83	2.77	2.71	2.67	2.60	2.53	2.46	2.42	2.38	2.34	2.30	2.25	2.21
14	4.60	3.74	3.34	3.11	2.96	2.85	2.76	2.70	2.65	2.60	2.53	2.46	2.39	2.35	2.31	2.27	2.22	2.18	2.13
15	4.54	3.68	3.29	3.06	2.90	2.79	2.71	2.64	2.59	2.54	2.48	2.40	2.33	2.29	2.25	2.20	2.16	2.11	2.07
16	4.49	3.63	3.24	3.01	2.85	2.74	2.66	2.59	2.54	2.49	2.42	2.35	2.28	2.24	2.19	2.15	2.11	2.06	2.01
17	4.45	3.59	3.20	2.96	2.81	2.70	2.61	2.55	2.49	2.45	2.38	2.31	2.23	2.19	2.15	2.10	2.06	2.01	1.96
18	4.41	3.55	3.16	2.93	2.77	2.66	2.58	2.51	2.46	2.41	2.34	2.27	2.19	2.15	2.11	2.06	2.02	1.97	1.92
19	4.38	3.52	3.13	2.90	2.74	2.63	2.54	2.48	2.42	2.38	2.31	2.23	2.16	2.11	2.07	2.03	1.98	1.93	1.88

(Continued)

C. Statistical tables

m \\ n	1	2	3	4	5	6	7	8	9	10	12	15	20	24	30	40	60	120	∞
20	4.35	3.49	3.10	2.87	2.71	2.60	2.51	2.45	2.39	2.35	2.28	2.20	2.12	2.08	2.04	1.99	1.95	1.90	1.84
21	4.32	3.47	3.07	2.84	2.68	2.57	2.49	2.42	2.37	2.32	2.25	2.18	2.10	2.05	2.01	1.96	1.92	1.87	1.81
22	4.30	3.44	3.05	2.82	2.66	2.55	2.46	2.40	2.34	2.30	2.23	2.15	2.07	2.03	1.98	1.94	1.89	1.84	1.78
23	4.28	3.42	3.03	2.80	2.64	2.53	2.44	2.37	2.32	2.27	2.20	2.13	2.05	2.01	1.96	1.91	1.86	1.81	1.76
24	4.26	3.40	3.01	2.78	2.62	2.51	2.42	2.36	2.30	2.25	2.18	2.11	2.03	1.98	1.94	1.89	1.84	1.79	1.73
25	4.24	3.39	2.99	2.76	2.60	2.49	2.40	2.34	2.28	2.24	2.16	2.09	2.01	1.96	1.92	1.87	1.82	1.77	1.71
26	4.23	3.37	2.98	2.74	2.59	2.47	2.39	2.32	2.27	2.22	2.15	2.07	1.99	1.95	1.90	1.85	1.80	1.75	1.69
27	4.21	3.35	2.96	2.73	2.57	2.46	2.37	2.31	2.25	2.20	2.13	2.06	1.97	1.93	1.88	1.84	1.79	1.73	1.67
28	4.20	3.34	2.95	2.71	2.56	2.45	2.36	2.29	2.24	2.19	2.12	2.04	1.96	1.91	1.87	1.82	1.77	1.71	1.65
29	4.18	3.33	2.93	2.70	2.55	2.43	2.35	2.28	2.22	2.18	2.10	2.03	1.94	1.90	1.85	1.81	1.75	1.70	1.64
30	4.17	3.32	2.92	2.69	2.53	2.42	2.33	2.27	2.21	2.16	2.09	2.01	1.93	1.89	1.84	1.79	1.74	1.68	1.62
40	4.08	3.23	2.84	2.61	2.45	2.34	2.25	2.18	2.12	2.08	2.00	1.92	1.84	1.79	1.74	1.69	1.64	1.58	1.51
60	4.00	3.15	2.76	2.53	2.37	2.25	2.17	2.10	2.04	1.99	1.92	1.84	1.75	1.70	1.65	1.59	1.53	1.47	1.39
120	3.92	3.07	2.68	2.45	2.29	2.18	2.09	2.02	1.96	1.91	1.83	1.75	1.66	1.61	1.55	1.50	1.43	1.35	1.25
∞	3.84	3.00	2.60	2.37	2.21	2.10	2.01	1.94	1.88	1.83	1.75	1.67	1.57	1.52	1.46	1.39	1.32	1.22	1.00

$\alpha = 0.975$

n

m	1	2	3	4	5	6	7	8	9	10	12	15	20	24	30	40	60	120	∞
1	647.79	799.50	864.16	899.58	921.85	937.11	948.22	956.66	963.28	968.63	976.71	984.87	993.10	997.25	1001.41	1005.60	1009.80	1014.02	1018.26
2	38.51	39.00	39.17	39.25	39.30	39.33	39.36	39.37	39.39	39.40	39.41	39.43	39.45	39.46	39.46	39.47	39.48	39.49	39.50
3	17.44	16.04	15.44	15.10	14.88	14.73	14.62	14.54	14.47	14.42	14.34	14.25	14.17	14.12	14.08	14.04	13.99	13.95	13.90
4	12.22	10.65	9.98	9.60	9.36	9.20	9.07	8.98	8.90	8.84	8.75	8.66	8.56	8.51	8.46	8.41	8.36	8.31	8.26
5	10.01	8.43	7.76	7.39	7.15	6.98	6.85	6.76	6.68	6.62	6.52	6.43	6.33	6.28	6.23	6.18	6.12	6.07	6.02
6	8.81	7.26	6.60	6.23	5.99	5.82	5.70	5.60	5.52	5.46	5.37	5.27	5.17	5.12	5.07	5.01	4.96	4.90	4.85
7	8.07	6.54	5.89	5.52	5.29	5.12	4.99	4.90	4.82	4.76	4.67	4.57	4.47	4.41	4.36	4.31	4.25	4.20	4.14
8	7.57	6.06	5.42	5.05	4.82	4.65	4.53	4.43	4.36	4.30	4.20	4.10	4.00	3.95	3.89	3.84	3.78	3.73	3.67
9	7.21	5.71	5.08	4.72	4.48	4.32	4.20	4.10	4.03	3.96	3.87	3.77	3.67	3.61	3.56	3.51	3.45	3.39	3.33
10	6.94	5.46	4.83	4.47	4.24	4.07	3.95	3.85	3.78	3.72	3.62	3.52	3.42	3.37	3.31	3.26	3.20	3.14	3.08
11	6.72	5.26	4.63	4.28	4.04	3.88	3.76	3.66	3.59	3.53	3.43	3.33	3.23	3.17	3.12	3.06	3.00	2.94	2.88
12	6.55	5.10	4.47	4.12	3.89	3.73	3.61	3.51	3.44	3.37	3.28	3.18	3.07	3.02	2.96	2.91	2.85	2.79	2.72
13	6.41	4.97	4.35	4.00	3.77	3.60	3.48	3.39	3.31	3.25	3.15	3.05	2.95	2.89	2.84	2.78	2.72	2.66	2.60
14	6.30	4.86	4.24	3.89	3.66	3.50	3.38	3.29	3.21	3.15	3.05	2.95	2.84	2.79	2.73	2.67	2.61	2.55	2.49
15	6.20	4.77	4.15	3.80	3.58	3.41	3.29	3.20	3.12	3.06	2.96	2.86	2.76	2.70	2.64	2.59	2.52	2.46	2.40
16	6.12	4.69	4.08	3.73	3.50	3.34	3.22	3.12	3.05	2.99	2.89	2.79	2.68	2.63	2.57	2.51	2.45	2.38	2.32
17	6.04	4.62	4.01	3.66	3.44	3.28	3.16	3.06	2.98	2.92	2.82	2.72	2.62	2.56	2.50	2.44	2.38	2.32	2.25
18	5.98	4.56	3.95	3.61	3.38	3.22	3.10	3.01	2.93	2.87	2.77	2.67	2.56	2.50	2.44	2.38	2.32	2.26	2.19

(Continued)

C. Statistical tables

m	1	2	3	4	5	6	7	8	9	10	12	15	20	24	30	40	60	120	∞
19	5.92	4.51	3.90	3.56	3.33	3.17	3.05	2.96	2.88	2.82	2.72	2.62	2.51	2.45	2.39	2.33	2.27	2.20	2.13
20	5.87	4.46	3.86	3.51	3.29	3.13	3.01	2.91	2.84	2.77	2.68	2.57	2.46	2.41	2.35	2.29	2.22	2.16	2.09
21	5.83	4.42	3.82	3.48	3.25	3.09	2.97	2.87	2.80	2.73	2.64	2.53	2.42	2.37	2.31	2.25	2.18	2.11	2.04
22	5.79	4.38	3.78	3.44	3.22	3.05	2.93	2.84	2.76	2.70	2.60	2.50	2.39	2.33	2.27	2.21	2.14	2.08	2.00
23	5.75	4.35	3.75	3.41	3.18	3.02	2.90	2.81	2.73	2.67	2.57	2.47	2.36	2.30	2.24	2.18	2.11	2.04	1.97
24	5.72	4.32	3.72	3.38	3.15	2.99	2.87	2.78	2.70	2.64	2.54	2.44	2.33	2.27	2.21	2.15	2.08	2.01	1.94
25	5.69	4.29	3.69	3.35	3.13	2.97	2.85	2.75	2.68	2.61	2.51	2.41	2.30	2.24	2.18	2.12	2.05	1.98	1.91
26	5.66	4.27	3.67	3.33	3.10	2.94	2.82	2.73	2.65	2.59	2.49	2.39	2.28	2.22	2.16	2.09	2.03	1.95	1.88
27	5.63	4.24	3.65	3.31	3.08	2.92	2.80	2.71	2.63	2.57	2.47	2.36	2.25	2.19	2.13	2.07	2.00	1.93	1.85
28	5.61	4.22	3.63	3.29	3.06	2.90	2.78	2.69	2.61	2.55	2.45	2.34	2.23	2.17	2.11	2.05	1.98	1.91	1.83
29	5.59	4.20	3.61	3.27	3.04	2.88	2.76	2.67	2.59	2.53	2.43	2.32	2.21	2.15	2.09	2.03	1.96	1.89	1.81
30	5.57	4.18	3.59	3.25	3.03	2.87	2.75	2.65	2.57	2.51	2.41	2.31	2.20	2.14	2.07	2.01	1.94	1.87	1.79
40	5.42	4.05	3.46	3.13	2.90	2.74	2.62	2.53	2.45	2.39	2.29	2.18	2.07	2.01	1.94	1.88	1.80	1.72	1.64
60	5.29	3.93	3.34	3.01	2.79	2.63	2.51	2.41	2.33	2.27	2.17	2.06	1.94	1.88	1.82	1.74	1.67	1.58	1.48
120	5.15	3.80	3.23	2.89	2.67	2.52	2.39	2.30	2.22	2.16	2.05	1.94	1.82	1.76	1.69	1.61	1.53	1.43	1.31
∞	5.02	3.69	3.12	2.79	2.57	2.41	2.29	2.19	2.11	2.05	1.94	1.83	1.71	1.64	1.57	1.48	1.39	1.27	1.00

n

C.7 Signed-rank test

The table gives critical values of w_+ for a one-tailed signed-rank test for samples of size n. For a two-tailed test, use the statistic w at a 2α value.

The data in this table was produced using the function qsignrank() from the R stats library in an RStudio script. Formulas for the calculation of this table are given in F. Wilcoxon, S.K. Katti, and R.A. Wilcox, 'Critical values and probability levels for the Wilcoxon rank-sum test and the Wilcoxon signed-rank test', *Selected Tables in Mathematical Statistics*, vol. 1 (1973), American Mathematical Society.

n	One-tailed $\alpha = 0.01$ Two-tailed $\alpha = 0.02$	One-tailed $\alpha = 0.025$ Two-tailed $\alpha = 0.05$	One-tailed $\alpha = 0.05$ Two-tailed $\alpha = 0.10$
5	0	0	1
6	0	1	3
7	1	3	4
8	2	4	6
9	4	6	9
10	6	9	11
11	8	11	14
12	10	14	18
13	13	18	22
14	16	22	26
15	20	26	31
16	24	30	36
17	28	35	42
18	33	41	48
19	38	47	54
20	44	53	61
21	50	59	68
22	56	66	76
23	63	74	84
24	70	82	92
25	77	90	101
26	85	99	111
27	93	108	120

(Continued)

n	One-tailed $\alpha = 0.01$ Two-tailed $\alpha = 0.02$	One-tailed $\alpha = 0.025$ Two-tailed $\alpha = 0.05$	One-tailed $\alpha = 0.05$ Two-tailed $\alpha = 0.10$
28	102	117	131
29	111	127	141
30	121	138	152

C.8   Rank-sum test

Values of $u'_{1,2}$ such that $P\left[u_{1,2} \leq u'_{1,2}\right] \leq \alpha$, for samples of size n_1 and $n_2 > n_1$. For a two-tailed test, use the statistic u at 2α value.

The data in these tables were calculated using the qwilcox() function from the stats library in the R programming language.

One-tailed test, $\alpha = 0.025$; or two-tailed test, $\alpha = 0.05$

n_1	n_2 3	4	5	6	7	8	9	10	11	12	13	14	15	16	17	18	19
2						1	1	1	1	2	2	2	2	2	3	3	3
3			1	2	2	3	3	4	4	5	5	6	6	7	7	8	8
4		1	2	3	4	5	5	6	7	8	9	10	11	12	12	13	14
5			3	4	6	7	8	9	10	12	13	14	15	16	18	19	20
6				6	7	9	11	12	14	15	17	18	20	22	23	25	26
7					9	11	13	15	17	19	21	23	25	27	29	31	33
8						14	16	18	20	23	25	27	30	32	35	37	39
9							18	21	24	27	29	32	35	38	40	43	46
10								24	27	30	34	37	40	43	46	49	53

One-tailed test, $\alpha = 0.05$; or two-tailed test, $\alpha = 0.10$

n_1	n_2 3	4	5	6	7	8	9	10	11	12	13	14	15	16	17	18	19	20
2			1	1	1	2	2	2	2	3	3	3	4	4	4	5	5	5
3		1	2	3	3	4	4	5	6	6	7	8	8	9	10	10	11	12
4		2	3	4	5	6	7	8	9	10	11	12	13	15	16	17	18	19

(Continued)

n_1	3	4	5	6	7	8	9	10	11	12	13	14	15	16	17	18	19	20
										n_2								
5			5	6	7	9	10	12	13	14	16	17	19	20	21	23	24	26
6				8	9	11	13	15	17	18	20	22	24	26	27	29	31	33
7					12	14	16	18	20	22	25	27	29	31	34	36	38	40
8						16	19	21	24	27	29	32	34	37	40	42	45	48
9							22	25	28	31	34	37	40	43	46	49	52	55
10								28	32	35	38	42	45	49	52	56	59	63

C.9   Run test

These tables give lower and upper critical values for r, the number of runs, for given values of n and $m \geq n$, the numbers of events of the two types. Results are given for significance levels $\alpha = 0.05$ and $\alpha = 0.025$ for one-tailed tests. The same results can be used for two-tailed tests at twice the significance, that is, $\alpha = 0.1$ and $\alpha = 0.05$, respectively.

In the upper critical value tables, it should be noted that many of the critical values satisfying the condition (11.28b) are the maximum possible number of runs. For $n = m$, this is $2n$, and for $n < m$, the maximum possible number of runs is $2n + 1$. In these cases, only the maximum possible number of runs is in the critical region.

In the lower critical value tables, a dash means that the value satisfying the condition (11.28a) is one. Since an outcome of only one run is impossible when both n and m are nonzero, there is no possible value that satisfies the required conditions.

These tables were created in the R programming language using the functions qruns() and pruns() in the library randtests.

Upper critical values for $\alpha = 0.05$

n	4	5	6	7	8	9	10	11	12	13	14	15	16	17	18	19	20
									m								
2	5	5	5	5	5	5	5	5	5	5	5	5	5	5	5	5	5
3	6	7	7	7	7	7	7	7	7	7	7	7	7	7	7	7	7
4	7	8	8	8	9	9	9	9	9	9	9	9	9	9	9	9	9
5		8	9	9	10	10	10	11	11	11	11	11	11	11	11	11	11
6			10	10	11	11	11	12	12	12	12	13	13	13	13	13	13

(Continued)

n	m																
	4	5	6	7	8	9	10	11	12	13	14	15	16	17	18	19	20
7				11	12	12	12	13	13	13	13	14	14	14	14	14	15
8					12	13	13	14	14	14	15	15	15	15	15	15	16
9						13	14	14	15	15	16	16	16	16	17	17	17
10							15	15	16	16	16	17	17	17	18	18	18
11								16	16	17	17	18	18	18	19	19	19
12									17	17	18	18	19	19	20	20	20
13										18	19	19	20	20	20	21	21
14											19	20	20	21	21	22	22
15												20	21	21	22	22	23
16													22	22	23	23	24
17														23	23	24	24
18															24	24	25
19																25	26
20																	26

Lower critical values for $\alpha = 0.05$

n	m																
	4	5	6	7	8	9	10	11	12	13	14	15	16	17	18	19	20
2	-	-	-	-	2	2	2	2	2	2	2	2	2	2	2	2	2
3	-	2	2	2	2	2	3	3	3	3	3	3	3	3	3	3	3
4	2	2	3	3	3	3	3	3	4	4	4	4	4	4	4	4	4
5		3	3	3	3	4	4	4	4	4	5	5	5	5	5	5	5
6			3	4	4	4	5	5	5	5	5	6	6	6	6	6	6
7				4	4	5	5	5	6	6	6	6	6	7	7	7	7
8					5	5	6	6	6	6	7	7	7	7	8	8	8
9						6	6	6	7	7	7	8	8	8	8	8	9
10							6	7	7	8	8	8	8	9	9	9	9
11								7	8	8	8	9	9	9	10	10	10
12									8	9	9	9	10	10	10	10	11
13										9	9	10	10	10	11	11	11

(Continued)

n	4	5	6	7	8	9	10	11	12	13	14	15	16	17	18	19	20
14											10	10	11	11	11	12	12
15												11	11	11	12	12	12
16													11	12	12	13	13
17														12	13	13	13
18															13	14	14
19																14	14
20																	15

Upper critical values for $\alpha = 0.025$

n	4	5	6	7	8	9	10	11	12	13	14	15	16	17	18	19	20
2	5	5	5	5	5	5	5	5	5	5	5	5	5	5	5	5	5
3	7	7	7	7	7	7	7	7	7	7	7	7	7	7	7	7	7
4	8	8	8	9	9	9	9	9	9	9	9	9	9	9	9	9	9
5	0	9	9	10	10	11	11	11	11	11	11	11	11	11	11	11	11
6			10	11	11	12	12	12	12	13	13	13	13	13	13	13	13
7				12	12	13	13	13	13	14	14	14	15	15	15	15	15
8					13	13	14	14	15	15	15	15	16	16	16	16	16
9						14	15	15	15	16	16	17	17	17	17	17	17
10							15	16	16	17	17	17	18	18	18	19	19
11								16	17	18	18	18	19	19	19	20	20
12									18	18	19	19	20	20	20	21	21
13										19	19	20	20	21	21	22	22
14											20	21	21	22	22	22	23
15												21	22	22	23	23	24
16													22	23	24	24	24
17														24	24	25	25
18															25	25	26

(Continued)

C. Statistical tables

n	4	5	6	7	8	9	10	11	12	13	14	15	16	17	18	19	20
																m	
19																26	26
20																	27

Lower critical values for $\alpha = 0.025$

n	4	5	6	7	8	9	10	11	12	13	14	15	16	17	18	19	20
																m	
2	-	-	-	-	-	-	-	-	2	2	2	2	2	2	2	2	2
3	-	-	2	2	2	2	2	2	2	2	3	3	3	3	3	3	3
4	-	2	2	2	3	3	3	3	3	3	3	3	4	4	4	4	4
5		2	3	3	3	3	3	4	4	4	4	4	4	4	5	5	5
6			3	3	3	4	4	4	4	5	5	5	5	5	5	6	6
7				3	4	4	5	5	5	5	5	6	6	6	6	6	6
8					4	5	5	5	6	6	6	6	6	7	7	7	7
9						5	5	6	6	6	7	7	7	7	8	8	8
10							6	6	7	7	7	7	8	8	8	8	9
11								7	7	7	8	8	8	9	9	9	9
12									7	8	8	8	9	9	9	10	10
13										8	9	9	9	10	10	10	10
14											9	9	10	10	10	11	11
15												10	10	11	11	11	12
16													11	11	11	12	12
17														11	12	12	13
18															12	13	13
19																13	13
20																	14

C.10 Rank correlation coefficient

Critical values of Spearman's rank correlation coefficient for specified values of the significance level α.

The data in this table was calculated using the qSpearman() function from the SuppDists library in the R programming language.

n	$\alpha = 0.05$	$\alpha = 0.025$	$\alpha = 0.01$
5	0.800	0.900	0.900
6	0.829	0.943	1.000
7	0.750	0.821	0.893
8	0.667	0.762	0.833
9	0.617	0.700	0.783
10	0.576	0.648	0.733
11	0.536	0.618	0.700
12	0.507	0.587	0.668
13	0.484	0.560	0.643
14	0.464	0.538	0.622
15	0.446	0.521	0.600
16	0.432	0.503	0.579
17	0.417	0.485	0.561
18	0.404	0.472	0.546
19	0.391	0.458	0.532
20	0.380	0.447	0.518
21	0.371	0.434	0.505
22	0.362	0.425	0.494
23	0.353	0.415	0.483
24	0.345	0.406	0.473
25	0.338	0.398	0.464
26	0.331	0.389	0.454
27	0.324	0.382	0.446
28	0.318	0.375	0.438
29	0.312	0.368	0.431
30	0.307	0.362	0.423

Chapter 1

1.1 By directly counting all the possible outcomes, the size of the sample space is 18.

1.3 The bins have widths 12.5: $0 \leq x < 12.5$, $12.5 \leq x < 25.0$, etc, with frequencies 0, 1, 5, 7, 9, 11, 4, 3 and cumulative frequencies 0, 1, 6, 13, 22, 33, 37, 40.

1.4 The area of the histogram is unity, so there are no data points outside the given bins. Then, taking the value of x_i for each bin to be the middle of that bin, $F_n(6) = 0.95$, the area of the first three bins.

1.5 Median = 59, $P_{0.67} = 67$.

1.6 Unbinned: $\bar{x} = 58.375$, $s = 18.62$. Binned: $\bar{x} = 58.75$, $s = 18.82$. Shepard's correction makes no change to the mean and reduces the standard deviation to 18.47. The percentage of unbinned data in the range $\bar{x} \pm 2s$ is 97.5%. For a normal distribution, this would be 95.5%.

1.10 The max height of the Gaussian is $1/\sigma\sqrt{2\pi}$, so the value of the random variable x at half the maximum height is the solution of $\exp\left[-(x-\mu)^2/2\sigma^2\right] = 1/2$. Taking the logarithm produces a quadratic equation for x. The half width is the distance between the two solutions, 2.35σ.

Chapter 2

2.1 (a) $\bar{A} \cap D = (6, 9)$ (b) $(B \cap \bar{C}) \cup A = (2, 3, 4, 5, 8, 9)$
(c) $(A \cap \bar{B}) \cap C = (4)$ (d) $A \cap (\bar{B} \cup \bar{D}) = (2, 4, 8)$

2.3 (a) Probability B wins both games = 9/100. (b) A wins at least one game = 3/4.

2.6 (a) $P[9 \cup C] = 4/13$ (b) Probability that three cards are of the same suit is 11869/270725 = 0.044.

2.8 Probability that the suspect did not commit the crime is 25%.

2.10 Probability of going from I to O in just two stages is 15/54.

2.12 The probability that the component was made by machine 2, given that it is defective, is 0.375.

Chapter 3

3.2 The probability that a given envelope will contain the correct letter is $1/n$. Letting $x_i = 1$ if the ith envelope contains the correct letter and zero; otherwise, the expected value of the sum of the x_i is 1. Thus, on average, one letter will arrive at its intended destination, independent of n.

3.5 (a) $P[x < y] = 0.25$ and (b) $P[x > 1, y < 2] = 0.367$.

3.7 $\sigma_h^2 = a^2 \sigma^2(X) + b^2 \sigma^2(Y) + 2ab \operatorname{cov}(X, Y)$.

3.9 (a) $P[x \leq 2, y = 1] = 19/100$, (b) $P[x > 2, y \leq 2] = 1/10$, (c) $P[x + y = 5] = 19/100$.

3.11 $E\left[X_1^2(X_2 - 2X_3)^2\right] = 16$.

3.12 $\mu_x = \mu_y = \int_0^1 \int_0^1 f(x, y)\, dx\, dy = 1$. Similarly, $\sigma_x^2 = \sigma_y^2 = 1$, and $\sigma_{xy}^2 = 0$.

Chapter 4

4.1 (a) Probability that the component will last 780 to 850 days is 0.8351.

(b) Minimum lifetime of the longest-lived 12% of the components is 823.5 days.

4.5 Using the binomial distribution with r as the number of hits, we need the value of n such that $P[r \geq 2; n, 0.95] \geq 0.999$. We find the minimum is $n = 4$ with $P[r \geq 2; 4, 0.95] = 0.9995$.

4.8 The expected cost of each method is found by using the exponential distribution

$$f(t, \lambda) = \lambda e^{-\lambda t}, \lambda > 0, t \geq 0,$$

where λ is the inverse mean lifetime. Thus

$$E[C_A] = (C + P) \int_0^{400} \frac{1}{250} \exp\left(\frac{-t}{250}\right) dt + C \int_{400}^{\infty} \frac{1}{250} \exp\left(\frac{-t}{250}\right) dt.$$

Carrying out the calculation gives $E[C_A] = C + 0.798P$. Similarly, for C_B, we have $E[C_B] = kC + 0.680P$. The supplier should use method A if $k > 1 + 0.118P/C$.

4.9 The lifetime $T = t_1 + t_2 + t_3$ has a gamma density with $\alpha = 3$ and $\lambda = 0.01$. The probability the system will operate for at least T hours is $e^{-0.01T}\left[1 + (0.01T) + (0.01T)^2/2\right]$.

4.10 As an incident particle either does or does not produce a new particle, the binomial distribution may be used. The probability that at least five particles are produced is 0.564. Using the Poisson approximation to the binomial, the result is 0.559.

4.12 Since p is small and the number of detectors is large, we can use the Poisson approximation. The result is that the probability of more than three detectors failing is 0.0091.

Chapter 5

5.2 Using the mean and variance of the uniform distribution and the central limit theorem, we have $E[\bar{x}_n] = \mu = I + 1/2$ and $\text{var}(\bar{x}_n) = \sigma^2/n = 1/12n$. Then an unbiased estimator for I is $\bar{x}_n - 1/2$ and $E\left[(\bar{x}_n - I)^2\right] = \text{var}(\bar{x}_n) + (1/2)^2 = 1/12n + 1/4$.

5.4 Let $x \equiv \bar{x}_1 - \bar{x}_2$. Then $\mu_x = \mu_1 - \mu_2 = 10$, and $\sigma_x^2 = \sigma_1^2/n_1 + \sigma_2^2/n_2 = 4$. Standardizing x as $z = (x - \mu)/\sigma = (x - 10)/2$, we find

$$P[x < 7.5] = P[z < -1.25] = 1 - N(1.25) = 0.1056.$$

5.5 Let $W_n = \left(\sum_{i=1}^{n} r_i\right) - R$. Then P_n, the probability of sustaining damage after n exposures, is $P_n = P[W_n \geq 0]$. The problem is to find n such that $P_n \leq 0.03$. Since the sum is normally distributed with mean $4n$ and variance $2n$, W_n has mean $(4n - 100)$ and variance $(2n + 25)$. Thus $z_n = (W_n - 4n + 100)/\sqrt{2n + 25}$ is $N(0, 1)$, and

$$P_n = P[W_n \geq 0] = P\left[z_n \geq \frac{100 - 4n}{(2n + 25)^{1/2}}\right].$$

with $P_n = 0.03, z_n = 1.88$. For smaller P_n, z_n must be larger. Thus, n must satisfy

$$\frac{(100 - 4n)}{(2n + 25)^{1/2}} \geq 1.88.$$

The largest integer value of n for which this is satisfied is $n = 21$, so any more than 21 exposures will result in permanent damage with a probability greater than 3%.

5.6 We need $P[\bar{x}_A \leq 49]$, where $\bar{x}_A$ is the mean of the sample A. Using the central limit theorem, the sample standard deviation is $\sigma/\sqrt{n} = 0.7$, and the standardized variable $z = (\bar{x}_A - 50)/0.7$ is $N(0, 1)$. Thus

$$P[\bar{x}_A \leq 49] = P[z \leq -1.43] = 1 - N[1.43] = 0.0764.$$

This indicates it is likely that plant A is producing power units of lower output than expected and further testing should be done to confirm if there is a problem with the manufacturing process in plant A.

5.7 We seek the value of n such that

$$P\left[|\bar{R}_n - \bar{R}| \geq \delta\right] = P\left[\bar{R}_n \leq \bar{R} - \delta\right] + P\left[\bar{R}_n \geq \bar{R} + \delta\right] \leq \alpha,$$

where $\alpha = 0.05$, and $\delta = 0.03\,\bar{R}$. From the central limit theorem, this is approximately

$$N\left\{\frac{(\bar{R} - \delta) - \bar{R}}{\sigma/\sqrt{n}}\right\} + \left[1 - N\left\{\frac{(\bar{R} + \delta) - \bar{R}}{\sigma/\sqrt{n}}\right\}\right] = 2\left[1 - N\left(\frac{\delta\sqrt{n}}{\sigma}\right)\right],$$

where N is the standard normal distribution function. The smallest value of n is found by setting this expression equal to α, which corresponds to $\delta\sqrt{n}/\sigma = 1.96$, with the result $n \approx 96$.

5.10 Let $x = G(y) = \int_{-\infty}^{y} g(y')dy'$ be the Breit-Wigner distribution function. The integration can be done by the substitution $u = 2(y-a)/\Gamma$, with $du = \frac{2}{\Gamma}dy$ with the result $x = \frac{1}{\pi}[\arctan u]_{-\infty}^{(2(y-a)/\Gamma}$. Inverting this, we have $y = a + \frac{\Gamma}{2}\tan\left\{\pi\left(x-\frac{1}{2}\right)\right\}$. Thus, as shown in Section 4.1.1, if the x's are random numbers in the range $[0,1]$, then y follows a Breit-Wigner distribution.

Chapter 6

6.1 The statistic $z = \sqrt{2\chi^2} - \sqrt{2n-1}$ for large n is a standardized normal variate. So $\chi^2 = \frac{1}{2}[z + \sqrt{2n-1}]^2$. From Table C.1 the 90th percentile of z is 1.28, so the 90th percentage point, z_{90}, is -1.28. and $\chi^2 = 82.3$.

6.3 The characteristic function of χ_j^2 with n_j degrees of freedom is $\phi_{\chi_j^2}(t) = (1-2it)^{-n_j/2}$. The characteristic function of a sum of independent variables is the product of the characteristic functions, so for two independent χ^2 variates χ_1^2 and χ_2^2 with n_1 and n_2 degrees of freedom, the *cf* of their sum is the product of their *cf*s:

$$\phi_{\chi_1^2+\chi_2^2}(t) = \phi_{\chi_1^2}(t)\,\phi_{\chi_2^2}(t) = (1-2it)^{-(n_1+n_2)/2},$$

which by the inversion theorem is the *cf* of a χ^2 variate with (n_1+n_2) degrees of freedom.

6.4 The distance d is given by $d^2 = \sum_{i=1}^{3} x_i^2$ and $z_i = x_i/3$ is a standard normal random variable. It follows that

$$P[d^2 < 36] = P\left[\sum_{i=1}^{3} z_i^2 < 4\right] = P[\chi^2 < 4], \text{ for 3 degrees of freedom.}$$

Then from Table C.4, $P[\chi^2 > 4] \approx 0.25$.

6.5 The t statistic for this problem is $t = \sqrt{20}(15-14.5)/\sqrt{2.31} = 1.47$. From Table C.5, the probability, for 19 degrees of freedom, of obtaining a value less than this is between 0.90 and 0.95 and so the data are compatible with the assumption.

6.6 If x_i is the error on the ith coordinate, the quantities $z_i = x_i/\sqrt{4/3}$ are distributed as $N(0,1)$. Assuming the fixed point is the desired position of the target, the distance D is given by

$$D^2 = \sum_{i=1}^{n} x_i^2,$$

and

$$P\left[D^2 > 16\right] = P\left[\sum_{i=1}^{n} z_i^2 > 12\right] = P\left[\chi_n^2 > 12\right] = 0.10.$$

Using Table C.4 gives $n = 7$.

6.11 From Table C.5, we find that for 24 degrees of freedom, $t_{0.05} = 1.711$, and hence the filling mechanism would be acceptable if the t-value lies between -1.711 and 1.711. Using the data given and the definition of t, we have

$$t = \frac{\bar{x} - \mu}{s/\sqrt{n}} = \frac{310 - 300}{20/\sqrt{25}} = 2.5,$$

which is outside the critical region. Thus either the barrels are being systematically overfilled, and/or the spread of volumes is greater than estimated by the sample standard deviation s.

Chapter 7

7.1 Taking the logarithm of both sides of (7.1) gives

$$\ln L(\mu, \sigma^2) = -n \ln\left[(2\pi\sigma^2)^{1/2}\right] - \frac{1}{2\sigma^2} \sum_{i=1}^{n} (x_i - \mu)^2.$$

Setting the derivatives with respect to σ^2 and μ to zero gives

$$\hat{\sigma}^2 = \frac{1}{n} \sum_{i=1}^{n} (x_i - \hat{\mu})^2,$$

and the result for $\hat{\mu}$ from Example 7.3. We know that the sample variance

$$s^2 = \frac{1}{n-1} \sum_{i=1}^{n} (x_i - \hat{\mu})^2$$

is an unbiased estimator for σ^2, and so it follows that $\hat{\sigma}^2$ is not an unbiased estimator and $E[\hat{\sigma}^2] = [(n-1)/n]\sigma^2$. Thus $\hat{\sigma}^2$ is only asymptotically unbiased.

7.3 Since events can only be found or missed, the distribution is binomial and $E_1 = \hat{p}$. To find $\hat{p}$ and its variance, we can use the result of Example 7.9, that is, we approximate p by its estimator $\hat{p}$:

$$\hat{p} = \frac{n_1 + n_2}{2n_2},$$

where we take n_2 as the larger of the two numbers of events found as an estimate of the total number of events in the data set. Then,

$$\sigma^2(E_1) \approx \frac{\dfrac{n_1 + n_2}{2n_2}\left(1 - \dfrac{n_1 + n_2}{2n_2}\right)}{2n_2} = \frac{(n_2^2 - n_1^2)}{4n_2^2}.$$

So we can write the result in the form

$$E_1 \pm \sigma(E_1) = \frac{n_1 + n_2}{2n_2} \pm \sqrt{\frac{(n_2^2 - n_1^2)}{4n_2^2}}.$$

7.4 From the Breit-Wigner density, the likelihood function is

$$L(E_0, \Gamma) = \left(\frac{2\Gamma}{\pi}\right)^n \prod_{i=1}^{n} \frac{1}{4(E_i - E_0)^2 + \Gamma^2},$$

and

$$\ln L(E_0, \Gamma) = n[\ln \Gamma + \ln 2 - \ln \pi] - \sum_{i=1}^{n} \ln\left[4(E_i - E_0)^2 + \Gamma^2\right].$$

The ML estimators are obtained by differentiating with respect to E_0 and Γ and setting the derivatives to zero.

Using $|E_i - E_0| \ll \Gamma$ and $\Gamma > 0$, we have approximately:

$$\hat{E}_0 = \frac{1}{n}\sum_{i=1}^{n} E_i = \bar{E}.$$

7.5 The likelihood function is

$$L(x; \theta) = \prod_{i=1}^{n} \frac{\theta}{x_i^{\theta+1}} = \frac{\theta^n}{\left(\displaystyle\prod_{i=1}^{n} x_i\right)^{\theta+1}} .$$

Taking the logarithm and setting the derivative with respect to θ to zero, we find

$$\widehat{\theta} = \frac{n}{\displaystyle\sum_{i=1}^{n} \ln(x_i)},$$

and using the data, the estimate is $\widehat{\theta} = 0.423$.

7.9 Using the Poisson probability,

$$\ln L(\lambda) = \sum_{i=1}^{n} \{ k_i \ln \lambda - \ln[(k_i)!] - \lambda \}$$

and

$$\frac{d \ln L(\lambda)}{d\lambda} = \frac{1}{\lambda} \sum_{i=1}^{n} (k_i - \lambda) = \frac{n}{\lambda} (\overline{k} - \lambda).$$

Setting $d \ln L(\lambda)/d\lambda = 0$ shows that the mean $\overline{k}$ is an estimator $\widehat{\lambda}$ for the parameter λ. Moreover, $\ln L(\lambda)$ may be written

$$\ln L(\lambda) = n\widehat{\lambda}\ln \lambda - \sum_{i=1}^{n} \{ \ln[(k_i)!] - \lambda \}$$

which is of the form (7.36) and thus $\widehat{\lambda} = \overline{k}$ is a minimum variance estimator for λ.

7.10 As described in Chapter 4.16, θ can be interpreted as the mean of the distribution and as shown in Problem 7.4, the ML estimator $\widehat{\theta}$ of the parameter θ is the mean of the sample. Thus, $\widehat{\theta}$ is an unbiased estimator. The MVB is given by (7.34), that is,

$$\mathrm{var}\left(\widehat{\theta}\right) = \left\{ E\left[-\frac{d^2 \ln L}{d\theta^2} \right] \right\}^{-1} = -\left\{ nE\left[\frac{d^2 \ln f}{d\theta^2} \right] \right\}^{-1} .$$

The right-hand side gives

$$-nE\left[\frac{d^2 \ln f}{d\theta^2}\right] = -\frac{n}{\pi} \int_{-\infty}^{\infty} \frac{2[(x-\theta)^2 - 1]}{[(x-\theta)^2 + 1]^3} dx.$$

and evaluating the integral gives var $\left(\widehat{\theta}\right) = 2/n$.

Chapter 8

8.3 Since the second experiment is independent of the first, the variance matrix for the two experiments, with $y_1 = \lambda_1, y_2 = \lambda_2$, and $y_3 = \lambda_2$, is

$$\mathbf{V} = \begin{pmatrix} 2.0 & -1.0 & 0 \\ -1.0 & 1.5 & 0 \\ 0 & 0 & 1.0 \end{pmatrix} \times 10^{-2} \text{ with } \mathbf{V}^{-1} = \frac{1}{4} \begin{pmatrix} 3 & 2 & 0 \\ 2 & 4 & 0 \\ 0 & 0 & 4 \end{pmatrix} \times 10^2.$$

and thus

$$\left(\mathbf{\Phi}^T \mathbf{V}^{-1} \mathbf{\Phi}\right)^{-1} = \frac{1}{5} \begin{pmatrix} 8 & -2 \\ -2 & 3 \end{pmatrix} \times 10^{-2}.$$

Then from (8.21)

$$\widehat{\lambda}_1 = 1.04 \quad \text{and} \quad \widehat{\lambda}_2 = -1.06$$

The associated error matrix follows from (8.28),

$$\mathbf{E} = \begin{pmatrix} 1.6 & -0.4 \\ -0.4 & 0.6 \end{pmatrix} \times 10^{-2}.$$

8.5 Chi-squared is $\chi^2 = \sum_{i=1}^{3} \left(\frac{x_i - h_i}{\sigma}\right)^2$, where the sides are subject to the constraint

$$\varphi(\mathbf{x}) = x_1^2 + x_2^2 - x_3^2 = 0$$

from the Pythagoras theorem. Using a Lagrange multiplier λ, the minimum χ^2 estimates for $x_i(i = 1, 2, 3)$ are

$$\frac{x_1}{h_1} = \frac{x_2}{h_2} = \frac{1}{1 + \lambda\sigma^2} \quad \text{and} \quad \frac{x_3}{h_3} = \frac{1}{1 - \lambda\sigma^2}.$$

Using the constraint to determine λ, we find $\lambda\sigma^2 = B - \sqrt{B^2 - 1}$, in which,

$$B = \frac{h_1^2 + h_2^2 + h_3^2}{h_1^2 + h_2^2 - h_3^2}.$$

With some algebra, the solution can be written in the form:

$$x_1 = \frac{h_1}{2}(1+A), x_2 = \frac{h_2}{2}(1+A), x_3 = \frac{h_3}{2}\left(1+\frac{1}{A}\right),$$

where

$$A = \frac{h_3}{\sqrt{h_1^2 + h_2^2}}.$$

8.7 The nth algebraic moment is given by the expectation values of x^n and so, noting that $\alpha e^{-\alpha x}$ is the exponential distribution, with expected value $1/\alpha$, we find $E[x] = e^{\alpha\beta}/\alpha$ and $E[x^2] = 2e^{\alpha\beta}/\alpha^2$. Equating these to the first and second sample moments m_1 and m_2 gives

$$\alpha = 2\frac{m_1}{m_2} \text{ and } \beta = \frac{m_2}{2m_1}\ln\left(2\frac{m_1}{m_2}\right).$$

8.9 Assuming the two experiments are independent, the variance matrix for the two determinations is

$$V = \begin{pmatrix} 1 & -1 & 0 & 0 \\ -1 & 2 & 0 & 0 \\ 0 & 0 & 1 & -1 \\ 0 & 0 & -1 & 3 \end{pmatrix} \text{ with } V^{-1} = \frac{1}{2}\begin{pmatrix} 4 & 2 & 0 & 0 \\ 2 & 2 & 0 & 0 \\ 0 & 0 & 3 & 1 \\ 0 & 0 & 1 & 1 \end{pmatrix}.$$

and thus

$$\left(\Phi^T V^{-1}\Phi\right)^{-1} = \frac{1}{6}\begin{pmatrix} 3 & -3 \\ -3 & 7 \end{pmatrix},$$

and from (8.21)

$$\hat{a} = 3.50 \quad \text{and} \quad \hat{b} = 12.84.$$

The associated error matrix follows from (8.28):

$$E = \begin{pmatrix} 0.5 & -0.5 \\ -0.5 & 1.167 \end{pmatrix}.$$

8.10 To convert the given function of particle counts per unit $\cos(\theta)$, we must divide by the normalization factor

$$N = \int_{-1}^{1} (a + b \cos^2 \theta) \, d \cos \theta = 2(a + b/3),$$

so the average value of $\cos^2 \theta$ is $\overline{\cos^2 \theta} = (5 + 3b/a)/5(3 + b/a)$ and hence $\frac{b}{a} = \frac{5(3\overline{\cos^2 \theta} - 1)}{3 - 5\overline{\cos^2 \theta}}$. The average value of $\cos^2 \theta$ from the data is

$$\overline{\cos^2 \theta} = \frac{1}{n} \sum_{i=1}^{n} \cos^2 \theta_i,$$

Finally, the MM estimator for b/a is

$$\frac{b}{a} = \frac{5 \left(3 \sum_{i=1}^{n} \cos^2 \theta_i - n \right)}{3n - 5 \sum_{i=1}^{n} \cos^2 \theta_i}.$$

Chapter 9

9.1 There are only two possible answers, yes or no, so the binomial distribution is appropriate. For $n = 2000$, we can use the normal approximation. Then, with $\hat{p} = \bar{r} = 0.35$ and $\sqrt{\frac{\hat{p}(1-\hat{p})}{n}} = 0.01067$, a 95% central confidence interval is $0.329 < p < 0.371$.

9.2 From the data, $\bar{R} = 10$ and using Table 9.1 a two-tailed 95% confidence interval is $8.61 \leq S \leq 11.39$ and (b) the lower and upper one-tailed 95% limits are $S > 8.84$ and $S < 11.16$.

9.7 From equations (8.87) the mean μ_1 and variance σ_1^2 of the posterior distribution are $\mu_1 = 781.6$ and $\sigma_1^2 = 8.26$, and a 95% credible interval is $780.5 < \mu < 782.7$. The related confidence interval is $774.1 < \mu < 785.9$, which is wider than the credible interval because we have not used any prior information.

9.9 From the pooled sample variance defined in (6.21), we have $S_p = 9.27$, which we use to construct the t statistic. From Appendix C.5, we have for a 95%, two-tailed, central confidence interval with 23 degrees of freedom, $t_a = 2.069$, so

$$2.17 < \mu_1 - \mu_2 < 17.83.$$

9.11 Following Example 9.3(b), for a 95% central two-tailed confidence interval and 9 degrees of freedom, $t_\alpha = 2.262$. Using all 10 data points, $T_\alpha = 0.321$ and $\bar{x} = 0.0083$. The confidence interval is then $-0.719 \le \mu \le 0.735$. If the negative data values are set to zero, we still have 10 data points, so t_α remains the same but $\bar{x} = 0.419$ and $T_\alpha = 0.189$. In this case the confidence interval is $-0.00921 \le \mu \le 0.847$, giving a small but nonzero possibility that the mass is negative. If we ignore negative data values and just use the five positive values, $t_\alpha = 2.776$, $\bar{x} = 0.838$, and $T_\alpha = 0.271$. In this case the confidence interval is $0.0855 \le \mu \le 1.59$.

Chapter 10

10.2 The received signal is distributed as $N(S,8)$ and we will test the null hypothesis $H_0 : S = 13$ against the alternative $H_a : S \ne 13$. The test statistic is $z = \sqrt{n}|\bar{x} - S|/\sigma$. If H_0 is true, $S = 13$ and $z = 2.24$. From Table C.1, $z_{0.025} = 1.96$ and since $z > z_{0.025}$, the hypothesis must be rejected at the 5% significance level.

10.3 We are testing $H_0 : \mu = 5.0$ against $H_a : \mu > 5.0$, that is, a one-tailed test, for the population of all resistors sold by the supplier. The data give $\bar{x} = 5.08$ and $s^2 = 0.08$, and the test statistic is $t = \sqrt{n}(\bar{x} - \mu)/s = 1.19$. Since $t_{0.1}(19) = 1.33$, $t < t_{0.1}(19)$, the supplier's claim cannot be rejected.

10.6 We are testing $H_0 : \mu_1 = \mu_2$ against $H_a : \mu_1 > \mu_2$. From the data $\bar{x}_1 = 75, s_1^2 = 9, \bar{x}_2 = 70, s_2^2 = 25$, and so the pooled sample variance is $s_p^2 = 13.2$, from which we can calculate the test statistic to be $t = 2.85$. From Table C.6, $t_{0.05}(19) = 1.73$, and since $t > 1.73$, the alternative hypothesis cannot be rejected at the 5% significance level.

10.8 This can be solved using ANOVA. From (10.39), $SSB = 3223.5, SSW = 3011.5$, and so $F = s_B^2/s_W^2 = 7.14$. Since $F_{0.1}(3,20) = 2.38$ and $F > 2.38$, the hypothesis cannot be accepted at this significance level.

10.10 In this problem $H_0 : \mu = \mu_0 = 100$ and the acceptance region is $-1.73 \le t \le 1.73$ where the test statistic t is:

$$t = \frac{\bar{x} - \mu_0}{\sqrt{s^2/20}} = -0.91,$$

which is in the acceptance region, so the null hypothesis is not rejected.

Chapter 11

11.1 The sample mean for the number of counts is 5.26 and the sample variance is 4.56. The probability π_i for each number of counts i using the Poisson distribution with $\lambda = 5$ is:

π_i	0.007	0.034	0.084	0.14	0.175	0.175	0.146	0.104	0.065	0.036	0.018	0.008	0.003

Using (11.6), $\chi^2 = 21.33$. Table C.4 does not enable the p-value to be found exactly, but since $\chi^2_{0.05} = 21.0$ and $\chi^2_{0.025} = 23.3$, the hypothesis is acceptable at a level somewhat more significant than 0.05.

11.2 The expected number of intervals e_i with i counts, using the Poisson probabilities multiplied by 500 and $\lambda = 5.26$, is:

e_i	2.6	13.7	35.9	63.0	82.9	87.2	76.4	57.4	37.8	22.1	11.6	5.6	2.4

Using (11.7), $\chi^2 = 10.51$ for 11 degrees of freedom, Table C.4 does not enable the p-value to be found exactly, but since $\chi^2_{0.5} = 10.3$ and $\chi^2_{0.25} = 13.7$, the hypothesis is acceptable at any reasonable significance level.

11.4 The statistic D from (11.10) using the exponential distribution function $F(x) = 1 - e^{-x/100}$ is $D = 0.330$ and from (11.11) $D^* = 1.09$. Thus from (11.12) the hypothesis would be accepted at a 10% significance level.

11.7 There are 8 values below 35 and 6 above. Using the normal approximation, $\mu = np = 7$ and $\sigma^2 = np(1-p) = 3.5$. Then the quantity $z = (6-7)/\sqrt{3.5} = -0.53$ is a standardized normal variable and because we are dealing with a two-tailed test, the p-value is $2N(-0.53) = 2[1 - N(0.53)] = 0.7019$.

11.10 The mean is 5.15, and assigning $+$ and $-$ signs to whether the value is greater than or less than the mean, we find the number of pluses is $n = 9$, the number of minuses is $m = 11$, and $r = 11$ runs. From Table C.9, the critical regions at a 10% significance level are $r \leq 6$ and $r \geq 15$, so the hypothesis is accepted.

Bibliography

The list below contains some relevant works at the introductory and intermediate level, together with a few classic books on the subject.

Probability theory

Cramér H: *The elements of probability theory and some of its applications*, 1955, John Wiley and Sons. A very readable introduction to the frequency theory of probability and probability distributions.

Cramér H: *Random variables and probability distributions*, ed 3, 1970, Cambridge University Press. A short, more advanced, discussion of probability distributions.

General probability and statistics

The following substantial volumes provide good general introductions to probability and statistics, some with solved examples and many additional problems.

Cramér H: *Mathematical methods of statistics*, ed 11, 1956, Princeton University Press. A classic book on the mathematical theory of statistics. First published in 1945, with many subsequent editions.

Deep R: *Probability and statistics*, 2006, Elsevier.

deGroot MH, Schervish MJ: *Probability and statistics*, ed 4, 2012, Pearson.

Dekking FM, et al.: *A modern introduction to probability and statistics*, 2005, Springer.

Kendall M, Stuart A: *The advanced theory of statistics*, ed 4, 1976, Griffen. An immense, definitive three-volume work (earlier editions were in one or two volumes) on most aspects of mathematical statistics. Written in an authoritative style, with many examples, both worked and as exercises.

Mood AM, Graybill FA: *Introduction to the theory of statistics*, 1963, McGraw-Hill. A good intermediate level book. Numerous problems but no solutions.

Walpole RE, et al.: *Probability and statistics*, ed 9, 1990, Maxwell Macmillan.

Books for physical scientists and engineers

Brandt S: *Data analysis*, ed 3, 1998, Springer. An extensive book on statistical and computational methods with a useful chapter on optimization methods.

Hayter A: *Probability and statistics for engineers and scientists*, ed 4, 2013, Brooks/Cole.

James FE: *Statistical methods in experimental physics*, ed 2, 2006, World Scientific. An advanced work with many practical applications, mainly taken from particle physics.

Ross SM: *Probability and statistics for engineers and scientists*, ed 5, 2014, Elsevier.

Shorter books for physical scientists

Barlow RJ: *Statistics*, 1989, John Wiley and Sons. A short informal guide. Contains some worked examples and problems with solutions.

Bevington PR, Robinson DK: *Data reduction and error analysis for the physical sciences*, ed 3, 1969, McGraw Hill. Mainly about the least-squares method of estimation. Contains some worked examples and problems with answers.

Cooper BE: *Statistics for experimentalists*, 1969, Pergamon Press. General overview with some emphasis on tests of significance and analysis of variance. Some problems with hints for their solution.

Cowan G: *Statistical data analysis*, 1998, Oxford University Press. Similar to the book by Barlow above, but more advanced. The examples are taken largely from particle physics. No solved problems.

Lyons L: *Statistics for nuclear and particle physicists*, 1986, Cambridge University Press. Mainly about errors and estimation methods, but with a short introduction to general statistics. Contains a long section on the Monte Carlo technique.

Lyons L: *Data analysis for physical science students*, 1991, Cambridge University Press. A very short practical guide to experimental errors, and estimation by the least-squares method.

Roe BP: *Probability and statistics in experimental physics*, 1992, Springer-Verlag. Similar in level to the book by James above; some problems and worked examples.

Bayesian statistics

Three books that give clear introductions to statistics from the Bayesian viewpoint, with worked examples.

Antelman G: *Elementary bayesian statistics*, 1997, Edward Elgar.
Berry DA: *Statistics. A bayesian perspective*, 1996, Wadsworth Publishing Co.
Lee M: *Bayesian statistics: An introduction*, 2012, ed 4, John Wiley and Sons.

Optimization theory

Box MJ, Davies D, Swann WH: *Non-linear optimization techniques*, 1969, Oliver and Boyd. A short, very clear introduction, without detailed proofs.
Bunday BD: *Basic optimization methods*, 1984, Edward Arnold. A short book covering constrained and unconstrained optimization, and containing proofs, exercises, and sample computer codes in BASIC.
Kowalik J, Osborne MR: *Methods for unconstrained optimization problems*, 1968, American Elsevier. A more detailed book than that of Box et al. Despite the title, it also discusses constrained problems.

Statistical tables

Beyer WH, editor: *Handbook of tables for probability and statistics*, 1966, Chemical Rubber Company. A large reference work.
Pearson ES, Hartley HO: *Biometrika tables for statisticians*, Vol. 1, 1958, Cambridge University Press. A short collection of the most useful tables.

Index